Biomechanics

PRINCIPLES
and APPLICATIONS

Edited by
DANIEL J. SCHNECK
JOSEPH D. BRONZINO

CRC PRESS

Boca Raton   London   New York   Washington, D.C.

Library of Congress Cataloging-in-Publication Data

Biomechanics : principles and applications / edited by Daniel Schneck and Joseph D. Bronzino.
    p. cm.
    Includes bibliographical references and index.
    ISBN 0-8493-1492-5 (alk. paper)
    1. Biomechanics. I. Schneck, Daniel J. II. Bronzino, Joseph D., 1937–

QH513 .B585 2002
571.4'3—dc21
                                     2002073353
                                            CIP

Visit the CRC Press Web site at www.crcpress.com

© 2003 by CRC Press LLC
This material was originally published in Vol. 1 of *The Biomedical Engineering Handbook*, 2nd ed.,
Joseph D. Bronzino, Ed., CRC Press, Boca Raton, FL, 2000.

No claim to original U.S. Government works
International Standard Book Number 0-8493-1492-5
Library of Congress Card Number 2002073353
Printed in the United States of America 1 2 3 4 5 6 7 8 9 0
Printed on acid-free paper

Preface

MECHANICS IS THE ENGINEERING SCIENCE that deals with studying, defining, and mathematically quantifying "interactions" that take place among "things" in our universe. Our ability to perceive the physical manifestation of such interactions is embedded in the concept of a *force,* and the "things" that transmit forces among themselves are classified for purposes of analysis as being *solid, fluid,* or some combination of the two. The distinction between solid behavior and fluid behavior has to do with whether or not the "thing" involved has disturbance-response characteristics that are time rate dependent. A constant force transmitted to a *solid* material will generally elicit a discrete, finite, time-independent deformation response, whereas the same force transmitted to a *fluid* will elicit a continuous, time-dependent response called *flow.* In general, whether or not a given material will behave as a solid or a fluid often depends on its thermodynamic state (i.e., its temperature, pressure, etc.). Moreover, for a given thermodynamic state, some "things" are solid-like when deformed at certain rates but show fluid behavior when disturbed at other rates, so they are appropriately called *viscoelastic,* which literally means "fluid-solid." Thus a more technical definition of *mechanics* is the science that deals with the action of forces on solids, fluids, and viscoelastic materials. *Bio*mechanics then deals with the time and space response characteristics of *biological* solids, fluids, and viscoelastic materials to imposed systems of internal and external forces.

The field of biomechanics has a long history. As early as the fourth century B.C., we find in the works of Aristotle (384–322 B.C.) attempts to describe through geometric analysis the mechanical action of muscles in producing locomotion of parts or all of the animal body. Nearly 2000 years later, in his famous anatomic drawings, Leonardo da Vinci (A.D. 1452–1519) sought to describe the mechanics of standing, walking up and down hill, rising from a sitting position, and jumping, and Galileo (A.D. 1564–1643) followed with some of the earliest attempts to mathematically analyze physiologic function. Because of his pioneering efforts in defining the anatomic circulation of blood, William Harvey (A.D. 1578–1657) is credited by many as being the father of modern-day biofluid mechanics, and Alfonso Borelli (A.D. 1608–1679) shares the same honor for contemporary biosolid mechanics because of his efforts to explore the amount of force produced by various muscles and his theorization that bones serve as levers that are operated and controlled by muscles. The early work of these pioneers of biomechanics was followed up by the likes of Sir Isaac Newton (A.D. 1642–1727), Daniel Bernoulli (A.D. 1700–1782), Jean L. M. Poiseuille (A.D. 1799–1869), Thomas Young (A.D. 1773–1829), Euler (whose work was published in 1862), and others of equal fame. To enumerate all their individual contributions would take up much more space than is available in this short introduction, but there is a point to be made if one takes a closer look.

In reviewing the preceding list of biomechanical scientists, it is interesting to observe that many of the earliest contributions to our ultimate understanding of the fundamental laws of *physics* and *engineering* (e.g., Bernoulli's equation of hydrodynamics, the famous Young's modulus in elasticity theory, Poiseuille flow, and so on) came from *physicians, physiologists,* and other health care practitioners seeking to study and explain *physiologic* structure and function. The irony in this is that as history has progressed, we have just about turned this situation completely around. That is, more recently, it has been *biomedical engineers* who have been making the greatest contributions to the advancement of the *medical* and *physiologic* sciences. These contributions will become more apparent in the chapters that follow that address the subjects of *biosolid* mechanics and *biofluid* mechanics as they pertain to various subsystems of the human body.

Since the physiologic organism is 60 to 75% fluid, it is not surprising that the subject of biofluid mechanics should be so extensive, including—but not limited to—lubrication of human synovial joints (Chapter 4), cardiac biodynamics (Chapter 11), mechanics of heart valves (Chapter 12), arterial macro-circulatory hemodynamics (Chapter 13), mechanics and transport in the microcirculation (Chapter 14), venous hemodynamics (Chapter 16), mechanics of the lymphatic system (Chapter 17), cochlear mechanics (Chapter 18), and vestibular mechanics (Chapter 19). The area of biosolid mechanics is somewhat more loosely defined—since all physiologic tissue is viscoelastic and not strictly solid in the engineering sense of the word. Also generally included under this heading are studies of the kinematics and kinetics of human posture and locomotion, i.e., *biodynamics,* so that under the generic section on biosolid mechanics in this *Handbook* you will find chapters addressing the mechanics of hard tissue (Chapter 1), the mechanics of blood vessels (Chapter 2) or, more generally, the mechanics of viscoelastic tissue, mechanics of joint articulating surface motion (Chapter 3), musculoskeletal soft tissue mechanics (Chapter 5), mechanics of the head/neck (Chapter 6), mechanics of the chest/abdomen (Chapter 7), the analysis of gait (Chapter 8), exercise physiology (Chapter 9), biomechanics and factors affecting mechanical work in humans (Chapter 10), and mechanics and deformability of hematocytes (blood cells) (Chapter 15). In all cases, the ultimate objectives of the science of biomechanics are generally twofold. First, biomechanics aims to understand fundamental aspects of physiologic function for purely medical purposes, and, second, it seeks to elucidate such function for mostly nonmedical applications.

In the first instance above, sophisticated techniques have been and continue to be developed to *monitor* physiologic function, to *process* the data thus accumulated, to formulate inductively *theories* that explain the data, and to extrapolate deductively, i.e., to *diagnose* why the human "engine" malfunctions as a result of disease (pathology), aging (gerontology), ordinary wear and tear from normal use (fatigue), and/or accidental impairment from extraordinary abuse (emergency medicine). In the above sense, engineers deal *directly* with *causation* as it relates to anatomic and physiologic malfunction. However, the work does not stop there, for it goes on to provide as well the foundation for the development of technologies to treat and maintain (*therapy*) the human organism in response to malfunction, and this involves biomechanical analyses that have as their ultimate objective an improved health care delivery system. Such improvement includes, but is not limited to, a much healthier *lifestyle* (exercise physiology and sports biomechanics), the ability to *repair* and/or *rehabilitate* body parts, and a technology to *support* ailing physiologic organs (orthotics) and/or, if it should become necessary, to *replace* them completely (with prosthetic parts). Nonmedical applications of biomechanics exploit essentially the same methods and technologies as do those oriented toward the delivery of health care, but in the former case, they involve mostly studies to define the response of the body to "unusual" environments—such as subgravity conditions, the aerospace milieu, and extremes of temperature, humidity, altitude, pressure, acceleration, deceleration, impact, shock and vibration, and so on. Additional applications include vehicular safety considerations, the mechanics of sports activity, the ability of the body to "tolerate" loading without failing, and the expansion of the envelope of human performance capabilities—for whatever purpose! And so, with this very brief introduction, let us take somewhat of a closer look at the subject of biomechanics.

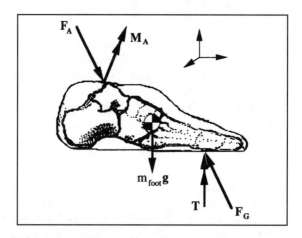

Free body diagram of the foot.

Contributors

Editors

Daniel J. Schneck
Virginia Polytechnic Institute
and State University
Blacksburg, Virginia

Joseph D. Bronzino
Trinity College
Hartford, Connecticut

Kai-Nan An
Biomechanics Laboratory
The Mayo Clinic
Rochester, Minnesota

Jeffrey T. Ellis
Georgia Institute of Technology
Atlanta, Georgia

Kenton R. Kaufman
Biomechanics Laboratory
The Mayo Clinic
Rochester, Minnesota

Gary J. Baker
Stanford University
Stanford, California

Michael J. Furey
Virginia Polytechnic Institute
and State University
Blacksburg, Virginia

Albert I. King
Wayne State University
Detroit, Michigan

Thomas J. Burkholder
Georgia Institute
of Technology
Atlanta, Georgia

Wallace Grant
Virginia Polytechnic Institute
and State University
Blacksburg, Virginia

Jack D. Lemmon
Georgia Institute of Technology
Atlanta, Georgia

Richard L. Lieber
University of California and
Veterans Administration
Medical Centers
San Diego, California

Thomas R. Canfield
Argonne National Laboratory
Argonne, Illinois

Alan R. Hargen
University of California
San Diego and NASA Ames
Research Center
San Diego, California

Roy B. Davis
Motion Analysis Laboratory
Shriners Hospitals for Children
Greenville, South Carolina

Andrew D. McCulloch
University of California
San Diego, California

Robert M. Hochmuth
Duke University
Durham, North Carolina

Peter A. DeLuca
Gait Analysis Laboratory
Connecticut Children's Medical
Center
Hartford, Connecticut

Sylvia Ounpuu
Gait Analysis Laboratory
Connecticut Children's Medical
Center
Hartford, Connecticut

Bernard F. Hurley
University of Maryland
College Park, Maryland

Philip B. Dobrin
Hines VA Hospital and Loyola
University Medical Center
Hines, Illinois

Arthur T. Johnson
University of Maryland
College Park, Maryland

Roland N. Pittman
Virginia Commonwealth
University
Richmond, Virginia

Cathryn R. Dooly
University of Maryland
College Park, Maryland

J. Lawrence Katz
Case Western Reserve University
Cleveland, Ohio

Aleksander S. Popel
The Johns Hopkins University
Baltimore, Maryland

Carl F. Rothe
Indiana University
Indianapolis, Indiana

Charles R. Steele
Stanford University
Stanford, California

Richard E. Waugh
University of Rochester
Rochester, New York

Geert Schmid-Schönbein
University of California
San Diego, California

Jason A. Tolomeo
Stanford University
Stanford, California

Ajit P. Yoganathan
Georgia Institute of Technology
Atlanta, Georgia

Artin A. Shoukas
The John Hopkins University
Baltimore, Maryland

David C. Viano
Wayne State University
Detroit, Michigan

Deborah E. Zetes-Tolomeo
Stanford University
Stanford, California

Contents

1

Mechanics of Hard Tissue

J. Lawrence Katz
*Case Western
Reserve University*

Hard tissue, mineralized tissue, and *calcified tissue* are often used as synonyms for bone when describing the structure and properties of bone or tooth. The *hard* is self-evident in comparison with all other mammalian tissues, which often are referred to as *soft tissues.* Use of the terms *mineralized* and *calcified* arises from the fact that, in addition to the principle protein, collagen, and other proteins, glycoproteins, and protein-polysaccherides, comprising about 50% of the volume, the major constituent of bone is a calcium phosphate (thus the term *calcified*) in the form of a crystalline carbonate *apatite* (similar to naturally occurring minerals, thus the term *mineralized*). Irrespective of its biological function, bone is one of the most interesting materials known in terms of structure–property relationships. Bone is an anisotropic, heterogeneous, inhomogeneous, nonlinear, thermorheologically complex viscoelastic material. It exhibits electromechanical effects, presumed to be due to streaming potentials, both *in vivo* and *in vitro* when wet. In the dry state, bone exhibits piezoelectric properties. Because of the complexity of the structure–property relationships in bone, and the space limitation for this chapter, it is necessary to concentrate on one aspect of the mechanics. Currey [1984] states unequivocally that he thinks, "the most important feature of bone material is its stiffness." This is, of course, the premiere consideration for the weight-bearing long bones. Thus, this chapter will concentrate on the elastic and viscoelastic properties of compact cortical bone and the elastic properties of trabecular bone as exemplar of mineralized tissue mechanics.

1.1 Structure of Bone

The complexity of bone's properties arises from the complexity in its structure. Thus it is important to have an understanding of the structure of mammalian bone in order to appreciate the related properties. Figure 1.1 is a diagram showing the structure of a human femur at different levels [Park, 1979]. For convenience, the structures shown in Fig. 1.1 will be grouped into four levels. A further subdivision of structural organization of mammalian bone is shown in Fig. 1.2 [Wainwright et al., 1982]. The individual figures within this diagram can be sorted into one of the appropriate levels of structure shown in Fig. 1.1 as described in the following. At the smallest unit of structure we have the *tropocollagen* molecule and

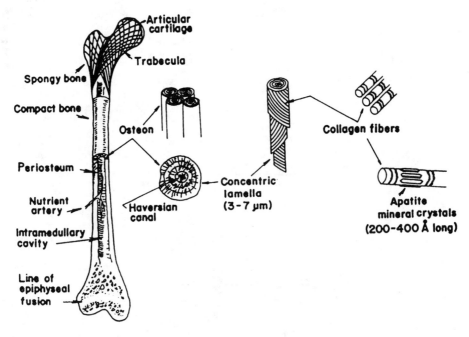

FIGURE 1.1   Hierarchical levels of structure in a human femur [Park, 1979]. (Courtesy of Plenum Press and Dr. J.B. Park.)

the associated apatite crystallites (abbreviated Ap). The former is approximately 1.5 by 280 nm, made up of three individual left-handed helical polypeptide (alpha) chains coiled into a right-handed triple helix. Ap crystallites have been found to be carbonate-substituted hydroxyapatite, generally thought to be nonstoichiometric. The crystallites appear to be about $4 \times 20 \times 60$ nm in size. This level is denoted the *molecular*. The next level we denote the *ultrastructural*. Here, the collagen and Ap are intimately associated and assembled into a microfibrilar composite, several of which are then assembled into fibers from approximately 3 to 5 μm thick. At the next level, the *microstructural*, these fibers are either randomly arranged (woven bone) or organized into concentric lamellar groups (*osteons*) or linear lamellar groups (*plexiform bone*). This is the level of structure we usually mean when we talk about bone *tissue* properties. In addition to the differences in lamellar organization at this level, there are also two different types of architectural structure. The dense type of bone found, for example, in the shafts of long bone is known as compact or *cortical bone*. A more porous or spongy type of bone is found, for example, at the articulating ends of long bones. This is called *cancellous bone*. It is important to note that the material and structural organization of collagen-Ap making up osteonic or *Haversian bone* and plexiform bone are the same as the material comprising cancellous bone.

Finally, we have the whole bone itself constructed of osteons and portions of older, partially destroyed osteons (called *interstitial lamellae*) in the case of humans or of osteons and/or plexiform bone in the case of mammals. This we denote the *macrostructural* level. The elastic properties of the whole bone results from the hierarchical contribution of each of these levels.

1.2   Composition of Bone

The composition of bone depends on a large number of factors: the species, which bone, the location from which the sample is taken, and the age, sex, and type of bone tissue, e.g., woven, cancellous, cortical. However, a rough estimate for overall composition by volume is one-third Ap, one-third collagen and other organic components, and one-third H_2O. Some data in the literature for the composition of adult human and bovine cortical bone are given in Table 1.1.

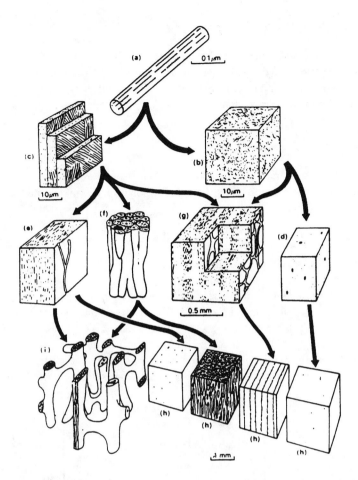

FIGURE 1.2 Diagram showing the structure of mammalian bone at different levels. Bone at the same level is drawn at the same magnification. The arrows show what types may contribute to structures at higher levels [Wainwright et al., 1982] (courtesy Princeton University Press). (a) Collagen fibril with associated mineral crystals. (b) Woven bone. The collagen fibrils are arranged more or less randomly. Osteocytes are not shown. (c) Lamellar bone. There are separate lamellae, and the collagen fibrils are arranged in "domains" of preferred fibrillar orientation in each lamella. Osteocytes are not shown. (d) Woven bone. Blood channels are shown as large black spots. At this level woven bone is indicated by light dotting. (e) Primary lamellar bone. At this level lamellar bone is indicated by fine dashes. (f) Haversian bone. A collection of Haversian systems, each with concentric lamellae round a central blood channel. The large black area represents the cavity formed as a cylinder of bone is eroded away. It will be filled in with concentric lamellae and form a new Haversian system. (g) Laminar bone. Two blood channel networks are exposed. Note how layers of woven and lamellar bone alternate. (h) Compact bone of the types shown at the lower levels. (i) Cancellous bone.

TABLE 1.1 Composition of Adult Human and Bovine Cortical Bone

Species	% H$_2$O	Ap	% Dry Weight Collagen	GAG[a]	Ref.
Bovine	9.1	76.4	21.5	N.D[b]	Herring [1977]
Human	7.3	67.2	21.2	0.34	Pellagrino and Blitz [1965]; Vejlens [1971]

[a] Glycosaminoglycan
[b] Not determined

1.3 Elastic Properties

Although bone is a viscoelastic material, at the quasi-static strain rates in mechanical testing and even at the ultrasonic frequencies used experimentally, it is a reasonable first approximation to model cortical bone as an anisotropic, linear elastic solid with Hooke's law as the appropriate constitutive equation. Tensor notation for the equation is written as:

$$\sigma_{ij} = C_{ijkl}\, e_{kl} \tag{1.1}$$

where σ_{ij} and ε_{kl} are the second-rank stress and infinitesimal second rank strain tensors, respectively, and C_{ijkl} is the fourth-rank elasticity tenor. Using the reduced notation, we can rewrite Eq. (1.1) as

$$\sigma_i = C_{ij}\epsilon_j \quad i, j = 1 \text{ to } 6 \tag{1.2}$$

where the C_{ij} are the stiffness coefficients (elastic constants). The inverse of the C_{ij}, the S_{ij}, are known as the *compliance coefficients.*

The anisotropy of cortical bone tissue has been described in two symmetry arrangements. Lang [1969], Katz and Ukraincik [1971], and Yoon and Katz [1976a,b] assumed bone to be *transversely isotropic* with the bone axis of symmetry (the 3 direction) as the unique axis of symmetry. Any small difference in elastic properties between the radial (1 direction) and transverse (2 direction) axes, due to the apparent gradient in porosity from the periosteal to the endosteal sides of bone, was deemed to be due essentially to the defect and did not alter the basic symmetry. For a transverse isotropic material, the stiffness matrix $[C_{ij}]$ is given by

$$[C_{ij}] = \begin{bmatrix} C_{11} & C_{12} & C_{13} & 0 & 0 & 0 \\ C_{12} & C_{11} & C_{13} & 0 & 0 & 0 \\ C_{13} & C_{13} & C_{33} & 0 & 0 & 0 \\ 0 & 0 & 0 & C_{44} & 0 & 0 \\ 0 & 0 & 0 & 0 & C_{44} & 0 \\ 0 & 0 & 0 & 0 & 0 & C_{66} \end{bmatrix} \tag{1.3}$$

where $C_{66} = 1/2\,(C_{11} - C_{12})$. Of the 12 nonzero coefficients, only 5 are independent.

However, Van Buskirk and Ashman [1981] used the small differences in elastic properties between the radial and tangential directions to postulate that bone is an *orthotropic* material; this requires that 9 of the 12 nonzero elastic constants be independent, that is,

$$[C_{ij}] = \begin{bmatrix} C_{11} & C_{12} & C_{13} & 0 & 0 & 0 \\ C_{12} & C_{22} & C_{23} & 0 & 0 & 0 \\ C_{13} & C_{23} & C_{33} & 0 & 0 & 0 \\ 0 & 0 & 0 & C_{44} & 0 & 0 \\ 0 & 0 & 0 & 0 & C_{55} & 0 \\ 0 & 0 & 0 & 0 & 0 & C_{66} \end{bmatrix} \tag{1.4}$$

Corresponding matrices can be written for the compliance coefficients, the S_{ij}, based on the inverse equation to Eq. (1.2):

$$\epsilon_i = S_{ij}\,\sigma_j \quad i, j = 1 \text{ to } 6 \tag{1.5}$$

where the S_{ij}th compliance is obtained by dividing the $[C_{ij}]$ stiffness matrix, minus the ith row and jth column, by the full $[C_{ij}]$ matrix and vice versa to obtain the C_{ij} in terms of the S_{ij}. Thus, although $S_{33} = 1/E_3$, where E_3 is Young's modulus in the bone axis direction, $E_3 \neq C_{33}$, since C_{33} and S_{33}, are not reciprocals of one another even for an isotropic material, let alone for transverse isotropy or orthotropic symmetry.

The relationship between the compliance matrix and the technical constants such as Young's modulus (Ei) shear modulus (Gi) and Poisson's ratio (v_{ij}) measured in mechanical tests such as uniaxial or pure shear is expressed in Eq. (1.6):

$$[S_{ij}] = \begin{bmatrix} \dfrac{1}{E_1} & \dfrac{-v_{21}}{E_2} & \dfrac{-v_{31}}{E_3} & 0 & 0 & 0 \\[2mm] \dfrac{-v_{12}}{E_1} & \dfrac{1}{E_2} & \dfrac{-v_{32}}{E_3} & 0 & 0 & 0 \\[2mm] \dfrac{-v_{13}}{E_1} & \dfrac{-v_{23}}{E_2} & \dfrac{1}{E_3} & 0 & 0 & 0 \\[2mm] 0 & 0 & 0 & \dfrac{1}{G_{31}} & 0 & 0 \\[2mm] 0 & 0 & 0 & 0 & \dfrac{1}{G_{31}} & 0 \\[2mm] 0 & 0 & 0 & 0 & 0 & \dfrac{1}{G_{12}} \end{bmatrix} \tag{1.6}$$

Again, for an orthotropic material, only 9 of the above 12 nonzero terms are independent, due to the symmetry of the S_{ij} tensor:

$$\frac{v_{12}}{E_1} = \frac{v_{21}}{E_2} \quad \frac{v_{13}}{E_1} = \frac{v_{31}}{E_3} \quad \frac{v_{23}}{E_2} = \frac{v_{32}}{E_3} \tag{1.7}$$

For the transverse isotropic case, Eq. (1.5) reduces to only 5 independent coefficients, since

$$E_1 = E_2 \quad v_{12} = v_{21} \quad v_{31} = v_{32} = v_{13} = v_{23}$$

$$G_{23} = G_{31} \quad G_{12} = \frac{E_1}{2(1+v_{12})} \tag{1.8}$$

In addition to the mechanical tests cited above, ultrasonic wave propagation techniques have been used to measure the anisotropic elastic properties of bone [Lang, 1969; Yoon and Katz, 1976a,b; Van Buskirk and Ashman, 1981]. This is possible, since combining Hooke's law with Newton's second law results in a wave equation which yields the following relationship involving the stiffness matrix:

$$\rho V^2 U_m = C_{mrns} N_r N_s U_n \tag{1.9}$$

where ρ is the density of the medium, V is the wave speed, and $\mathbf{U}$ and $\mathbf{N}$ are unit vectors along the particle displacement and wave propagation directions, respectively, so that U_m, N_r, etc. are direction cosines.

Thus to find the five transverse isotropic elastic constants, at least five independent measurements are required, e.g., a dilatational longitudinal wave in the 2 and 1(2) directions, a transverse wave in the 13 (23) and 12 planes, etc. The technical moduli must then be calculated from the full set of C_{ij}. For

TABLE 1.2   Elastic Stiffness Coefficients for Various Human and Bovine Bones[a]

Experiment (Bone Type)	C_{11} (GPa)	C_{22} (GPa)	C_{33} (GPa)	C_{44} (GPa)	C_{55} (GPa)	C_{66} (GPa)	C_{12} (GPa)	C_{13} (GPa)	C_{23} (GPa)
Van Buskirk and Ashman [1981] (bovine femur)	14.1	18.4	25.0	7.00	6.30	5.28	6.34	4.84	6.94
Knets [1978] (human tibia)	11.6	14.4	22.5	4.91	3.56	2.41	7.95	6.10	6.92
Van Buskirk and Ashman [1981] (human femur)	20.0	21.7	30.0	6.56	5.85	4.74	10.9	11.5	11.5
Maharidge [1984] (bovine femur haversian)	21.2	21.0	29.0	6.30	6.30	5.40	11.7	12.7	11.1
Maharidge [1984] (bovine femur plexiform)	22.4	25.0	35.0	8.20	7.10	6.10	14.0	15.8	13.6

[a] All measurements made with ultrasound except for Knets [1978] mechanical tests.

improved statistics, redundant measurements should be made. Correspondingly, for orthotropic symmetry, enough independent measurements must be made to obtain all 9 C_{ij}; again, redundancy in measurements is a suggested approach.

One major advantage of the ultrasonic measurements over mechanical testing is that the former can be done with specimens too small for the latter technique. Second, the reproducibility of measurements using the former technique is greater than for the latter. Still a third advantage is that the full set of either five or nine coefficients can be measured on one specimen, a procedure not possible with the latter techniques. Thus, at present, most of the studies of elastic anisotropy in both human and other mammalian bone are done using ultrasonic techniques. In addition to the bulk wave type measurements described above, it is possible to obtain Young's modulus directly. This is accomplished by using samples of small cross sections with transducers of low frequency so that the wavelength of the sound is much larger than the specimen size. In this case, an extensional longitudinal (bar) wave is propagated (which experimentally is analogous to a uniaxial mechanical test experiment), yielding

$$V^2 = \frac{E}{\rho} \tag{1.10}$$

This technique was used successfully to show that bovine plexiform bone was definitely orthotropic while bovine Haversian bone could be treated as transversely isotropic [Lipson and Katz, 1984]. The results were subsequently confirmed using bulk wave propagation techniques with considerable redundancy [Maharidge, 1984].

Table 1.2 lists the C_{ij} (in GPa) for human (Haversian) bone and bovine (both Haversian and plexiform) bone. With the exception of Knet's [1978] measurements, which were made using quasi-static mechanical testing, all the other measurements were made using bulk ultrasonic wave propagation.

In Maharidge's study [1984], both types of tissue specimens, Haversian and plexiform, were obtained from different aspects of the same level of an adult bovine femur. Thus the differences in C_{ij} reported between the two types of bone tissue are hypothesized to be due essentially to the differences in microstructural organization (Fig. 1.3) [Wainwright et al., 1982]. The textural symmetry at this level of structure has dimensions comparable to those of the ultrasound wavelengths used in the experiment, and the molecular and ultrastructural levels of organization in both types of tissues are essentially identical. Note that while C_{11} almost equals C_{22} and that C_{44} and C_{55} are equal for bovine Haversian bone, C_{11} and C_{22} and C_{44} and C_{55} differ by 11.6 and 13.4%, respectively, for bovine plexiform bone. Similarly, although C_{66} and ½ $(C_{11} - C_{12})$ differ by 12.0% for the Haversian bone, they differ by 31.1% for plexiform bone. Only the differences between C_{13} and C_{23} are somewhat comparable: 12.6% for Haversian bone and 13.9% for plexiform. These results reinforce the importance of modeling bone as a hierarchical ensemble in order to understand the basis for bone's elastic properties as a composite material–structure system in

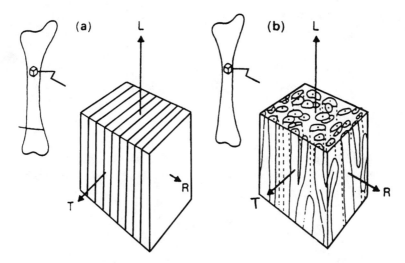

FIGURE 1.3 Diagram showing how laminar (plexiform) bone (a) differs more between radial and tangential directions (R and T) than does Haversian bone (b). The arrows are vectors representing the various directions [Wainwright et al., 1982]. (Courtesy Princeton University Press.)

which the collagen-Ap components define the material composite property. When this material property is entered into calculations based on the microtextural arrangement, the overall anisotropic elastic anisotropy can be modeled.

The human femur data [Van Buskirk and Ashman, 1981] support this description of bone tissue. Although they measured all nine individual C_{ij}, treating the femur as an orthotropic material, their results are consistent with a near transverse isotropic symmetry. However, their nine C_{ij} for bovine femoral bone clearly shows the influence of the orthotropic microtextural symmetry of the tissue's plexiform structure.

The data of Knets [1978] on human tibia are difficult to analyze. This could be due to the possibility of significant systematic errors due to mechanical testing on a large number of small specimens from a multitude of different positions in the tibia.

The variations in bone's elastic properties cited earlier above due to location is appropriately illustrated in Table 1.3, where the mean values and standard deviations (all in GPa) for all g orthotropic C_{ij} are given for bovine cortical bone at each aspect over the entire length of bone.

Since the C_{ij} are simply related to the "technical" elastic moduli, such as Young's modulus (E), shear modulus (G), bulk modulus (K), and others, it is possible to describe the moduli along any given direction. The full equations for the most general anisotropy are too long to present here. However, they

TABLE 1.3 Mean Values and Standard Deviations for the C_{ij} Measured by Van Buskirk and Ashman [1981] at Each Aspect over the Entire Length of Bone (all values in GPa)

	Anterior	Medial	Posterior	Lateral
C_{11}	18.7 ± 1.7	20.9 ± 0.8	20.1 ± 1.0	20.6 ± 1.6
C_{22}	20.4 ± 1.2	22.3 ± 1.0	22.2 ± 1.3	22.0 ± 1.0
C_{33}	28.6 ± 1.9	30.1 ± 2.3	30.8 ± 1.0	30.5 ± 1.1
C_{44}	6.73 ± 0.68	6.45 ± 0.35	6.78 ± 1.0	6.27 ± 0.28
C_{55}	5.55 ± 0.41	6.04 ± 0.51	5.93 ± 0.28	5.68 ± 0.29
C_{66}	4.34 ± 0.33	4.87 ± 0.35	5.10 ± 0.45	4.63 ± 0.36
C_{12}	11.2 ± 2.0	11.2 ± 1.1	10.4 ± 1.0	10.8 ± 1.7
C_{13}	11.2 ± 1.1	11.2 ± 2.4	11.6 ± 1.7	11.7 ± 1.8
C_{23}	10.4 ± 1.4	11.5 ± 1.0	12.5 ± 1.7	11.8 ± 1.1

can be found in Yoon and Katz [1976a]. Presented below are the simplified equations for the case of transverse isotropy. Young's modulus is

$$\frac{1}{E(\gamma_3)} = S'_{33} = \left(1 - \gamma_3^2\right) 2 S_{11} + \gamma_3^4 \, S_{33}$$
$$+ \gamma_3^2 \left(1 - \gamma_3^2\right)\left(2 S_{13} + S_{44}\right) \tag{1.11}$$

where $\gamma_3 = \cos \phi$, and ϕ is the angle made with respect to the bone (3) axis.

The shear modulus (rigidity modulus or torsional modulus for a circular cylinder) is

$$\frac{1}{G(\gamma_3)} = \frac{1}{2}\left(S'_{44} + S'_{55}\right) = S_{44} + \left(S_{11} - S_{12}\right) - \frac{1}{2} S_{44}\left(1 - \gamma_3^2\right)$$
$$+ 2\left(S_{11} + S_{33} - 2 S_{13} - S_{44}\right) \gamma_3^2 \left(1 - \gamma_3^2\right) \tag{1.12}$$

where, again $\gamma_3 = \cos \phi$.

The bulk modulus (reciprocal of the volume compressibility) is

$$\frac{1}{K} = S_{33} + 2\left(S_{11} + S_{12} + 2 S_{13}\right) = \frac{C_{11} + C_{12} + 2 C_{33} - 4 C_{13}}{C_{33}\left(C_{11} + C_{12}\right) - 2 C_{13}^2} \tag{1.13}$$

Conversion of Eqs. (1.11) and (1.12) from S_{ij} to C_{ij} can be done by using the following transformation equations:

$$S_{11} = \frac{C_{22} \, C_{33} - C_{23}^2}{\Delta} \quad S_{22} = \frac{C_{33} \, C_{11} - C_{13}^2}{\Delta}$$

$$S_{33} = \frac{C_{11} \, C_{22} - C_{12}^2}{\Delta} \quad S_{12} = \frac{C_{13} \, C_{23} - C_{12} \, C_{33}}{\Delta}$$

$$S_{13} = \frac{C_{12} \, C_{23} - C_{13} \, C_{22}}{\Delta} \quad S_{23} = \frac{C_{12} \, C_{13} - C_{23} \, C_{11}}{\Delta} \tag{1.14}$$

$$S_{44} = \frac{1}{C_{44}} \quad S_{55} = \frac{1}{C_{55}} \quad S_{66} = \frac{1}{C_{66}}$$

where

$$\Delta = \begin{vmatrix} C_{11} & C_{12} & C_{13} \\ C_{12} & C_{22} & C_{23} \\ C_{13} & C_{23} & C_{33} \end{vmatrix} = C_{11} \, C_{22} \, C_{33} + 2 \, C_{12} \, C_{23} \, C_{13} - \left(C_{11} \, C_{23}^2 + C_{22} \, C_{13}^2 + C_{33} \, C_{12}^2\right) \tag{1.15}$$

In addition to data on the elastic properties of cortical bone presented above, there is also available a considerable set of data on the mechanical properties of cancellous (trabecullar) bone including measurements of the elastic properties of single trabeculae. Indeed as early as 1993, Keaveny and Hayes (1993)

TABLE 1.4 Elastic Moduli of Trabecular Bone Material Measured by Different Experimental Methods

Study	Method	Average Modulus	(GPa)
Townsend et al. [1975]	Buckling	11.4	(Wet)
	Buckling	14.1	(Dry)
Ryan and Williams [1989]	Uniaxial tension	0.760	
Choi et al. [1992]	Four-point bending	5.72	
Ashman and Rho [1988]	Ultrasound	13.0	(Human)
	Ultrasound	10.9	(Bovine)
Rho et al. [1993]	Ultrasound	14.8	
	Tensile test	10.4	
Rho et al. [1999]	Nanoindentation	19.4	(Longitudinal)
	Nanoindentation	15.0	(Transverse)
Turner et al. [1999]	Acoustic microscopy	17.5	
	Nanoindentation	18.1	
Bumrerraj [1999]	Acoustic microscopy	17.4	

presented an analysis of 20 years of studies on the mechanical properties of trabecular bone. Most of the earlier studies used mechanical testing of bulk specimens of a size reflecting a cellular solid, i.e., of the order of cubic mm or larger. These studies showed that both the modulus and strength of trabecular bone are strongly correlated to the apparent density, where apparent density, ρ_a, is defined as the product of individual trabeculae density, ρ_t, and the volume fraction of bone in the bulk specimen, V_f, and is given by $\rho_a = \rho_t V_f$.

Elastic moduli, E, from these measurements generally ranged from approximately 10 MPa to the order of 1 GPa depending on the apparent density and could be correlated to the apparent density in g/cc by a power law relationship, $E = 6.13 P_a^{144}$, calculated for 165 specimens with an $r^2 = 0.62$ [Keaveny and Hayes, 1993].

With the introduction of micromechanical modeling of bone, it became apparent that in addition to knowing the bulk properties of trabecular bone it was necessary to determine the elastic properties of the individual trabeculae. Several different experimental techniques have been used for these studies. Individual trabeculae have been machined and measured in buckling, yielding a modulus of 11.4 GPa (wet) and 14.1 GPa (dry) [Townsend et al., 1975], as well as by other mechanical testing methods providing average values of the elastic modulus ranging from less than 1 GPa to about 8 GPa (Table 1.4). Ultrasound measurements [Ashman and Rho, 1988; Rho et al., 1993] have yielded values commensurate with the measurements of Townsend et al. (1975) (Table 1.4). More recently, acoustic microscopy and nanoindentation have been used, yielding values significantly higher than those cited above. Rho et al. [1999] using nanoindentation obtained average values of modulus ranging from 15.0 to 19.4 GPa depending on orientation, as compared to 22.4 GPa for osteons and 25.7 GPa for the interstitial lamellae in cortical bone (Table 1.4). Turner et al. (1999) compared nanoindentation and acoustic microscopy at 50 MHz on the same specimens of trabecular and cortical bone from a common human donor. While the nanoindentation resulted in Young's moduli greater than those measured by acoustic microscopy by 4 to 14%, the anisotropy ratio of longitudinal modulus to transverse modulus for cortical bone was similar for both modes of measurement; the trabecular values are given in Table 1.4. Acoustic microscopy at 400 MHz has also been used to measure the moduli of both human trabecular and cortical bone [Bumrerraj, 1999], yielding results comparable to those of Turner et al. (1999) for both types of bone (Table 1.4).

These recent studies provide a framework for micromechanical analyses using material properties measured on the microstructural level. They also point to using nano-scale measurements, such as those provided by atomic force microscopy (AFM), to analyze the mechanics of bone on the smallest unit of structure shown in Figure 1.1.

1.4 Characterizing Elastic Anisotropy

Having a full set of five or nine C_{ij} does permit describing the anisotropy of that particular specimen of bone, but there is no simple way of comparing the relative anisotropy between different specimens of the same bone or between different species or between experimenters' measurements by trying to relate individual C_{ij} between sets of measurements. Adapting a method from crystal physics [Chung and Buessem, 1968], Katz and Meunier [1987] presented a description for obtaining two scalar quantities defining the compressive and shear anisotropy for bone with transverse isotropic symmetry. Later, they developed a similar pair of scalar quantities for bone exhibiting orthotropic symmetry [Katz and Meunier, 1990]. For both cases, the percentage compressive (Ac^*) and shear (As^*) elastic anisotropy are given, respectively, by

$$Ac^*\left(\%\right) = 100\,\frac{K^V - K_R}{K^V + K_R}$$

$$As^*\left(\%\right) = 100\,\frac{G_V - G_R}{G^V + G_R} \tag{1.16}$$

where K^V and K_R are the Voigt (uniform strain across an interface) and Reuss (uniform stress across an interface) bulk moduli, respectively, and G^V and G_R are the Voigt and Reuss shear moduli, respectively. The equations for K^V, K_R, G^V, and G_R are provided for both transverse isotropy and orthotropic symmetry in the Appendix to this chapter.

Table 1.5 lists the values of K^V, K_R, G^V, G_R, Ac^*, and As^* for the five experiments whose C_{ij} are given in Table 1.2.

It is interesting to note that Haversian bones, whether human or bovine, have both their compressive and shear anisotropy factors considerably lower than the respective values for plexiform bone. Thus, not only is plexiform bone both stiffer and more rigid than Haversian bone, it is also more anisotropic. The higher values of Ac^* and As^*, especially the latter at 7.88% for the Knets [1978] mechanical testing data on human Haversian bone, supports the possibility of the systematic errors in such measurements suggested above.

1.5 Modeling Elastic Behavior

Currey [1964] first presented some preliminary ideas of modeling bone as a composite material composed of a simple linear superposition of collagen and Ap. He followed this later [1969] with an attempt to take into account the orientation of the Ap crystallites using a model proposed by Cox [1952] for fiber-reinforced composites. Katz [1971a] and Piekarski [1973] independently showed that the use of Voigt and Reuss or even Hashin–Shtrikman [1963] composite modeling showed the limitations of using linear combinations of either elastic moduli or elastic compliances. The failure of all these early models could be traced to the fact that they were based only on considerations of material properties. This is comparable to trying to determine the properties of an Eiffel Tower built using a composite material by simply

TABLE 1.5 Values of K^V, K_R, G^V, and G_R (all in GPa), and Ac^* and As^* (%) for the Bone Specimens Given in Table 1.2

Experiments (Bone Type)	K^V	K_R	G^V	G_R	Ac^*	As^*
Van Buskirk and Ashman [1981] (bovine femur)	10.4	9.87	6.34	6.07	2.68	2.19
Knets [1978] (human tibia)	10.1	9.52	4.01	3.43	2.68	7.88
Van Buskirk and Ashman [1981] (human femur)	15.5	15.0	5.95	5.74	1.59	1.82
Maharidge [1984] (bovine femur Haversian)	15.8	15.5	5.98	5.82	1.11	1.37
Maharidge [1984] (bovine femur plexiform)	18.8	18.1	6.88	6.50	1.84	2.85

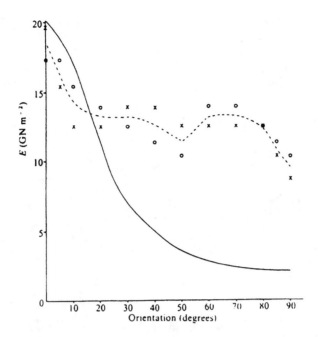

FIGURE 1.4   Variation in Young's modulus of bovine femur specimens (*E*) with the orientation of specimen axis to the long axis of the bone, for wet (o) and dry (x) conditions compared with the theoretical curve (————) predicted from a fiber-reinforced composite model [Bonfield and Grynpas, 1977]. (Courtesy *Nature* 270:453, 1977. © Macmillan Magazines Ltd.)

modeling the composite material properties without considering void spaces and the interconnectivity of the structure [Lakes, 1993]. In neither case is the complexity of the structural organization involved. This consideration of hierarchical organization clearly must be introduced into the modeling.

Katz in a number of papers [1971b, 1976] and meeting presentations put forth the hypothesis that Haversian bone should be modeled as a hierarchical composite, eventually adapting a hollow fiber composite model by Hashin and Rosen [1964]. Bonfield and Grynpas [1977] used extensional (longitudinal) ultrasonic wave propagation in both wet and dry bovine femoral cortical bone specimens oriented at angles of 5, 10, 20, 40, 50, 70, 80, and 85 degrees with respect to the long bone axis. They compared their experimental results for Young's moduli with the theoretical curve predicted by Currey's model [1969]; this is shown in Fig. 1.4. The lack of agreement led them to "conclude, therefore that an alternative model is required to account for the dependence of Young's modulus on orientation" [Bonfield and Grynpas, 1977]. Katz [1980, 1981], applying his hierarchical material-structure composite model, showed that the data in Fig. 1.4 could be explained by considering different amounts of Ap crystallites aligned parallel to the long bone axis; this is shown in Fig. 1.5. This early attempt at hierarchical micromechanical modeling is now being extended with more sophisticated modeling using either finite-element micromechanical computations [Hogan, 1992] or homogenization theory [Crolet et al., 1993]. Further improvements will come by including more definitive information on the structural organization of collagen and Ap at the molecular-ultrastructural level [Wagner and Weiner, 1992; Weiner and Traub, 1989].

1.6   Viscoelastic Properties

As stated earlier, bone (along with all other biologic tissues) is a viscoelastic material. Clearly, for such materials, Hooke's law for linear elastic materials must be replaced by a constitutive equation which includes the time dependency of the material properties. The behavior of an anisotropic linear viscoelastic material may be described by using the *Boltzmann superposition integral* as a constitutive equation:

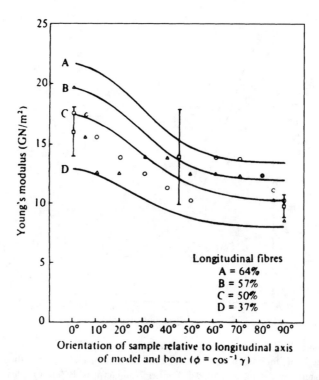

FIGURE 1.5   Comparison of predictions of Katz two-level composite model with the experimental data of Bonfield and Grynpas. Each curve represents a different lamellar configuration within a single osteon, with longitudinal fibers A, 64%; B, 57%; C, 50%; D, 37%; and the rest of the fibers assumed horizontal. (From Katz JL, *Mechanical Properties of Bone*, AMD, Vol. 45, New York, American Society of Mechanical Engineers, 1981. With permission.)

$$\sigma_{ij}\big(t\big) = \int_{-\infty}^{t} C_{ijkl}\big(t-\tau\big)\frac{d\epsilon_{kl}\big(\tau\big)}{d\tau}\,d\tau \qquad (1.17)$$

where $\sigma_{ij}(t)$ and $\epsilon_{kl}(\tau)$ are the time-dependent second rank stress and strain tensors, respectively, and $C_{ijkl}(t-\tau)$ is the fourth-rank relaxation modulus tensor. This tensor has 36 independent elements for the lowest symmetry case and 12 nonzero independent elements for an orthotropic solid. Again, as for linear elasticity, a reduced notation is used, i.e., $11 \to 1$, $22 \to 2$, $33 \to 3$, $23 \to 4$, $31 \to 5$, and $12 \to 6$. If we apply Eq. (1.17) to the case of an orthotropic material, e.g., plexiform bone, in uniaxial tension (compression) in the 1 direction [Lakes and Katz, 1974], in this case using the reduced notation, we obtain

$$\sigma_{1}\big(t\big) = \int_{-\infty}^{t}\left[C_{11}\big(t-\tau\big)\frac{d\epsilon_{1}\big(\tau\big)}{d\tau} + C_{12}\big(t-\tau\big)\frac{d\epsilon_{2}\big(\tau\big)}{d\tau} + C_{13}\big(t-\tau\big)\frac{d\epsilon_{3}\big(\tau\big)}{d\tau}\right] d\tau \qquad (1.18)$$

$$\sigma_{2}\big(t\big) = \int_{-\infty}^{t}\left[C_{21}\big(t-\tau\big)\frac{d\epsilon_{1}\big(\tau\big)}{d\tau} + C_{22}\big(t-\tau\big)\frac{d\epsilon_{2}\big(\tau\big)}{d\tau} + C_{23}\big(t-\tau\big)\frac{d\epsilon_{3}\big(\tau\big)}{d\tau}\right] = 0 \qquad (1.19)$$

for all t, and

$$\sigma_3(t) = \int_{-\infty}^{t} \left[C_{31}(t-\tau) \frac{d\epsilon_1(\tau)}{d\tau} + C_{32}(t-\tau) \frac{d\epsilon_2(\tau)}{d\tau} + C_{33}(t-\tau) \frac{d\epsilon_3(\tau)}{d\tau} \right] d\tau = 0 \qquad (1.20)$$

for all *t*.

Having the integrands vanish provides an obvious solution to Eqs. (1.19) and (1.20). Solving them simultaneously for $\frac{[d\epsilon_2(\tau)]}{d\tau}$ and $\frac{[d\epsilon_3(\tau)]}{d\tau}$ and substituting these values in Eq. (1.17) yields

$$\sigma_1(t) = \int_{-\infty}^{t} E_1(t-\tau) \frac{d\epsilon_1(\tau)}{d\tau} d\tau \qquad (1.21)$$

where, if for convenience we adopt the notation $C_{ij} \equiv C_{ij}(t-\tau)$, then Young's modulus is given by

$$E_1(t-\tau) = C_{11} + C_{12} \frac{\left[C_{31} - \left(C_{21}C_{33}/C_{23} \right) \right]}{\left[\left(C_{21}C_{33}/C_{23} \right) - C_{32} \right]} + C_{13} \frac{\left[C_{21} - \left(C_{31}C_{22}/C_{32} \right) \right]}{\left[\left(C_{22}C_{33}/C_{32} \right) - C_{23} \right]} \qquad (1.22)$$

In this case of uniaxial tension (compression), only nine independent orthotropic tensor components are involved, the three shear components being equal to zero. Still, this time-dependent Young's modulus is a rather complex function. As in the linear elastic case, the inverse form of the Boltzmann integral can be used; this would constitute the compliance formulation.

If we consider the bone being driven by a strain at a frequency ω, with a corresponding sinusoidal stress lagging by an angle δ, then the complex Young's modulus $E^*(\omega)$ may be expressed as

$$E^*(\omega) = E'(\omega) + iE''(\omega) \qquad (1.23)$$

where $E'(\omega)$, which represents the stress–strain ratio in phase with the strain, is known as the *storage modulus*, and $E''(\omega)$, which represents the stress–strain ratio 90° out of phase with the strain, is known as the *loss modulus*. The ratio of the loss modulus to the storage modulus is then equal to tan δ. Usually, data are presented by a graph of the storage modulus along with a graph of tan δ, both against frequency. For a more complete development of the values of $E'(\omega)$ and $E''(\omega)$, as well as for the derivation of other viscoelastic technical moduli, see Lakes and Katz [1974]. For a similar development of the shear storage and loss moduli, see Cowin [1989].

Thus, for a more complete understanding of bone's response to applied loads, it is important to know its rheologic properties. There have been a number of early studies of the viscoelastic properties of various long bones [Sedlin, 1965; Smith and Keiper, 1965; Laird and Kingsbury, 1973; Lugassy, 1968; Black and Korostoff, 1973]. However, none of these was performed over a wide enough range of frequency (or time) to completely define the viscoelastic properties measured, e.g., creep or stress relaxation. Thus it is not possible to mathematically transform one property into any other to compare results of three different experiments on different bones [Lakes and Katz, 1974].

In the first experiments over an extended frequency range, the biaxial viscoelastic as well as uniaxial viscoelastic properties of wet cortical human and bovine femoral bone were measured using both dynamic and stress relaxation techniques over eight decades of frequency (time) [Lakes et al., 1979]. The results of these experiments showed that bone was both nonlinear and thermorheologically complex, i.e., time–temperature superposition could not be used to extend the range of viscoelastic measurements. A nonlinear constitutive equation was developed based on these measurements [Lakes and Katz, 1979a].

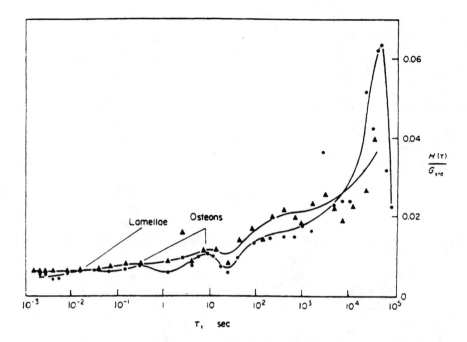

FIGURE 1.6   Comparison of relaxation spectra for wet human bone, specimens 5 and 6 [Lakes et al., 1979] in simple torsion; $T = 37°C$. First approximation from relaxation and dynamic data. ● Human tibial bone, specimen 6. ▲ Human tibial bone, specimen 5, $G_{std} = G(10 \text{ s})$. $G_{std}(5) = G(10 \text{ s})$. $G_{std}(5) = 0.590 \times 10^6 \text{ lb/in.}^2$. $G_{std}(6) \times 0.602 \times 10^6 \text{ lb/in.}^2$. (Courtesy *Journal of Biomechanics*, Pergamon Press.)

In addition, relaxation spectrums for both human and bovine cortical bone were obtained; Fig. 1.6 shows the former [Lakes and Katz, 1979b]. The contributions of several mechanisms to the loss tangent of cortical bone is shown in Fig. 1.7 [Lakes and Katz, 1979b]. It is interesting to note that almost all the major loss mechanisms occur at frequencies (times) at or close to those in which there are "bumps," indicating possible strain energy dissipation, on the relaxation spectra shown on Fig. 1.6. An extensive review of the viscoelastic properties of bone can be found in the CRC publication *Natural and Living Biomaterials* [Lakes and Katz, 1984].

Following on Katz's [1976, 1980] adaptation of the Hashin-Rosen hollow fiber composite model [1964], Gottesman and Hashin [1979] presented a viscoelastic calculation using the same major assumptions.

1.7   Related Research

As stated earlier, this chapter has concentrated on the elastic and viscoelastic properties of compact cortical bone and the elastic properties of trabecular bone. At present there is considerable research activity on the fracture properties of the bone. Professor William Bonfield and his associates at Queen Mary and Westfield College, University of London and Professor Dwight Davy and his colleagues at Case Western Reserve University are among those who publish regularly in this area. Review of the literature is necessary in order to become acquainted with the state of bone fracture mechanics.

An excellent introductory monograph which provides a fascinating insight into the structure-property relationships in bones including aspects of the two areas discussed immediately above is Professor John Currey's *The Mechanical Adaptations of Bones*, published in 1984 by Princeton University Press.

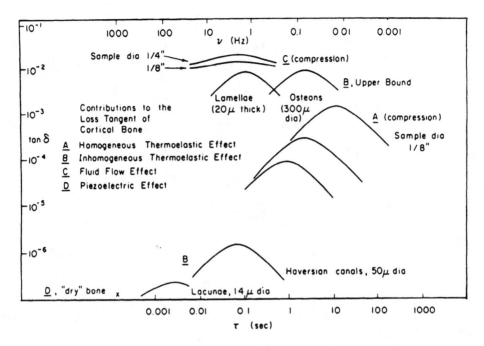

FIGURE 1.7 Contributions of several relaxation mechanisms to the loss tangent of cortical bone. A: Homogeneous thermoelastic effect. B: Inhomogeneous thermoelastic effect. C: Fluid flow effect. D: Piezoelectric effect [Lakes and Katz, 1984]. (Courtesy CRC Press.)

Defining Terms

Apatite: Calcium phosphate compound, stoichiometric chemical formula $Ca_5(PO_4)_3 \cdot X$, where X is OH^- (hydroxyapatite), F^- (fluorapatite), Cl^- (chlorapatite), etc. There are two molecules in the basic crystal unit cell.

Cancellous bone: Also known as *porous, spongy, trabecular bone.* Found in the regions of the articulating ends of tubular bones, in vertebrae, ribs, etc.

Cortical bone: The dense compact bone found throughout the shafts of long bones such as the femur, tibia, etc. also found in the outer portions of other bones in the body.

Haversian bone: Also called *osteonic.* The form of bone found in adult humans and mature mammals, consisting mainly of concentric lamellar structures, surrounding a central canal called the *Haversian canal,* plus lamellar remnants of older Haversian systems (osteons) called *interstitial lamellae.*

Interstitial lamellae: See **Haversian bone** above.

Orthotropic: The symmetrical arrangement of structure in which there are three distinct orthogonal axes of symmetry. In crystals this symmetry is called *orthothombic.*

Osteons: See **Haversian bone** above.

Plexiform: Also called *laminar.* The form of parallel lamellar bone found in younger, immature non-human mammals.

Transverse isotropy: The symmetry arrangement of structure in which there is a unique axis perpendicular to a plane in which the other two axes are equivalent. The long bone direction is chosen as the unique axis. In crystals this symmetry is called *hexagonal.*

References

Ashman RB, Rho JY. 1988. Elastic modulus of trabecular bone material. *J Biomech* 21:177.

Black J, Korostoff E. 1973. Dynamic mechanical properties of viable human cortical bone. *J Biomech* 6:435.

Bonfield W, Grynpas MD. 1977. Anisotropy of Young's modulus of bone. *Nature, London* 270:453.

Bumrerraj S. 1999. Scanning Acoustic Microscopy Studies of Human Cortical and Trabecular Bone, M.S. (BME) project (Katz, JL, advisor), Case Western Reserve University, Cleveland, OH.

Choi K, Goldstein SA. 1992. A comparison of the fatigue behavior of human trabecular and cortical bone tissue. *J Biomech* 25:1371.

Chung DH, Buessem WR. 1968. In Vahldiek, FW and Mersol, SA (Eds.), *Anisotropy in Single-Crystal Refractory Compounds*, Vol. 2, p. 217. New York, Plenum Press.

Cowin SC. 1989. *Bone Mechanics.* Boca Raton, FL, CRC Press.

Cox HL. 1952. The elasticity and strength of paper and other fibrous materials. *Br Appl Phys* 3:72.

Crolet JM, Aoubiza B, Meunier A. 1993. Compact bone: numerical simulation of mechanical character-istics. *J Biomech* 26:(6)677.

Currey JD. 1964. Three analogies to explain the mechanical properties of bone. *Biorheology* (2):1.

Currey JD. 1969. The relationship between the stiffness and the mineral content of bone. *J Biomech* (2):477.

Currey J. 1984. *The Mechanical Adaptations of Bones.* Princeton, NJ, Princeton University Press.

Gottesman T, Hashin Z. 1979. Analysis of viscoelastic behavior of bones on the basis of microstructure. *J Biomech* 13:89.

Hashin Z, Rosen BW. 1964. The elastic moduli of fiber reinforced materials. *J Appl Mech* (31):223.

Hashin Z, Shtrikman S. 1963. A variational approach to the theory of elastic behavior of multiphase materials. *J Mech Phys Solids* (11):127.

Hastings GW, Ducheyne P (Eds.). 1984. *Natural and Living Biomaterials*, Boca Raton, FL, CRC Press.

Herring GM. 1977. Methods for the study of the glycoproteins and proteoglycans of bone using bacterial collagenase. Determination of bone sialoprotein and chondroitin sulphate. *Calcif Tiss Res* (24):29.

Hogan HA. 1992. Micromechanics modeling of Haversian cortical bone properties. *J Biomech* 25(5):549.

Katz JL. 1971a. Hard tissue as a composite material: I. Bounds on the elastic behavior. *J Biomech* 4:455.

Katz JL. 1971b. Elastic properties of calcified tissues. *Isr J Med Sci* 7:439.

Katz JL. 1976. Hierarchical modeling of compact haversian bone as a fiber reinforced material. In Mates, RE and Smith, CR (Eds.), *Advances in Bioengineering*, pp. 17–18. New York, American Society of Mechanical Engineers.

Katz JL. 1980. Anisotropy of Young's modulus of bone. *Nature* 283:106.

Katz JL. 1981. Composite material models for cortical bone. In Cowin SC (Ed.), *Mechanical Properties of Bone*, Vol. 45, pp. 171–184. New York, American Society of Mechanical Engineers.

Katz JL, Meunier A. 1987. The elastic anisotropy of bone. *J Biomech* 20:1063.

Katz JL, Meunier A. 1990. A generalized method for characterizing elastic anisotropy in solid living tissues. *J Mater Sci Mater Med* 1:1.

Katz JL, Ukraincik K. 1971. On the anisotropic elastic properties of hydroxyapatite. *J Biomech* 4:221.

Katz JL, Ukraincik K. 1972. A fiber-reinforced model for compact haversian bone. Program and Abstracts of the 16th Annual Meeting of the Biophysical Society, 28a FPM-C15, Toronto.

Keaveny TM, Hayes WC. 1993. A 20-year perspective on the mechanical properties of trabecular bone. *J Biomech Eng* 115:535.

Knets IV. 1978. *Mekhanika Polimerov* 13:434.

Laird GW, Kingsbury HB. 1973. Complex viscoelastic moduli of bovine bone. *J Biomech* 6:59.

Lakes RS. 1993. Materials with structural hierarchy. *Nature* 361:511.

Lakes RS, Katz JL. 1974. Interrelationships among the viscoelastic function for anisotropic solids: appli-cation to calcified tissues and related systems. *J Biomech* 7:259.

Lakes RS, Katz JL. 1979a. Viscoelastic properties and behavior of cortical bone. Part II. Relaxation mechanisms. *J Biomech* 12:679.

Lakes RS, Katz JL. 1979b. Viscoelastic properties of wet cortical bone: III. A nonlinear constitutive equation. *J Biomech* 12:689.

Lakes RS, Katz JL. 1984. Viscoelastic properties of bone. In Hastings, GW and Ducheyne, P (Eds.), *Natural and Living Tissues*, pp 1–87. Boca Raton, FL, CRC Press.

Lakes RS, Katz JL, Sternstein SS. 1979. Viscoelastic properties of wet cortical bone: I. Torsional and biaxial studies. *J Biomech* 12:657.

Lang SB. 1969. Elastic coefficients of animal bone. *Science* 165:287.

Lipson SF, Katz JL. 1984. The relationship between elastic properties and microstructure of bovine cortical bone. *J Biomech* 4:231.

Lugassy AA. 1968. Mechanical and Viscoelastic Properties of Bone and Dentin in Compression, thesis, Metallurgy and Materials Science, University of Pennsylvania.

Maharidge R. 1984. Ultrasonic Properties and Microstructure of Bovine Bone and Haversian Bovine Bone Modeling, thesis, Rensselaer Polytechnic Institute, Troy, NY.

Park JB. 1979. *Biomaterials: An Introduction.* New York, Plenum Press.

Pellegrino ED, Biltz RM. 1965. The composition of human bone in uremia. *Medicine* 44:397.

Piekarski K. 1973. Analysis of bone as a composite material. *Int J Eng Sci* 10:557.

Reuss A. 1929. Berechnung der Fliessgrenze von Mischkristallen auf Grund der Plastizitatsbedingung für Einkristalle, A. *Zeitschrift für Angewandte Mathematik und Mechanik* 9:49–58.

Rho JY, Ashman RB, Turner CH. 1993. Young's modulus of trabecular and cortical bone material; ultrasonic and microtensile measurements. *J Biomech* 26:111.

Rho JY, Roy ME, Tsui TY, Pharr GM. 1999. Elastic properties of microstructural components of human bone tissue as measured by indentation. *J Biomed Mater Res* 45:48.

Ryan SD, Williams JL. 1989. Tensile testing of rodlike trabeculae excised from bovine femoral bone. *J Biomech* 22:351.

Sedlin E. 1965. A rheological model for cortical bone. *Acta Orthop Scand* 36(suppl 83).

Smith R, Keiper D. 1965. Dynamic measurement of viscoelastic properties of bone. *Am J Med Elec* 4:156.

Townsend PR, Rose RM, Radin EL. 1975. Buckling studies of single human trabeculae. *J Biomech* 8:199.

Turner CH, Rho JY, Takano Y, Tsui TY, Pharr GM. 1999. The elastic properties of trabecular and cortical bone tissues are simular: results from two microscopic measurement techniques. *J Biomech* 32:437.

Van Buskirk WC, Ashman RB. 1981. The elastic moduli of bone. In Cowin, SC (Ed.), *Mechanical Properties of Bone*, AMD Vol. 45, pp. 131–143. New York, American Society of Mechanical Engineers.

Vejlens L. 1971. Glycosaminoglycans of human bone tissue: I. Pattern of compact bone in relation to age. *Calcif Tiss Res* 7:175.

Voigt W. 1966. *Lehrbuch der Kristallphysik*, Teubner, Leipzig 1910; reprinted (1928) with an additional appendix. Leipzig, Teubner, New York, Johnson Reprint.

Wagner HD, Weiner S. 1992. On the relationship between the microstructure of bone and its mechanical stiffness. *J Biomech* 25:1311.

Wainwright SA, Briggs WD, Currey JD, Gosline JM. 1982. *Mechanical Design in Organisms*. Princeton, NJ, Princeton University Press.

Weiner S, Traub W. 1989. Crystal size and organization in bone. *Conn Tissue Res* 21:259.

Yoon HS, Katz JL. 1976a. Ultrasonic wave propagation in human cortical bone: I. Theoretical considerations of hexagonal symmetry. *J Biomech* 9:407.

Yoon HS, Katz JL. 1976b. Ultrasonic wave propagation in human cortical bone: II. Measurements of elastic properties and microhardness. *J Biomech* 9:459.

Further Information

Several societies both in the United States and abroad hold annual meetings during which many presentations, both oral and poster, deal with hard tissue biomechanics. In the United States these societies include the Orthopaedic Research Society, the American Society of Mechanical Engineers, the Biomaterials Society, the American Society of Biomechanics, the Biomedical Engineering Society, and the Society for Bone and Mineral Research. In Europe there are alternate year meetings of the European Society of Biomechanics and the European Society of Biomaterials. Every four years there is a World Congress of Biomechanics; every three years there is a World Congress of Biomaterials. All of these meetings result in documented proceedings; some with extended papers in book form.

The two principal journals in which bone mechanics papers appear frequently are the *Journal of Biomechanics* published by Elsevier and the *Journal of Biomechanical Engineering* published by the American Society of Mechanical Engineers. Other society journals which periodically publish papers in the field are the *Journal of Orthopaedic Research* published for the Orthopaedic Research Society, the *Annals of Biomedical Engineering* published for the Biomedical Engineering Society, and the *Journal of Bone and Joint Surgery* (both American and English issues) for the American Academy of Orthopaedic Surgeons and the British Organization, respectively. Additional papers in the field may be found in the journal *Bone and Calcified Tissue International.*

The 1984 CRC volume, *Natural and Living Biomaterials* (Hastings, G.W. and Ducheyne, P., Eds.) provides a good historical introduction to the field. A more advanced book is *Bone Mechanics* (Cowin, S.C., 1989); the second edition was published by CRC Press in 2001.

Many of the biomaterials journals and society meetings will have occasional papers dealing with hard tissue mechanics, especially those dealing with implant–bone interactions.

Appendix

The Voigt and Reuss moduli for both transverse isotropic and orthotropic symmetry are given below:

Voigt Transverse Isotropic

$$K^V = \frac{2(C_{11}+C_{12})+4(C_{13}+C_{33})}{9}$$

$$G^V = \frac{(C_{11}+C_{12})-4C_{13}+2C_{33}+12(C_{44}+C_{66})}{30}$$

(1.A1)

Reuss Transverse Isotropic

$$K_R = \frac{C_{33}(C_{11}+C_{12})-2C_{13}^2}{(C_{11}+C_{12}-4C_{13}+2C_{33})}$$

$$G_R = \frac{5\left[C_{33}(C_{11}+C_{12})-2C_{13}^2\right]C_{44}C_{66}}{2\left\{\left[C_{33}(C_{11}+C_{12})-2C_{13}^2\right](C_{44}+C_{66})+\left[C_{44}C_{66}(2C_{11}+C_{12})+4C_{13}+C_{33}\right]/3\right\}}$$

(1.A2)

Voigt Orthotropic

$$K^V = \frac{C_{11}+C_{22}+C_{33}+2(C_{12}+C_{13}+C_{23})}{9}$$

$$G^V = \frac{\left[C_{11}+C_{22}+C_{33}+3(C_{44}+C_{55}+C_{66})-(C_{12}+C_{13}+C_{23})\right]}{15}$$

(1.A3)

Reuss Orthotropic

$$K_R = \frac{\Delta}{C_{11}C_{22}+C_{22}C_{33}+C_{33}C_{11}} - 2(C_{11}C_{23}+C_{22}C_{13}+C_{33}C_{12})$$

$$+2(C_{12}C_{23}+C_{23}C_{13}+C_{13}C_{12})-(C_{12}^2+C_{13}^2+C_{23}^2)$$

$$G_R = 15\Big/\Big(4\big\{(C_{11}C_{22}+C_{22}C_{33}+C_{33}C_{11}+C_{11}C_{23}+C_{22}C_{13}+C_{33}C_{12})$$

$$-\left[C_{12}(C_{12}+C_{23})+C_{23}(C_{23}+C_{13})+C_{13}(C_{13}+C_{12})\right]\big\}\Big/\Delta$$

$$+3(1/C_{44}+1/C_{55}+1/C_{66})\Big)$$

(1.A4)

where Δ is given in Eq. (1.15).

2

Mechanics of Blood Vessels

Thomas R. Canfield
Argonne National Laboratory

Philip B. Dobrin
Hines VA Hospital and Loyola University Medical Center

2.1 Assumptions

This chapter is concerned with the mechanical behavior of blood vessels under static loading conditions and the methods required to analyze this behavior. The assumptions underlying this discussion are for *ideal* blood vessels that are at least regionally homogeneous, incompressible, elastic, and cylindrically orthotropic. Although physiologic systems are *nonideal*, much understanding of vascular mechanics has been gained through the use of methods based upon these ideal assumptions.

Homogeneity of the Vessel Wall

On visual inspection, blood vessels appear to be fairly homogeneous and distinct from surrounding connective tissue. The inhomogeneity of the vascular wall is realized when one examines the tissue under a low-power microscope, where one can easily identify two distinct structures: the media and adventitia. For this reason the assumption of vessel wall homogeneity is applied cautiously. Such an assumption may be valid only within distinct macroscopic structures. However, few investigators have incorporated macroscopic inhomogeneity into studies of vascular mechanics [17].

Incompressibility of the Vessel Wall

Experimental measurement of wall compressibility of 0.06% at 270 cm of H_2O indicates that the vessel can be considered incompressible when subjected to physiologic pressure and load [2]. In terms of the mechanical behavior of blood vessels, this is small relative to the large magnitude of the distortional strains that occur when blood vessels are deformed under the same conditions. Therefore, vascular

Work sponsored by the U.S. Department of Energy Order Contract W-31-109-Eng-38.

compressibility may be important to understanding other physiologic processes related to blood vessels, such as the transport of interstitial fluid.

Inelasticity of the Vessel Wall

That blood vessel walls exhibit inelastic behavior such as length-tension and pressure-diameter hysteresis, stress relaxation, and creep has been reported extensively [1, 10]. However, blood vessels are able to maintain stability and contain the pressure and flow of blood under a variety of physiologic conditions. These conditions are dynamic but slowly varying with a large static component.

Residual Stress and Strain

Blood vessels are known to retract both longitudinally and circumferentially after excision. This retraction is caused by the relief of distending forces resulting from internal pressure and longitudinal tractions. The magnitude of retraction is influenced by several factors. Among these factors are growth, aging, and hypertension. Circumferential retraction of medium-caliber blood vessels, such as the carotid, iliac, and bracheal arteries, can exceed 70% following reduction of internal blood pressure to zero. In the case of the carotid artery, the amount of longitudinal retraction tends to increase during growth and to decrease in subsequent aging [5]. It would seem reasonable to assume that blood vessels are in a nearly stress-free state when they are fully retracted and free of external loads. This configuration also seems to be a reasonable choice for the reference configuration. However, this ignores residual stress and strain effects that have been the subject of current research [4, 11–14, 16].

Blood vessels are formed in a dynamic environment which gives rise to imbalances between the forces that tend to extend the diameter and length and the internal forces that tend to resist the extension. This imbalance is thought to stimulate the growth of elastin and collagen and to effectively reduce the stresses in the underlying tissue. Under these conditions it is not surprising that a residual stress state exists when the vessel is fully retracted and free of external tractions. This process has been called *remodeling* [11]. Striking evidence of this remodeling is found when a cylindrical slice of the fully retracted blood vessel is cut longitudinally through the wall. The cylinder springs open, releasing bending stresses kept in balance by the cylindrical geometry [16].

2.2   Vascular Anatomy

A blood vessel can be divided anatomically into three distinct cylindrical sections when viewed under the optical microscope. Starting at the inside of the vessel, they are the intima, the media, and the adventitia. These structures have distinct functions in terms of the blood vessel physiology and mechanical properties.

The intima consists of a thin monolayer of endothelial cells that line the inner surface of the blood vessel. The endothelial cells have little influence on blood vessel mechanics but do play an important role in hemodynamics and transport phenomena. Because of their anatomical location, these cells are subjected to large variations in stress and strain as a result of pulsatile changes in blood pressure and flow.

The media represents the major portion of the vessel wall and provides most of the mechanical strength necessary to sustain structural integrity. The media is organized into alternating layers of interconnected smooth muscle cells and elastic lamellae. There is evidence of collagen throughout the media. These small collagen fibers are found within the bands of smooth muscle and may participate in the transfer of forces between the smooth muscle cells and the elastic lamellae. The elastic lamellae are composed principally of the fiberous protein elastin. The number of elastic lamellae depends upon the wall thickness and the anatomical location [18]. In the case of the canine carotid, the elastic lamellae account for a major component of the static structural response of the blood vessel [6]. This response is modulated by the smooth-muscle cells, which have the ability to actively change the mechanical characteristics of the wall [7].

The adventitia consists of loose, more disorganized fiberous connective tissue, which may have less influence on mechanics.

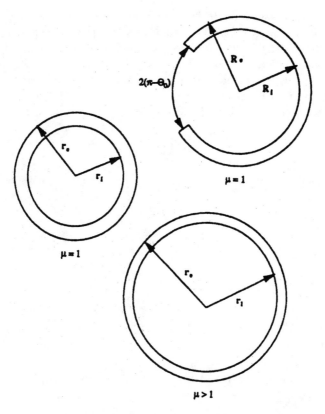

FIGURE 2.1 Cylindrical geometry of a blood vessel: top: stress-free reference configuration; middle: fully retracted vessel free of external traction; bottom: vessel *in situ* under longitudinal tether and internal pressurization.

2.3 Axisymmetric Deformation

In the following discussion we will concern ourselves with deformation of cylindrical tubes (see Fig. 2.1). Blood vessels tend to be nearly cylindrical *in situ* and tend to remain cylindrical when a cylindrical section is excised and studied *in vitro*. Only when the vessel is dissected further does the geometry begin to deviate from cylindrical. For this deformation there is a unique coordinate mapping:

$$\left(R, \Theta, Z\right) \rightarrow \left(r, \theta, z\right) \tag{2.1}$$

where the undeformed coordinates are given by (R, Θ, Z) and the deformed coordinates are given by (r, θ, z). The deformation is given by a set of restricted functions

$$r = r\left(R\right) \tag{2.2}$$

$$\theta = \beta\Theta \tag{2.3}$$

$$z = \mu Z + C_1 \tag{2.4}$$

where the constants μ and β have been introduced to account for a uniform longitudinal strain and a symmetric residual strain that are both independent of the coordinate Θ.

If $\beta = 1$, there is no residual strain. If $\beta \neq 1$, residual stresses and strains are present. If $\beta > 1$, a longitudinal cut through the wall will cause the blood vessel to open up, and the new cross-section will form a *c*-shaped section of an annulus with larger internal and external radii. If $\beta < 1$, the cylindrical shape is unstable, but a thin section will tend to overlap itself. In Choung and Fung's formulation, $\beta = \pi/\Theta_o$, where the angle Θ_o is half the angle spanned by the open annular section [4].

For cylindrical blood vessels there are two assumed constraints. The first assumption is that the longitudinal strain is uniform through the wall and therefore

$$\lambda_z = \mu = \text{a constant} \tag{2.5}$$

for any cylindrical configuration. Given this, the principal stretch ratios are computed from the above function as

$$\lambda_r = \frac{dr}{dR} \tag{2.6}$$

$$\lambda_\theta = \beta \frac{r}{R} \tag{2.7}$$

$$\lambda_z = \mu \tag{2.8}$$

The second assumption is wall incompressibility, which can be expressed by

$$\lambda_r \lambda_\theta \lambda_z \equiv 1 \tag{2.9}$$

or

$$\beta\mu \frac{r}{R} \frac{dr}{dR} = 1 \tag{2.10}$$

and therefore

$$r\,dr = \frac{1}{\beta\mu} R\,dR \tag{2.11}$$

Integration of this expression yields the solution

$$r^2 = \frac{1}{\beta\mu} R^2 + c_2 \tag{2.12}$$

where

$$c_2 = r_e^2 - \frac{1}{\beta\mu} R_e^2 \tag{2.13}$$

As a result, the principal stretch ratios can be expressed in terms of R as follows:

$$\lambda_r = \frac{R}{\sqrt{\beta\mu\left(R^2 + \beta\mu c_2\right)}} \tag{2.14}$$

$$\lambda_\theta = \sqrt{\frac{1}{\beta\mu} + \frac{c_2}{R^2}} \tag{2.15}$$

2.4 Experimental Measurements

The basic experimental setup required to measure the mechanical properties of blood vessels *in vitro* is described in [7]. It consists of a temperature-regulated bath of physiologic saline solution to maintain immersed cylindrical blood vessel segments, devices to measure diameter, an apparatus to hold the vessel at a constant longitudinal extension and to measure longitudinal distending force, and a system to deliver and control the internal pressure of the vessel with 100% oxygen. Typical data obtained from this type of experiment are shown in Figs. 2.2 and 2.3.

2.5 Equilibrium

When blood vessels are excised, they retract both longitudinally and circumferentially. Restoration to natural dimensions requires the application of internal pressure, p_i, and a longitudinal tether force, F_T. The internal pressure and longitudinal tether are balanced by the development of forces within the vessel wall. The internal pressure is balanced in the circumferential direction by a wall tension, T. The longitudinal tether force and pressure are balanced by the retractive force of the wall, F_R:

$$T = p_i r_i \tag{2.16}$$

$$F_R = F_T + p_i \pi r_i^2 \tag{2.17}$$

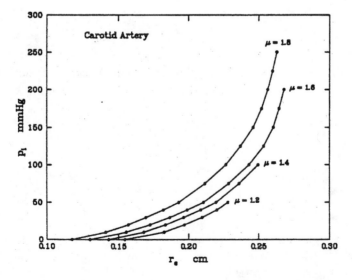

FIGURE 2.2 Pressure-radius curves for the canine carotid artery at various degrees of longitudinal extension.

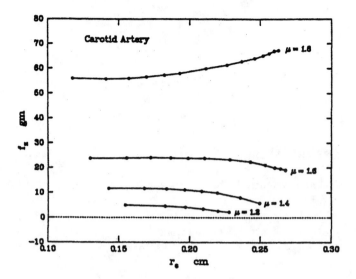

FIGURE 2.3 Longitudinal distending force as a function of radius at various degrees of longitudinal extension.

The first equation is the familiar law of Laplace for a cylindrical tube with internal radius r_i. It indicates that the force due to internal pressure, p_i, must be balanced by a tensile force (per unit length), T, within the wall. This tension is the integral of the circumferentially directed force intensity (or stress, σ_θ) across the wall:

$$T = \int_{r_i}^{r_e} \sigma_\theta dr = \overline{\sigma}_\theta h \qquad (2.18)$$

where $\overline{\sigma}_\theta$ is the mean value of the circumferential stress and h is the wall thickness. Similarly, the longitudinal tether force, F_T, and extending force due to internal pressure are balanced by a retractive internal force, F_R, due to axial stress, σ_z, in the blood vessel wall:

$$F_R = 2\pi \int_{r_i}^{r_e} \sigma_z r dr = \overline{\sigma}_z \pi h (r_e + r_i) \qquad (2.19)$$

where $\overline{\sigma}_z$ is the mean value of this longitudinal stress. The mean stresses are calculated from the above equation as

$$\overline{\sigma}_\theta = p_i \frac{r_i}{h} \qquad (2.20)$$

$$\overline{\sigma}_z = \frac{F_T}{\pi h (r_e + r_i)} + \frac{p_i}{2} \frac{r_i}{h} \qquad (2.21)$$

The mean stresses are a fairly good approximation for thin-walled tubes where the variations through the wall are small. However, the range of applicability of the thin-wall assumption depends upon the material properties and geometry. In a linear elastic material, the variation in σ_θ is less than 5% for $r/h > 20$. When the material is nonlinear or the deformation is large, the variations in stress can be more severe (see Fig. 2.10).

The stress distribution is determined by solving the equilibrium equation,

$$\frac{1}{r}\frac{d}{dr}\left(r\sigma_r\right)-\frac{\sigma_\theta}{r}=0 \tag{2.22}$$

This equation governs how the two stresses are related and must change in the cylindrical geometry. For uniform extension and internal pressurization, the stresses must be functions of a single radial coordinate, r, subject to the two boundary conditions for the radial stress:

$$\sigma_r\left(r_i,\mu\right)=-p_i \tag{2.23}$$

$$\sigma_r\left(r_e,\mu\right)=0 \tag{2.24}$$

2.6 Strain Energy Density Functions

Blood vessels are able to maintain their structural stability and contain steady oscillating internal pressures. This property suggests a strong elastic component, which has been called the *pseudoelasticity* [10]. This elastic response can be characterized by a single potential function called the *strain energy density*. It is a scalar function of the strains that determines the amount of stored elastic energy per unit volume. In the case of a cylindrically orthotropic tube of incompressible material, the strain energy density can be written in the following functional form:

$$W = W^*\left(\lambda_r, \lambda_\theta, \lambda_z\right)+\lambda_r\lambda_\theta\lambda_z p \tag{2.25}$$

where p is a scalar function of position, R. The stresses are computed from the strain energy by the following:

$$\sigma_i=\lambda_i\frac{\partial W^*}{\partial\lambda_i}+p \tag{2.26}$$

We make the following transformation [3]:

$$\lambda = \frac{\beta r}{\sqrt{\beta\mu\left(r^2-c_2\right)}} \tag{2.27}$$

which upon differentiation gives

$$r\frac{d\lambda}{dr}=\beta^{-1}\left(\beta\lambda-\mu\lambda^3\right) \tag{2.28}$$

After these expressions and the stresses in terms of the strain energy density function are introduced into the equilibrium equation, we obtain an ordinary differential equation for p:

$$\frac{dp}{d\lambda}=\frac{\beta\ W^*_{,\lambda_\theta}-W^*_{,\lambda_r}}{\beta\lambda=\mu\lambda^3}-\frac{dW^*_{,\lambda_r}}{d\lambda} \tag{2.29}$$

subject to the boundary conditions:

$$p(R_i) = p_i \tag{2.30}$$

$$p(R_e) = 0 \tag{2.31}$$

Isotropic Blood Vessels

A blood vessel generally exhibits anisotropic behavior when subjected to large variations in internal pressure and distending force. When the degree of anisotropy is small, the blood vessel may be treated as isotropic. For isotropic materials it is convenient to introduce the strain invariants:

$$I_1 = \lambda_r^2 + \lambda_\theta^2 + \lambda_z^2 \tag{2.32}$$

$$I_2 = \lambda_r^2 \lambda_\theta^2 + \lambda_\theta^2 \lambda_z^2 + \lambda_z^2 \lambda_r^2 \tag{2.33}$$

$$I_3 = \lambda_r^2 \lambda_\theta^2 \lambda_z^2 \tag{2.34}$$

These are measures of strain that are independent of the choice of coordinates. If the material is incompressible

$$I_3 = j^2 \equiv 1 \tag{2.35}$$

and the strain energy density is a function of the first two invariants, then

$$W = W(I_1, I_2). \tag{2.36}$$

The least complex form for an incompressible material is the first-order polynomial, which was first proposed by Mooney to characterize rubber:

$$W^* = \frac{G}{2}\left[(I_1 - 3) + k(I_2 - 3)\right] \tag{2.37}$$

It involves only two elastic constants. A special case, where $k = 0$, is the neo-Hookean material, which can be derived from thermodynamics principles for a simple solid. Exact solutions can be obtained for the cylindrical deformation of a thick-walled tube. In the case where there is no residual strain, we have the following:

$$p = -G(1 + k\mu^2)\left[\frac{\log \lambda}{\mu} + \frac{1}{2\mu^2 \lambda^2}\right] + c_0 \tag{2.38}$$

$$\sigma_r = G\left[\frac{1}{\lambda^2 \mu^2} + k\left(\frac{1}{\mu^2} + \frac{1}{\lambda^2}\right)\right] + p \tag{2.39}$$

$$\sigma_\theta = G\left[\lambda^2 + k\left(\frac{1}{\mu^2} + \lambda^2 \mu^2\right)\right] + p \tag{2.40}$$

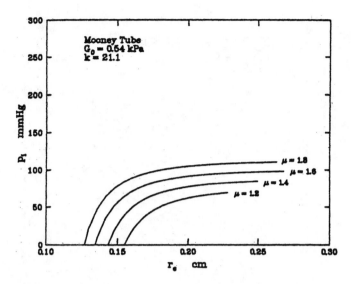

FIGURE 2.4   Pressure-radius curves for a Mooney–Rivlin tube with the approximate dimensions of the carotid.

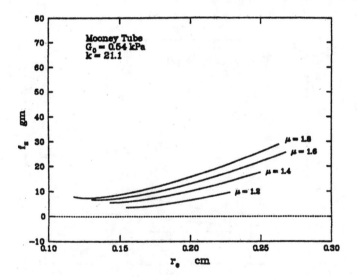

FIGURE 2.5   Longitudinal distending force as a function of radius for the Mooney–Rivlin tube.

$$\sigma_z = G\left[\mu^2 + k\left(\lambda^2\mu^2 + \frac{1}{\lambda^2}\right)\right] + p \tag{2.41}$$

However, these equations predict stress softening for a vessel subjected to internal pressurization at fixed lengths, rather than the stress stiffening observed in experimental studies on arteries and veins (see Figs. 2.4 and 2.5).

An alternative isotropic strain energy density function which can predict the appropriate type of stress stiffening for blood vessels is an exponential where the arguments is a polynomial of the strain invariants. The first-order form is given by

$$W^* = \frac{G_0}{2k_1} \exp\left[k_1\left(I_1 - 3\right) + k_2\left(I_2 - 3\right)\right] \tag{2.42}$$

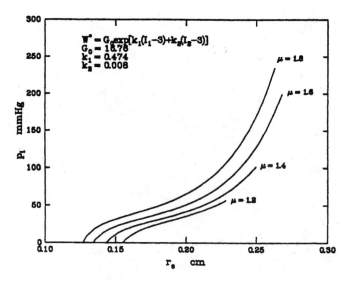

FIGURE 2.6 Pressure-radius curves for tube with the approximate dimensions of the carotid calculated using an isotropic exponential strain energy density function.

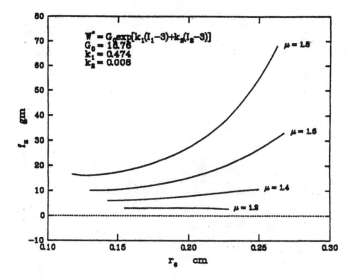

FIGURE 2.7 Longitudinal distending force as a function of radius for the isotropic tube.

This requires the determination of only two independent elastic constants. The third, G_0, is introduced to facilitate scaling of the argument of the exponent (see Figs. 2.6 and 2.7). This exponential form is attractive for several reasons. It is a natural extension of the observation that biologic tissue stiffness is proportional to the load in simple elongation. This stress stiffening has been attributed to a statistical recruitment and alignment of tangled and disorganized long chains of proteins. The exponential forms resemble statistical distributions derived from these same arguments.

Anisotropic Blood Vessels

Studies of the orthotropic behavior of blood vessels may employ polynomial or exponential strain energy density functions that include all strain terms or extension ratios. In particular, the strain energy density function can be of the form:

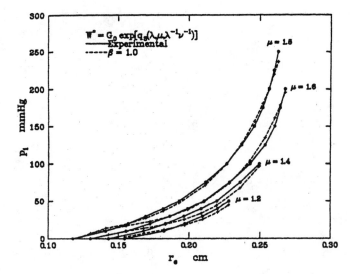

FIGURE 2.8 Pressure-radius curves for a fully orthotropic vessel calculated with an exponential strain energy density function.

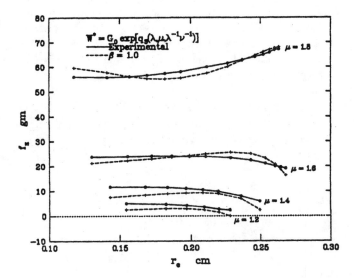

FIGURE 2.9 Longitudinal distending force as a function of radius for the orthotropic vessel.

$$W^* = q_n\left(\lambda_r, \lambda_\theta, \lambda_z\right) \tag{2.43}$$

or

$$W^* = e^{q_n\left(\lambda_r, \lambda_\theta, \lambda_z\right)} \tag{2.44}$$

where q_n is a polynomial of order n. Since the material is incompressible, the explicit dependence upon λ_r can be eliminated either by substituting $\lambda_r = \lambda_\theta^{-1} \lambda_z^{-1}$ or by assuming that the wall is thin and hence that the contribution of these terms is small. Figures 2.8 and 2.9 illustrate how well the experimental data can be fitted to an exponential strain density function whose argument is a polynomial of order $n = 3$.

Care must be taken to formulate expressions that will lead to stresses that behave properly. For this reason it is convenient to formulate the strain energy density in terms of the Lagrangian strains:

$$e_i = 1/2\left(\lambda_i^2 - 1\right) \tag{2.45}$$

and in this case we can consider polynomials of the lagrangian strains, $q_n(e_r, e_\theta, e_z)$.

Vaishnav et al. [15] proposed using a polynomial of the form:

$$W^* = \sum_{i=2}^{n} \sum_{j=0}^{i} a_{ij-i} e_\theta^{i-j} e_z^{j} \tag{2.46}$$

to approximate the behavior of the canine aorta. They found better correlation with order-three polynomials over order-two, but order-four polynomials did not warrant the additional work.

Later, Fung et al. [10] found very good correlation with an expression of the form:

$$W = \frac{C}{2} \exp\left[a_1\left(e_\theta^2 - e_z^{*2}\right) + a_2\left(e_z^2 - e_z^{*2}\right) + 2a_4\left(e_\theta e_z - e_\theta^* e_z^*\right)\right] \tag{2.47}$$

for the canine carotid artery, where e_θ^* and e_z^* are the strains in a reference configuration at *in situ* length and pressure. Why should this work? One answer appears to be related to residual stresses and strains.

When residual stresses are ignored, large-deformation analysis of thick-walled blood vessels predicts steep distributions in σ_θ and σ_z through the vessel wall, with the highest stresses at the interior. This prediction is considered significant because high tensions in the inner wall could inhibit vascularization and oxygen transport to vascular tissue.

When residual stresses are considered, the stress distributions flatten considerably and become almost uniform at *in situ* length and pressure. Figure 2.10 shows the radial stress distributions computed for a vessel with $\beta = 1$ and $\beta = 1.11$. Takamizawa and Hayashi have even considered the case where the strain distribution is uniform *in situ* [13]. The physiologic implications are that vascular tissue is in a constant

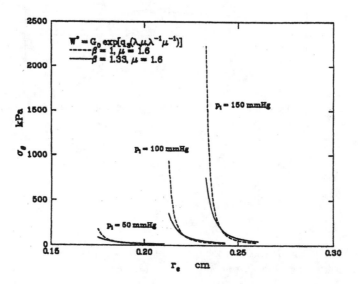

FIGURE 2.10 Stress distributions through the wall at various pressures for the orthotropic vessel.

state of flux. New tissue is synthesized in a state of stress that allows it to redistribute the internal loads more uniformly. There probably is no stress-free reference state [8, 11, 12]. Continuous dissection of the tissue into smaller and smaller pieces would continue to relieve residual stresses and strains [14].

References

1. Bergel DH. 1961. The static elastic properties of the arterial wall. *J Physiol* 156:445.
2. Carew TE, Vaishnav RN, Patel DJ. 1968. Compressibility of the arterial walls. *Circ Res* 23:61.
3. Chu BM, Oka S. 1973. Influence of longitudinal tethering on the tension in thick-walled blood vessels in equilibrium. *Biorheology* 10:517.
4. Choung CJ, Fung YC. 1986. On residual stresses in arteries. *J Biomed Eng* 108:189.
5. Dobrin PB. 1978. Mechanical properties of arteries. *Physiol Rev* 58:397.
6. Dobrin PB, Canfield TR. 1984. Elastase, collagenase, and the biaxial elastic properties of dog carotid artery. *Am J Physiol* 2547:H124.
7. Dobrin PB, Rovick AA. 1969. Influence of vascular smooth muscle on contractile mechanics and elasticity of arteries. *Am J Physiol* 217:1644.
8. Dobrin PD, Canfield T, Sinha S. 1975. Development of longitudinal retraction of carotid arteries in neonatal dogs. *Experientia* 31:1295.
9. Doyle JM, Dobrin PB. 1971. Finite deformation of the relaxed and contracted dog carotid artery. *Microvasc Res* 3:400.
10. Fung YC, Fronek K, Patitucci P. 1979. Pseudoelasticity of arteries and the choice of its mathematical expression. *Am J Physiol* 237:H620.
11. Fung YC, Liu SQ, Zhou JB. 1993. Remodeling of the constitutive equation while a blood vessel remodels itself under strain. *J Biomech Eng* 115:453.
12. Rachev A, Greenwald S, Kane T, Moore J, Meister J-J. 1994. Effects of age-related changes in the residual strains on the stress distribution in the arterial wall. In J Vossoughi (Ed.), *Proceedings of the Thirteenth Society of Biomedical Engineering Recent Developments*, pp. 409–412, Washington, D.C., University of District of Columbia.
13. Takamizawa K, Hayashi K. 1987. Strain energy density function and the uniform strain hypothesis for arterial mechanics. *J Biomech* 20:7.
14. Vassoughi J. 1992. Longitudinal residual strain in arteries. Proc of the 11th South Biomed Engrg Conf, Memphis, TN.
15. Vaishnav RN, Young JT, Janicki JS, Patel DJ. 1972. Nonlinear anisotropic elastic properties of the canine aorta. *Biophys J* 12:1008.
16. Vaishnav RN, Vassoughi J. 1983. Estimation of residual stresses in aortic segments. In CW Hall (Ed.), *Biomedical Engineering*, II, *Recent Developments*, pp. 330–333, New York, Pergamon Press.
17. Von Maltzahn W-W, Desdo D, Wiemier W. 1981. Elastic properties of arteries: a nonlinear two-layer cylindrical model. *J Biomech* 4:389.
18. Wolinsky H, Glagov S. 1969. Comparison of abdominal and thoracic aortic media structure in mammals. *Circ Res* 25:677.

3

Joint-Articulating Surface Motion

Kenton R. Kaufman
Biomechanics Laboratory,
The Mayo Clinic

Kai-Nan An
Biomechanics Laboratory,
The Mayo Clinic

Knowledge of joint-articulating surface motion is essential for design of prosthetic devices to restore function; assessment of joint wear, stability, and degeneration; and determination of proper diagnosis and surgical treatment of joint disease. In general, kinematic analysis of human movement can be arranged into two separate categories: (1) gross movement of the limb segments interconnected by joints, or (2) detailed analysis of joint articulating surface motion which is described in this chapter. Gross movement is the relative three-dimensional joint rotation as described by adopting the Eulerian angle system. Movement of this type is described in Chapter 8: Analysis of Gait. In general, the three-dimensional unconstrained rotation and translation of an articulating joint can be described utilizing the concept of the screw displacement axis. The most commonly used analytic method for the description of 6-degree-of-freedom displacement of a rigid body is the screw displacement axis [Kinzel et al. 1972; Spoor and Veldpaus, 1980; Woltring et al. 1985].

Various degrees of simplification have been used for kinematic modeling of joints. A hinged joint is the simplest and most common model used to simulate an anatomic joint in planar motion about a single axis embedded in the fixed segment. Experimental methods have been developed for determination

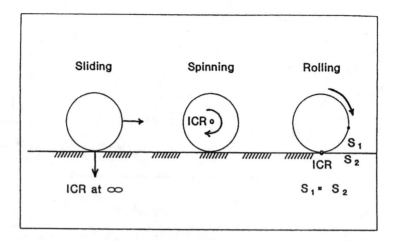

FIGURE 3.1   Three types of articulating surface motion in human joints.

of the instantaneous center of rotation for planar motion. The *instantaneous center of rotation* is defined as the point of zero velocity. For a true hinged motion, the instantaneous center of rotation will be a fixed point throughout the movement. Otherwise, loci of the instantaneous center of rotation or centrodes will exist. The center of curvature has also been used to define joint anatomy. The *center of curvature* is defined as the geometric center of coordinates of the articulating surface.

For more general planar motion of an articulating surface, the term *sliding, rolling,* and *spinning* are commonly used (Fig. 3.1). Sliding (gliding) motion is defined as the pure translation of a moving segment against the surface of a fixed segment. The contact point of the moving segment does not change, while the contact point of the fixed segment has a constantly changing contact point. If the surface of the fixed segment is flat, the instantaneous center of rotation is located at infinity. Otherwise, it is located at the center of curvature of the fixed surface. Spinning motion (rotation) is the exact opposite of sliding motion. In this case, the moving segment rotates, and the contact points on the fixed surface does not change. The instantaneous center of rotation is located at the center of curvature of the spinning body that is undergoing pure rotation. Rolling motion occurs between moving and fixed segments where the contact points in each surface are constantly changing and the arc lengths of contact are equal on each segment. The instantaneous center of rolling motion is located at the contact point. Most planar motion of anatomic joints can be described by using any two of these three basic descriptions.

In this chapter, various aspects of joint-articulating motion are covered. Topics include the anatomical characteristics, joint contact, and axes of rotation. Joints of both the upper and lower extremity are discussed.

3.1   Ankle

The ankle joint is composed of two joints: the talocrural (ankle) joint and the talocalcaneal (subtalar joint). The talocrural joint is formed by the articulation of the distal tibia and fibula with the trochlea of the talus. The talocalcaneal joint is formed by the articulation of the talus with the calcaneus.

Geometry of the Articulating Surfaces

The upper articular surface of the talus is wedge-shaped, its width diminishing from front to back. The talus can be represented by a conical surface. The wedge shape of the talus is about 25% wider in front than behind with an average difference of 2.4 mm ± 1.3 mm and a maximal difference of 6 mm [Inman, 1976].

TABLE 3.1 Talocalcaneal (Ankle) Joint Contact Area

Investigators	Plantarflexion	Neutral	Dorsiflexion
Ramsey and Hamilton [1976]		4.40 ± 1.21	
Kimizuka et al. [1980]		4.83	
Libotte et al. [1982]	5.01 (30°)	5.41	3.60 (30°)
Paar et al. [1983]	4.15 (10°)	4.15	3.63 (10°)
Macko et al. [1991]	3.81 ± 0.93 (15°)	5.2 ± 0.94	5.40 ± 0.74 (10°)
Driscoll et al. [1994]	2.70 ± 0.41 (20°)	3.27 ± 0.32	2.84 ± 0.43 (20°)
Hartford et al. [1995]		3.37 ± 0.52	
Pereira et al. [1996]	1.49 (20°)	1.67	1.47 (10°)

Note: The contact area is expressed in square centimeters.

Joint Contact

The talocrural joint contact area varies with flexion of the ankle (Table 3.1). During plantarflexion, such as would occur during the early stance phase of gait, the contact area is limited and the joint is incongruous. As the position of the joint progresses from neutral to dorsiflexion, as would occur during the midstance of gait, the contact area increases and the joint becomes more stable. The area of the subtalar articulation is smaller than that of the talocrural joint. The contact area of the subtalar joint is 0.89 ± 0.21 cm^2 for the posterior facet and 0.28 ± 15 cm^2 for the anterior and middle facets [Wang et al., 1994]. The total contact area (1.18 ± 0.35 cm^2) is only 12.7% of the whole subtalar articulation area (9.31 ± 0.66 cm^2) [Wang et al., 1994]. The contact area/joint area ratio increases with increases in applied load (Fig. 3.2).

Axes of Rotation

Joint motion of the talocrural joint has been studied to define the axes of rotation and their location with respect to specific anatomic landmarks (Table 3.2). The axis of motion of the talocrural joint essentially passes through the inferior tibia at the fibular and tibial malleoli (Fig. 3.3). Three types of motion have been used to describe the axes of rotation: fixed, quasi-instantaneous, and instantaneous axes. The motion that occurs in the ankle joints consists of dorsiflexion and plantarflexion. Minimal or no transverse rotation takes place within the talocrural joint. The motion in the talocrural joint is intimately related to the motion in the talocalcaneal joint which is described next.

The motion axes of the talocalcaneal joint have been described by several authors (Table 3.3). The axis of motion in the talocalcaneal joint passes from the anterior medial superior aspect of the navicular bone to the posterior lateral inferior aspect of the calcaneus (Fig. 3.4). The motion that occurs in the talocalcaneal joint consists of inversion and eversion.

3.2 Knee

The knee is the intermediate joint of the lower limb. It is composed of the distal femur and proximal tibia. It is the largest and most complex joint in the body. The knee joint is composed of the tibiofemoral articulation and the patellofemoral articulation.

Geometry of the Articulating Surfaces

The shape of the articular surfaces of the proximal tibia and distal femur must fulfill the requirement that they move in contact with one another. The profile of the femoral condyles varies with the condyle examined (Fig. 3.5 and Table 3.4). The tibial plateau widths are greater than the corresponding widths of the femoral condyles (Fig. 3.6 and Table 3.5). However, the tibial plateau depths are less than those

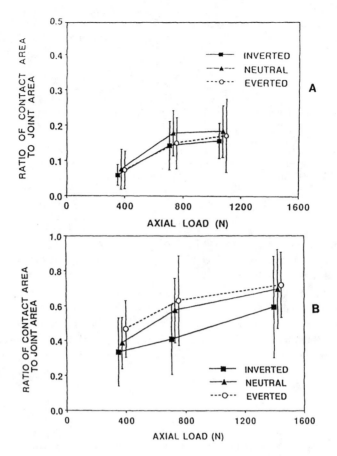

FIGURE 3.2 Ratio of total contact area to joint area in the (A) anterior/middle facet and (B) posterior facet of the subtalar joint as a function of applied axial load for three different positions of the foot. (*Source:* Wagner UA, Sangeorzan BJ, Harrington RM, Tencer AF. 1992. Contact characteristics of the subtalar joint: load distribution between the anterior and posterior facets. *J Orthop Res* 10:535. With permission.)

of the femoral condyle distances. The medial condyle of the tibia is concave superiorly (the center of curvature lies above the tibial surface) with a radius of curvature of 80 mm [Kapandji, 1987]. The lateral condyle is convex superiorly (the center of curvature lies below the tibial surface) with a radius of curvature of 70 mm [Kapandji, 1987]. The shape of the femoral surfaces is complementary to the shape of the tibial plateaus. The shape of the posterior femoral condyles may be approximated by spherical surfaces (Table 3.4).

The geometry of the patellofemoral articular surfaces remains relatively constant as the knee flexes. The knee sulcus angle changes only ±3.4° from 15 to 75° of knee flexion (Fig. 3.7). The mean depth index varies by only ±4% over the same flexion range (Fig. 3.7). Similarly, the medial and lateral patellar facet angles (Fig. 3.8) change by less than a degree throughout the entire knee flexion range (Table 3.6). However, there is a significant difference between the magnitude of the medial and lateral patellar facet angles.

Joint Contact

The mechanism of movement between the femur and tibia is a combination of rolling and gliding. Backward movement of the femur on the tibia during flexion has long been observed in the human knee. The magnitude of the rolling and gliding changes through the range of flexion. The tibial–femoral contact

TABLE 3.2   Axis of Rotation for the Ankle

Investigator	Axis[a]	Position
Elftman [1945]	Fix.	67.6° ± 7.4° with respect to sagittal plane
Isman and Inman [1969]	Fix.	8 mm anterior, 3 mm inferior to the distal tip of the lateral malleolus; 1 mm posterior, 5 mm inferior to the distal tip of the medial malleolus
Inman and Mann [1979]	Fix.	79° (68–88°) with respect to the sagittal plane
Allard et al. [1987]	Fix.	95.4° ± 6.6° with respect to the frontal plane, 77.7° ± 12.3° with respect to the sagittal plane, and 17.9° ± 4.5° with respect to the transverse plane
Singh et al. [1992]	Fix.	3.0 mm anterior, 2.5 mm inferior to distal tip of lateral malleolus, 2.2 mm posterior, 10 mm inferior to distal tip of medial malleolus
Sammarco et al. [1973]	Ins.	Inside and outside the body of the talus
D'Ambrosia et al. [1976]	Ins.	No consistent pattern
Parlasca et al. [1979]	Ins.	96% within 12 mm of a point 20 mm below the articular surface of the tibia along the long axis.
Van Langelaan [1983]	Ins.	At an approximate right angle to the longitudinal direction of the foot, passing through the corpus tali, with a direction from anterolaterosuperior to posteromedioinferior
Barnett and Napier [1952]	Q-I	Dorsiflexion: down and lateral Plantarflexion: down and medial
Hicks [1953]	Q-I	Dorsiflexion: 5 mm inferior to tip of lateral malleolus to 15 mm anterior to tip of medial malleolus Plantarflexion: 5 mm superior to tip of lateral malleolus to 15 mm anterior, 10 mm inferior to tip of medial malleolus

[a] Fix. = fixed axis of rotation; Ins. = instantaneous axis of rotation; Q-I = quasi-instantaneous axis of rotation

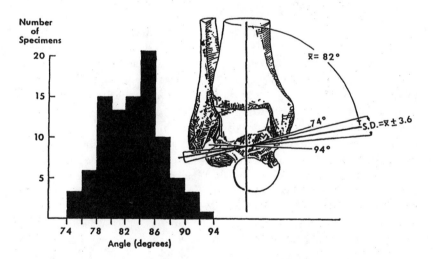

FIGURE 3.3   Variations in angle between middle of tibia and empirical axis of ankle. The histogram reveals a considerable spread of individual values. (*Source:* Inman VT. 1976. *The Joints of the Ankle*, Baltimore, Williams & Wilkins. With permission.)

point has been shown to move posteriorly as the knee is flexed, reflecting the coupling of anterior/ posterior motion with flexion/extension (Fig. 3.9). During flexion, the weight-bearing surfaces move backward on the tibial plateaus and become progressively smaller (Table 3.7).

It has been shown that in an intact knee at full extension the center of pressure is approximately 25 mm from the anterior edge of the knee joint line [Andriacchi et al., 1986]. This net contact point moves posteriorly with flexion to approximately 38.5 mm from the anterior edge of the knee joint. Similar displacements have been noted in other studies (Table 3.8).

TABLE 3.3 Axis of Rotation for the Talocalcaneal (Subtalar) Joint

Investigator	Axis[a]	Position
Manter [1941]	Fix.	16° (8–24°) with respect to sagittal plane, and 42° (29–47°) with respect to transverse plane
Shephard [1951]	Fix.	Tuberosity of the calcaneus to the neck of the talus
Hicks [1953]	Fix.	Posterolateral corner of the heel to superomedial aspect of the neck of the talus
Root et al. [1966]	Fix.	17° (8–29°) with respect to sagittal plane, and 41° (22–55°) with respect to transverse plane
Isman and Inman [1969]	Fix.	23° ± 11° with respect to sagittal plane, and 41° ± 9° with respect to transverse plane
Kirby [1947]	Fix.	Extends from the posterolateral heel, posteriorly, to the first intermetatarsal space, anteriorly
Rastegar et al. [1980]	Ins.	Instant centers of rotation pathways in posterolateral quadrant of the distal articulating tibial surface, varying with applied load
Van Langelaan [1983]	Ins.	A bundle of axes that make an acute angle with the longitudinal direction of the foot passing through the tarsal canal having a direction from anteromediosuperior to posterolateroinferior
Engsberg [1987]	Ins.	A bundle of axes with a direction from anteromediosuperior to posterolateroinferior

[a] Fix. = fixed axis of rotation; Ins. = instantaneous axis of rotation

The patellofemoral contact area is smaller than the tibiofemoral contact area (Table 3.9). As the knee joint moves from extension to flexion, a band of contact moves upward over the patellar surface (Fig. 3.10). As knee flexion increases, not only does the contact area move superiorly, but it also becomes larger. At 90° of knee flexion, the contact area has reached the upper level of the patella. As the knee continues to flex, the contact area is divided into separate medial and lateral zones.

Axes of Rotation

The tibiofemoral joint is mainly a joint with two degrees of freedom. The first degree of freedom allows movements of flexion and extension in the sagittal plane. The axis of rotation lies perpendicular to the sagittal plane and intersects the femoral condyles. Both fixed axes and screw axes have been calculated (Fig. 3.11). In Fig. 3.11, the optimal axes are fixed axes, whereas the screw axis is an instantaneous axis. The symmetric optimal axis is constrained such that the axis is the same for both the right and left knee. The screw axis may sometimes coincide with the optimal axis but not always, depending upon the motions of the knee joint. The second degree of freedom is the axial rotation around the long axis of the tibia. Rotation of the leg around its long axis can only be performed with the knee flexed. There is also an automatic axial rotation which is involuntarily linked to flexion and extension. When the knee is flexed, the tibia internally rotates. Conversely, when the knee is extended, the tibia externally rotates.

During knee flexion, the patella makes a rolling/gliding motion along the femoral articulating surface. Throughout the entire flexion range, the gliding motion is clockwise (Fig. 3.12). In contrast, the direction of the rolling motion is counter-clockwise between 0° and 90° and clockwise between 90° and 120° (Fig. 3.12). The mean amount of patellar gliding for all knees is approximately 6.5 mm per 10° of flexion between 0° and 80° and 4.5 mm per 10° of flexion between 80° and 120°. The relationship between the angle of flexion and the mean rolling/gliding ratio for all knees is shown in Fig. 3.13. Between 80° and 90° of knee flexion, the rolling motion of the articulating surface comes to a standstill and then changes direction. The reversal in movement occurs at the flexion angle where the quadriceps tendon first contacts the femoral groove.

3.3 Hip

The hip joint is composed of the head of the femur and the acetabulum of the pelvis. The hip joint is one of the most stable joints in the body. The stability is provided by the rigid ball-and-socket configuration.

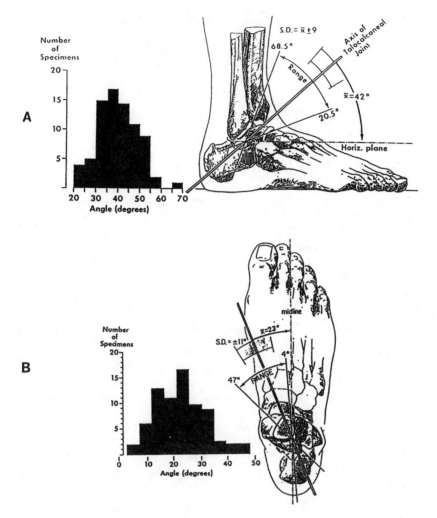

FIGURE 3.4 (A) Variations in inclination of axis of subtalar joint as projected upon the sagittal plane. The distribution of the measurements on the individual specimens is shown in the histogram. The single observation of an angle of almost 70° was present in a markedly cavus foot. (B) Variations in position of subtalar axis as projected onto the transverse plane. The angle was measured between the axis and the midline of the foot. The extent of individual variation is shown on the sketch and revealed in the histogram. (*Source:* Inman VT. 1976. *The Joints of the Ankle*, Baltimore, Williams & Wilkins. With permission.)

Geometry of the Articulating Surfaces

The femoral head is spherical in its articular portion which forms two-thirds of a sphere. The diameter of the femoral head is smaller for females than for males (Table 3.10). In the normal hip, the center of the femoral head coincides exactly with the center of the acetabulum. The rounded part of the femoral head is spheroidal rather than spherical because the uppermost part is flattened slightly. This causes the load to be distributed in a ringlike pattern around the superior pole. The geometrical center of the femoral head is traversed by the three axes of the joint, the horizontal axis, the vertical axis, and the anterior/posterior axis. The head is supported by the neck of the femur, which joins the shaft. The axis of the femoral neck is obliquely set and runs superiorly, medially, and anteriorly. The angle of inclination of the femoral neck to the shaft in the frontal plane is the neck-shaft angle (Fig. 3.14). In most adults,

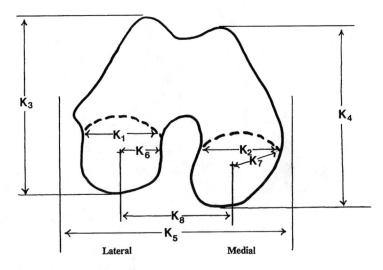

FIGURE 3.5 Geometry of distal femur. The distances are defined in Table 3.4.

TABLE 3.4 Geometry of the Distal Femur

	Condyle					
	Lateral		Medial		Overall	
Parameter	Symbol	Distance (mm)	Symbol	Distance (mm)	Symbol	Distance (mm)
Medial/lateral distance	K1	31 ± 2.3 (male) 28 ± 1.8 (female)	K2	32 ± 31 (male) 27 ± 3.1 (female)		
Anterior/posterior distance	K3	72 ± 4.0 (male) 65 ± 3.7 (female)	K4	70 ± 4.3 (male) 63 ± 4.5 (female)		
Posterior femoral condyle spherical radii	K6	19.2 ± 1.7	K7	20.8 ± 2.4		
Epicondylar width					K5	90 ± 6 (male) 80 ± 6 (female)
Medial/lateral spacing of center of spherical surfaces					K8	45.9 ± 3.4

Note: See Fig. 3.5 for location of measurements.
Sources: Yoshioka Y, Siu D, Cooke TDV. 1987. The anatomy of functional axes of the femur. *J Bone Joint Surg* 69A(6):873–880. Kurosawa H, Walker PS, Abe S, Garg A, Hunter T. 1985. Geometry and motion of the knee for implant and orthotic design. *J Biomech* 18(7):487.

this angle is about 130° (Table 3.10). An angle exceeding 130° is known as *coxa valga;* an angle less than 130° is known as *coxa vara*. The femoral neck forms an acute angle with the transverse axis of the femoral condyles. This angle faces medially and anteriorly and is called the *angle of anteversion* (Fig. 3.15). In the adult, this angle averages about 7.5° (Table 3.10).

The acetabulum receives the femoral head and lies on the lateral aspect of the hip. The acetabulum of the adult is a hemispherical socket. Its cartilage area is approximately 16 cm² [Von Lanz and Wauchsmuth, 1938]. Together with the labrum, the acetabulum covers slightly more than 50% of the femoral head [Tönnis, 1987]. Only the sides of the acetabulum are lined by articular cartilage, which is interrupted inferiorly by the deep acetabular notch. The central part of the cavity is deeper than the articular cartilage and is nonarticular. This part is called the *acetabular fossae* and is separated from the interface of the pelvic bone by a thin plate of bone.

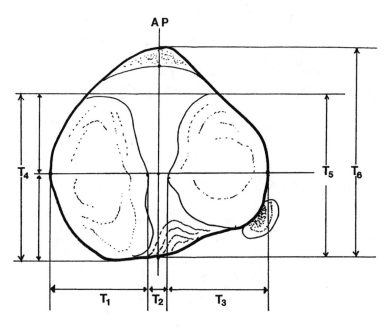

FIGURE 3.6 Contour of the tibial plateau (transverse plane). The distances are defined in Table 3.5.

TABLE 3.5 Geometry of the Proximal Tibia

Parameter	Symbol	All Limbs	Male	Female
Tibial plateau with widths (mm)				
Medial plateau	T_1	32 ± 3.8	34 ± 3.9	30 ± 22
Lateral plateau	T_3	33 ± 2.6	35 ± 1.9	31 ± 1.7
Overall width	$T_1 + T_2 + T_3$	76 ± 6.2	81 ± 4.5	73 ± 4.5
Tibial plateau depths (mm)				
AP depth, medial	T_4	48 ± 5.0	52 ± 3.4	45 ± 4.1
AP depth, lateral	T_5	42 ± 3.7	45 ± 3.1	40 ± 2.3
Interspinous width (mm)	T_2	12 ± 1.7	12 ± 0.9	12 ± 2.2
Intercondylar depth (mm)	T_6	48 ± 5.9	52 ± 5.7	45 ± 3.9

Source: Yoshioka Y, Siu D, Scudamore RA, Cooke TDV. 1989. Tibial anatomy in functional axes. *J Orthop Res* 7:132.

Joint Contact

Miyanaga et al. [1984] studied the deformation of the hip joint under loading, the contact area between the articular surfaces, and the contact pressures. They found that at loads up to 1000 N, pressure was distributed largely to the anterior and posterior parts of the lunate surface with very little pressure applied to the central portion of the roof itself. As the load increased, the contact area enlarged to include the outer and inner edges of the lunate surface (Fig. 3.16). However, the highest pressures were still measured anteriorly and posteriorly. Of five hip joints studied, only one had a pressure maximum at the zenith or central part of the acetabulum.

Davy et al. [1989] utilized a telemetered total hip prosthesis to measure forces across the hip after total hip arthroplasty. The orientation of the resultant joint contact force varies over a relatively limited range during the weight-load-bearing portions of gait. Generally, the joint contact force on the ball of the hip prosthesis is located in the anterior/superior region. A three-dimensional plot of the resultant joint force during the gait cycle, with crutches, is shown in Fig. 3.17.

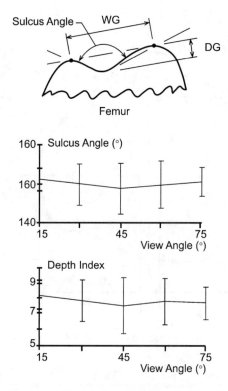

FIGURE 3.7   The trochlear geometry indices. The sulcus angle is the angle formed by the lines drawn from the top of the medial and lateral condyles to the deepest point of the sulcus. The depth index is the ratio of the width of the groove (WG) to the depth (DG). Mean and SD; $n = 12$. (*Source:* Farahmand F. et al. 1998. Quantitative study of the quadriceps muscles and trochlear groove geometry related to instability of the patellofemoral joint, *J Orthop Res* 16(1):140.)

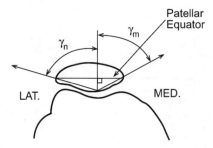

FIGURE 3.8   Medial (γ_m) and lateral (γ_n) patellar facet angles. (*Source:* Ahmed AM, Burke DL, Hyder A. 1987. Force analysis of the patellar mechanism. *J Orthop Res* 5:69–85.)

TABLE 3.6   Patellar Facet Angles

Facet Angle	Knee Flexion Angle				
	0°	30°	60°	90°	120°
γ_n (deg)	60.88	60.96	61.43	61.30	60.34
	3.89[a]	4.70	4.12	4.18	4.51
γ_m (deg)	67.76	68.05	69.36	68.39	68.20
	4.15	3.97	3.63	4.01	3.67

[a] SD

Source: Ahmed AM, Burke DL, Hyder A. 1987. Force analysis of the patellar mechanism. *J Orthop Res* 5:69–85.

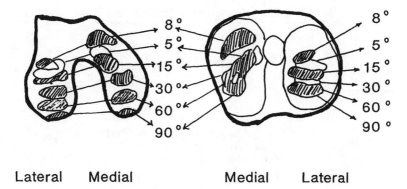

Lateral   Medial   Medial   Lateral

FIGURE 3.9 Tibio-femoral contact area as a function of knee flexion angle. (*Source:* Iseki F, Tomatsu T. 1976. The biomechanics of the knee joint with special reference to the contact area. *Keio J Med* 25:37. With permission.)

TABLE 3.7 Tibiofemoral Contact Area

Knee Flexion (deg)	Contact Area (cm²)
−5	20.2
5	19.8
15	19.2
25	18.2
35	14.0
45	13.4
55	11.8
65	13.6
75	11.4
85	12.1

Source: Maquet PG, Vandberg AJ, Simonet JC. 1975. Femorotibial weight bearing areas: experimental determination. *J Bone Joint Surg* 57A(6):766.

TABLE 3.8 Posterior Displacement of the Femur Relative to the Tibia

Author	Condition	A/P Displacement (mm)
Kurosawa [1985]	*In vitro*	14.8
Andriacchi [1986]	*In vitro*	13.5
Draganich [1987]	*In vitro*	13.5
Nahass [1991]	*In vivo* (walking)	12.5
	In vivo (stairs)	13.9

Axes of Rotation

The human hip is a modified spherical (ball-and-socket) joint. Thus, the hip possesses three degrees of freedom of motion with three correspondingly arranged, mutually perpendicular axes that intersect at the geometric center of rotation of the spherical head. The transverse axis lies in the frontal plane and controls movements of flexion and extension. An anterior/posterior axis lies in the sagittal plane and controls movements of adduction and abduction. A vertical axis which coincides with the long axis of the limb when the hip joint is in the neutral position controls movements of internal and external rotation. Surface motion in the hip joint can be considered as spinning of the femoral head on the

TABLE 3.9   Patellofemoral Contact Area

Knee Flexion (degrees)	Contact Area (cm²)
20	2.6 ± 0.4
30	3.1 ± 0.3
60	3.9 ± 0.6
90	4.1 ± 1.2
120	4.6 ± 0.7

Source: Hubert HH, Hayes WC. 1984. Patello-femoral contact pressures: the influence of Q-angle and tendofemoral contact. *J Bone Joint Surg* 66A(5):715–725.

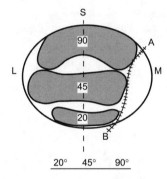

FIGURE 3.10   Diagrammatic representation of patella contact areas for varying degrees of knee flexion. (*Source:* Goodfellow J, Hungerford DS, Zindel M. 1976. Patellofemoral joint mechanics and pathology. *J Bone Joint Surg* 58B(3):288.)

acetabulum. The pivoting of the bone socket in three planes around the center of rotation in the femoral head produces the spinning of the joint surfaces.

3.4 Shoulder

The shoulder represents the group of structures connecting the arm to the thorax. The combined movements of four distinct articulations—glenohumeral, acromioclavicular, sternoclavicular, and scapu-lothoracic—allow the arm to be positioned in space.

Geometry of the Articulating Surfaces

The articular surface of the humerus is approximately one-third of a sphere (Fig. 3.18). The articular surface is oriented with an upward tilt of approximately 45° and is retroverted approximately 30° with respect to the condylar line of the distal humerus [Morrey and An, 1990]. The average radius of curvature of the humeral head in the coronal plane is 24.0 ± 2.1 mm [Iannotti et al., 1992]. The radius of curvature

LATERAL   POSTERIOR   MEDIAL

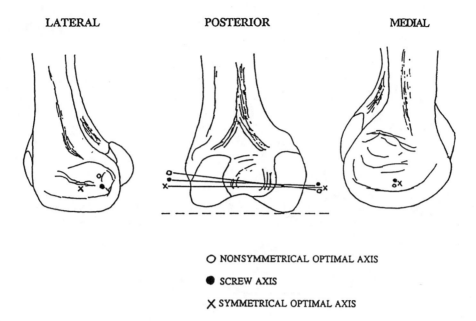

○ NONSYMMETRICAL OPTIMAL AXIS

● SCREW AXIS

✗ SYMMETRICAL OPTIMAL AXIS

FIGURE 3.11 Approximate location of the optimal axis (case 1—nonsymmetric, case 3—symmetric), and the screw axis (case 2) on the medial and lateral condyles of the femur of a human subject for the range of motion of 0–90° flexion (standing to sitting, respectively). (*Source:* Lewis JL, Lew WD. 1978. A method for locating an optimal "fixed" axis of rotation for the human knee joint. *J Biomech Eng* 100:187. With permission.)

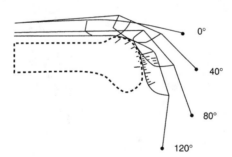

FIGURE 3.12 Position of patellar ligament, patella, and quadriceps tendon and location of the contact points as a function of the knee flexion angle. (*Source:* van Eijden TMGJ, Kouwenhoven E, Verburg J, et al. 1986. A mathematical model of the patellofemoral joint. *J Biomech* 19(3):227.)

in the anteroposterior and axillary-lateral view is similar, measuring 13.1 ± 1.3 mm and 22.9 ± 2.9 mm, respectively [McPherson et al., 1997]. The humeral articulating surface is spherical in the center. However, the peripheral radius is 2 mm less in the axial plane than in the coronal plane. Thus the peripheral contour of the articular surface is elliptical with a ratio of 0.92 [Iannotti et al., 1992]. The major axis is superior to inferior and the minor axis is anterior to posterior [McPherson et al., 1997]. More recently, the three-dimensional geometry of the proximal humerus has been studied extensively. The articular surface, which is part of a sphere, varies individually in its orientation with respect to inclination and retroversion, and it has variable medial and posterior offsets [Boileau and Walch, 1997]. These findings have great impact in implant design and placement in order to restore soft-tissue function.

The glenoid fossa consists of a small, pear-shaped, cartilage-covered bony depression that measures 39.0 ± 3.5 mm in the superior/inferior direction and 29.0 ± 3.2 mm in the anterior/posterior direction [Iannotti et al., 1992]. The anterior/posterior dimension of the glenoid is pear-shaped with the lower

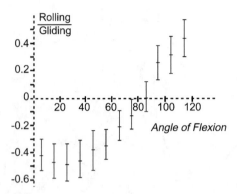

FIGURE 3.13   Calculated rolling/gliding ratio for the patellofemoral joint as a function of the knee flexion angle. (*Source:* van Eijden TMGJ, Kouwenhoven E, Verburg J, et al. 1986. A mathematical model of the patellofemoral joint. *J Biomech* 19(3):226.)

TABLE 3.10   Geometry of the Proximal Femur

Parameter	Females	Males
Femoral head diameter (mm)	45.0 ± 3.0	52.0 ± 3.3
Neck shaft angle (degrees)	133 ± 6.6	129 ± 7.3
Anteversion (degrees)	8 ± 10	7.0 ± 6.8

Source: Yoshioka Y, Siu D, Cooke TDV. 1987. The anatomy and functional axes of the femur. *J Bone Joint Surg* 69A(6):873.

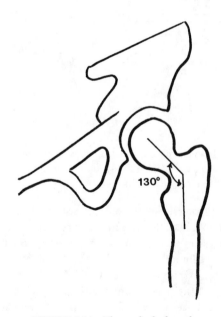

FIGURE 3.14   The neck-shaft angle.

half being larger than the top half. The ratio of the lower half to the top half is 1:0.80 ± 0.01 [Iannotti et al., 1992]. The glenoid radius of curvature is 32.2 ± 7.6 mm in the anteroposterior view and 40.6 ± 14.0 mm in the axillary-lateral view [McPherson et al., 1997]. The glenoid is therefore more curved

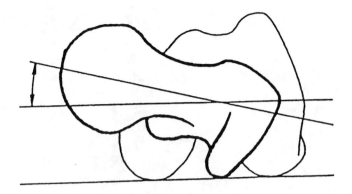

FIGURE 3.15 The normal anteversion angle formed by a line tangent to the femoral condyles and the femoral neck axis, as displayed in the superior view.

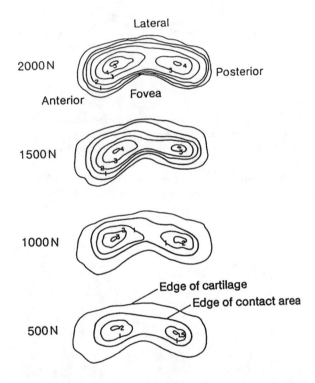

FIGURE 3.16 Pressure distribution and contact area of hip joint. The pressure is distributed largely to the anterior and posterior parts of the lunate surface. As the load increased, the contact area increased. (*Source:* Miyanaga Y, Fukubayashi T, Kurosawa H. 1984. Contact study of the hip joint: load deformation pattern, contact area, and contact pressure. *Arch Orth Trauma Surg* 103:13. With permission.)

superior to inferior (coronal plane) and relatively flatter in an anterior to posterior direction (sagittal plane). Glenoid depth is 5.0 ± 1.1 mm in the anteroposterior view and 2.9 ± 1.0 mm in the axillary-lateral [McPherson et al., 1997], again confirming that the glenoid is more curved superior to inferior. In the coronal plane the articular surface of the glenoid comprises an arc of approximately 75° and in the transverse plane the arc of curvature of the glenoid is about 50° [Morrey and An, 1990]. The glenoid has a slight upward tilt of about 5° [Basmajian and Bazant, 1959] with respect to the medial border of the scapula (Fig. 3.19) and is retroverted a mean of approximately 7° [Saha, 1971]. The relationship of the

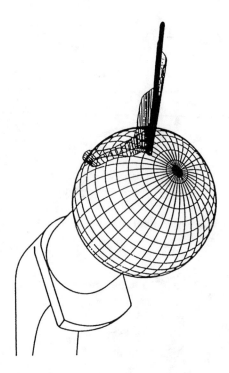

FIGURE 3.17 Scaled three-dimensional plot of resultant force during the gait cycle with crutches. The lengths of the lines indicate the magnitude of force. Radial line segments are drawn at equal increments of time, so the distance between the segments indicates the rate at which the orientation of the force was changing. For higher amplitudes of force during stance phase, line segments in close proximity indicate that the orientation of the force was changing relatively little with the cone angle between 30 and 40° and the polar angle between −25 and −15°. (*Source:* Davy DT, Kotzar DM, Brown RH, et al. 1989. Telemetric force measurements across the hip after total arthroplasty. *J Bone Joint Surg* 70A(1):45. With permission.)

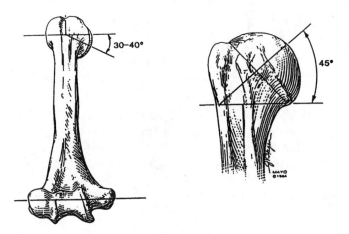

FIGURE 3.18 The two-dimensional orientation of the articular surface of the humerus with respect to the bicondylar axis. (By permission of the Mayo Foundation.)

dimension of the humeral head to the glenoid head is approximately 0.8 in the coronal plane and 0.6 in the horizontal or transverse plane [Saha, 1971]. The surface area of the glenoid fossa is only one-third to one-fourth that of the humeral head [Kent, 1971]. The arcs of articular cartilage on the humeral head

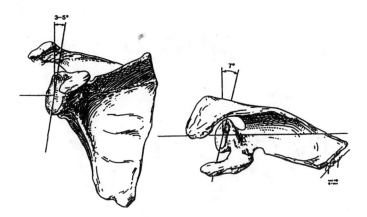

FIGURE 3.19 The glenoid faces slightly superior and posterior (retroverted) with respect to the body of the scapula. (By permission of the Mayo Foundation.)

and glenoid in the frontal and axial planes were measured [Jobe and Iannotti, 1995]. In the coronal plane, the humeral heads had an arc of 159° covered by 96° of glenoid, leaving 63° of cartilage uncovered. In the transverse plane, the humeral arc of 160° is opposed by 74° of glenoid, leaving 86° uncovered.

Joint Contact

The degree of conformity and constraint between the humeral head and glenoid has been represented by conformity index (radius of head/radius of glenoid) and constraint index (arc of enclosure/360) [McPherson, 1997]. Based on the study of 93 cadaveric specimens, the mean conformity index was 0.72 in the coronal and 0.63 in the sagittal plane. There was more constraint to the glenoid in the coronal vs. sagittal plane (0.18 vs. 0.13). These anatomic features help prevent superior–inferior translation of the humeral head but allow translation in the sagittal plane. Joint contact areas of the glenohumeral joint tend to be greater at mid-elevation positions than at either of the extremes of joint position (Table 3.11). These results suggest that the glenohumeral surface is maximum at these more functional positions, thus distributing joint load over a larger region in a more stable configuration. The contact point moves

TABLE 3.11 Glenohumeral Contact Areas

Elevation angle (°)	Contact areas at SR (cm²)	Contact areas at 20° internal to SR (cm²)
0	0.87 ± 1.01	1.70 ± 1.68
30	2.09 ± 1.54	2.44 ± 2.15
60	3.48 ± 1.69	4.56 ± 1.84
90	4.95 ± 2.15	3.92 ± 2.10
120	5.07 ± 2.35	4.84 ± 1.84
150	3.52 ± 2.29	2.33 ± 1.47
180	2.59 ± 2.90	2.51 ± NA

Note: SR = starting external rotation which allowed the shoulder to reach maximal elevation in the scapular plane (≈40° ± 8°); NA = not applicable.

Source: Soslowsky LJ, Flatow EL, Bigliani LU, Pablak RJ, Mow VC, Athesian GA. 1992. Quantitation of in situ contact areas at the glenohumeral joint: a biomechanical study. *J Orthop Res* 10(4):524. With permission.

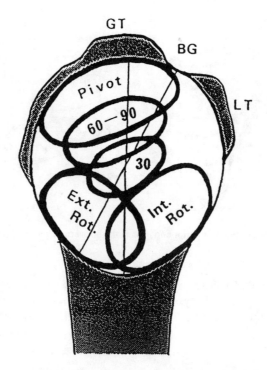

FIGURE 3.20   Humeral contact positions as a function of glenohumeral motion and positions. (*Source:* Morrey BF, An KN. 1990. Biomechanics of the shoulder. In CA Rockwood and FA Matsen (Eds.), *The Shoulder,* pp. 208–245, Philadelphia, Saunders. With permission.)

forward and inferior during internal rotation (Fig. 3.20). With external rotation, the contact is posterior/inferior. With elevation, the contact area moves superiorly. Lippitt and associates [1998] calculated the stability ratio, which is defined as a force necessary to translate the humeral head from the glenoid fossa divided by the compressive load times 100. The stability ratios were in the range of 50 to 60% in the superior–inferior direction and 30 to 40% in the anterior–posterior direction. After the labrum was removed, the ratio decreased by approximately 20%. Joint conformity was found to have significant influence on translations of humeral head during active positioning by muscles [Karduna et al., 1996].

Axes of Rotation

The shoulder complex consists of four distinct articulations: the glenohumeral joint, the acromioclavicular joint, the sternoclavicular joint, and the scapulothoracic articulation. The wide range of motion of the shoulder (exceeding a hemisphere) is the result of synchronous, simultaneous contributions from each joint. The most important function of the shoulder is arm elevation. Several investigators have attempted to relate glenohumeral and scapulothoracic motion during arm elevation in various planes (Table 3.12). About two-thirds of the motion takes place in the glenohumeral joint and about one-third in the scapulothoracic articulation, resulting in a 2:1 ratio.

Surface motion at the glenohumeral joint is primarily rotational. The center of rotation of the glenohumeral joint has been defined as a locus of points situated within 6.0 ± 1.8 mm of the geometric center of the humeral head [Poppen and Walker, 1976]. However, the motion is not purely rotational. The humeral head displaces, with respect to the glenoid. From 0–30°, and often from 30–60°, the humeral head moves upward in the glenoid fossa by about 3 mm, indicating that rolling and/or gliding has taken place. Thereafter, the humeral head has only about 1 mm of additional excursion. During arm elevation in the scapular plane, the scapula moves in relation to the thorax [Poppen and Walker, 1976]. From 0–30° the

TABLE 3.12 Arm Elevation: Glenohumeral/
Scapulothoracic Rotation

Investigator	Glenohumeral/Scapulothoracic Motion Ratio
Inman et al. [1994]	2:1
Freedman and Munro [1966]	1.35:1
Doody et al. [1970]	1.74:1
Poppen and Walker [1976]	4.3:1 (<24° elevation)
	1.25:1 (>24° elevation)
Saha [1971]	2.3:1 (30–135° elevation)

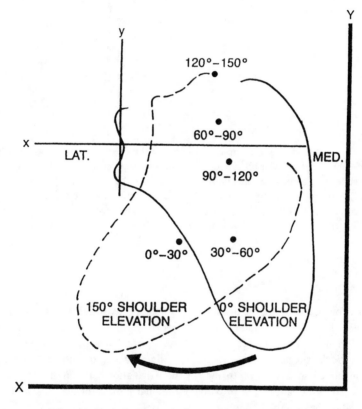

FIGURE 3.21 Rotation of the scapula on the thorax in the scapular plane. Instant centers of rotation (solid dots) are shown for each 30° interval of motion during shoulder elevation in the scapular plane from zero to 150°. The *x* and *y* axes are fixed in the scapula, whereas the *X* and *Y* axes are fixed in the thorax. From zero to 30° in the scapula rotated about its lower midportion; from 60° onward, rotation took place about the glenoid area, resulting in a medial and upward displacement of the glenoid face and a large lateral displacement of the inferior tip of the scapula. (*Source:* Poppen NK, Walker PS. 1976. Normal and abnormal motion of the shoulder. *J Bone Joint Surg* 58A:195. With permission.)

scapula rotates about its lower mid portion, and then from 60° onward the center of rotation shifts toward the glenoid, resulting in a large lateral displacement of the inferior tip of the scapula (Fig. 3.21). The center of rotation of the scapula for arm elevation is situated at the tip of the acromion as viewed from the edge on (Fig. 3.22). The mean amount of scapular twisting at maximum arm elevation is 40°. The superior tip of the scapula moves away from the thorax, and the inferior tip moves toward it.

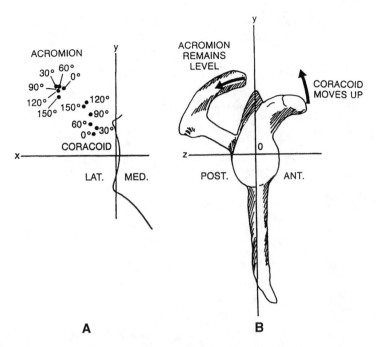

FIGURE 3.22 (A) A plot of the tips of the acromion and coracoid process on roentgenograms taken at successive intervals of arm elevation in the scapular plane shows upward movement of the coracoid and only a slight shift in the acromion relative to the glenoid face. This finding demonstrates twisting, or external rotation, of the scapula about the *x* axis. (B) A lateral view of the scapula during this motion would show the coracoid process moving upward while the acromion remains on the same horizontal plane as the glenoid. (*Source:* Poppen NK, Walker PS. 1976. Normal and abnormal motion of the shoulder. *J Bone Joint Surg* 58A:195. With permission.)

3.5 Elbow

The bony structures of the elbow are the distal end of the humerus and the proximal ends of the radius and ulna. The elbow joint complex allows two degrees of freedom in motion: flexion/extension and pronation/supination. The elbow joint complex is three separate synovial articulations. The humeral-ulnar joint is the articulation between the trochlea of the distal radius and the trochlear fossa of the proximal ulna. The humero-radial joint is formed by the articulation between the capitulum of the distal humerus and the head of the radius. The proximal radioulnar joint is formed by the head of the radius and the radial notch of the proximal ulna.

Geometry of the Articulating Surfaces

The curved, articulating portions of the trochlea and capitulum are approximately circular in a cross-section. The radius of the capitulum is larger than the central trochlear groove (Table 3.13). The centers of curvature of the trochlea and capitulum lie in a straight line located on a plane that slopes at 45–50° anterior and distal to the transepicondylar line and is inclined at 2.5° from the horizontal transverse plane [Shiba et al., 1988]. The curves of the ulnar articulations form two surfaces (coronoid and olecranon) with centers on a line parallel to the transepicondylar line but are distinct from it [Shiba et al., 1988]. The carrying angle is an angle made by the intersection of the longitudinal axis of the humerus and the forearm in the frontal plane with the elbow in an extended position. The carrying angle is contributed to, in part, by the oblique axis of the distal humerus and, in part, by the shape of the proximal ulna (Fig. 3.23).

TABLE 3.13 Elbow Joint Geometry

Parameter	Size (mm)
Capitulum radius	10.6 ± 1.1
Lateral trochlear flange radius	10.8 ± 1.0
Central trochlear groove radius	8.8 ± 0.4
Medial trochlear groove radius	13.2 ± 1.4
Distal location of flexion/extension axis from transepicondylar line:	
Lateral	6.8 ± 0.2
Medial	8.7 ± 0.6

Source: Shiba R, Sorbie C, Siu DW, Bryant JT, Cooke TDV, Weavers HW. 1988. Geometry of the humeral-ulnar joint. *J Orthop Res* 6:897.

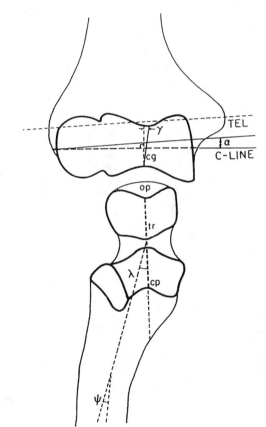

FIGURE 3.23 Components contributing to the carrying angles: $\alpha + \lambda + \psi$. Key: α, angle between C-line and TEL; γ, inclination of central groove (cg); λ, angle between trochlear notch (tn); ψ, reverse angulation of shaft of ulna; TLE, transepicondylar line; C-line, line joining centers of curvature of the trochlea and capitellum; cg, central groove; op, olecranon process; tr, trochlear ridge; cp, coronoid process. $\alpha = 2.5 \pm 0.0$; $\lambda = 17.5 \pm 5.0$ (females) and 12.0 ± 7.0 (males); $\psi = -6.5 \pm 0.7$ (females) and -9.5 ± 3.5 (males). (*Source:* Shiba R, Sorbie C, Siu DW, Bryant JT, Cooke TDV, Weavers HW. 1988. Geometry of the humero-ulnar joint. *J Orthop Res* 6:897. With permission.)

Joint Contact

The contact area on the articular surfaces of the elbow joint depends on the joint position and the loading conditions. Increasing the magnitude of the load not only increases the size of the contact area but shifts the locations as well (Fig. 3.25). As the axial loading is increased, there is an increased lateralization of the articular contact [Stormont et al., 1985]. The area of contact, expressed as a percentage of the total articulating surface area, is given in Table 3.14. Based on a finite element model of the humero-ulnar joint, Merz et al. [1997] demonstrated that the humero-ulnar joint incongruity brings about a bicentric distribution of contact pressure, a tensile stress exists in the notch that is the same order of magnitude as the compressive stress [Merz, 1997].

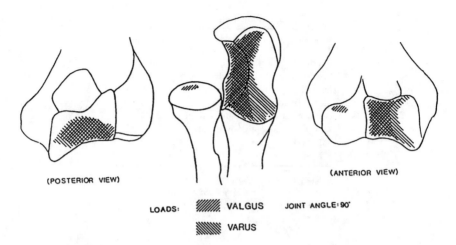

(POSTERIOR VIEW)   (ANTERIOR VIEW)

LOADS: ▨ VALGUS   JOINT ANGLE: 90°

▨ VARUS

FIGURE 3.24 Contact of the ulnohumeral joint with varus and valgus loads and the elbow at 90°. Notice only minimal radiohumeral contact in this loading condition. (*Source:* Stormont TJ, An KN, Morrey BF, Chae EY. 1985. In elbow joint contact study: comparison of techniques. *J Biomech* 18(5):329. Reprinted with permission of Elsevier.)

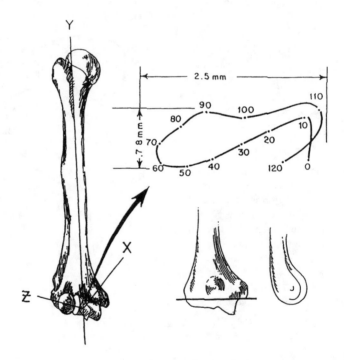

FIGURE 3.25 Very small locus of instant center of rotation for the elbow joint demonstrates that the axis may be replicated by a single line drawn from the inferior aspect of the medial epicondyle through the center of the lateral epicondyle, which is in the center of the lateral projected curvature of the trochlea and capitellum. (*Source:* Modified from Morrey BF, Chao EYS. 1976. In passive motion of the elbow joint: a biomechanical analysis. *J Bone Joint Surg* 58A(4):501. Reprinted with permission of the *Journal of Bone and Joint Surgery.*)

Axes of Rotation

The axes of flexion and extension can be approximated by a line passing through the center of the trochlea, bisecting the angle formed by the longitudinal axes of the humerus and the ulna [Morrey and Chao, 1976].

TABLE 3.14   Elbow Joint Contact Area

Position	Total Articulating Surface Area of Ulna and Radial Head (mm^2)	Contact Area (%)
Full extension	1598 ± 103	8.1 ± 2.7
90° flexion	1750 ± 123	10.8 ± 2.0
Full flexion	1594 ± 120	9.5 ± 2.1

Source: Goel VK, Singh D, Bijlani V. 1982. Contact areas in human elbow joints. *J Biomech Eng* 104:169.

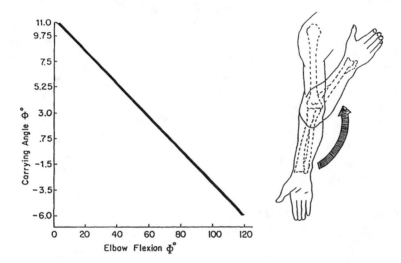

FIGURE 3.26   During elbow flexion and extension, a linear change in the carrying angle is demonstrated, typically going from valgus in extension to varus in flexion. (*Source:* Morrey BF, Chao EYS. 1976. In passive motion of the elbow joint: a biomechanical analysis. *J Bone Joint Surg* 58A(4):501. Reprinted with permission of the *Journal of Bone and Joint Surgery.*)

The instant centers of flexion and extension vary within 2–3 mm of this axis (Fig. 3.25). With the elbow fully extended and the forearm fully supinated, the longitudinal axes of humerus and ulna normally intersect at a valgus angle referred to as the *carrying angle*. In adults, this angle is usually 10–15° and normally is greater on average in women [Zuckerman and Matsen, 1989]. As the elbow flexes, the carrying angle varies as a function of flexion (Fig. 3.26). In extension there is a valgus angulation of 10°; at full flexion there is a varus angulation of 8° [Morrey and Chao, 1976]. More recently, the three-dimensional kinematics of the ulno-humeral joint under simulated active elbow joint flexion-extension was obtained by using an electromagnetic tracking device [Tanaka et al., 1998]. The optimal axis to best represent flexion–extension motion was found to be close to the line joining the centers of the capitellum and the trochlear groove. Furthermore, the joint laxity under valgus–varus stress was also examined. With the weight of the forearm as the stress, a maximum of 7.6° valgus–varus and 5.3° of axial rotation laxity were observed.

3.6   Wrist

The wrist functions by allowing changes of orientation of the hand relative to the forearm. The wrist joint complex consists of multiple articulations of eight carpal bones with the distal radius, the structures

TABLE 3.15   Changes of Wrist Geometry with Grasp

	Resting	Grasp	Analysis of Variance (p = level)
Distal radioulnar joint space (mm)	1.6 ± 0.3	1.8 ± 0.6	0.06
Ulnar variance (mm)	−0.2 ± 1.6	0.7 ± 1.8	0.003
Lunate, uncovered length (mm)	6.0 ± 1.9	7.6 ± 2.6	0.0008
Capitate length (mm)	21.5 ± 2.2	20.8 ± 2.3	0.0002
Carpal height (mm)	33.4 ± 3.4	31.7 ± 3.4	0.0001
Carpal ulnar distance (mm)	15.8 ± 4.0	15.8 ± 3.0	NS
Carpal radial distance (mm)	19.4 ± 1.8	19.7 ± 1.8	NS
Third metacarpal length (mm)	63.8 ± 5.8	62.6 ± 5.5	NS
Carpal height ratio	52.4 ± 3.3	50.6 ± 4.1	0.02
Carpal ulnar ratio	24.9 ± 5.9	25.4 ± 5.3	NS
Lunate uncovering index	36.7 ± 12.1	45.3 ± 14.2	0.002
Carpal radial ratio	30.6 ± 2.4	31.6 ± 2.3	NS
Radius—third metacarpal angle (degrees)	−0.3 ± 9.2	−3.1 ± 12.8	NS
Radius—capitate angle (degrees)	0.4 ± 15.4	−3.8 ± 22.2	NS

Note: 15 normal subjects with forearm in neutral position and elbow at 90° flexion.
Source: Schuind FA, Linscheid RL, An KN, Chao EYS. 1992. Changes in wrist and forearm configuration with grasp and isometric contraction of elbow flexors. *J Hand Surg* 17A:698.

of the ulnocarpal space, the metacarpals, and each other. This collection of bones and soft tissues is capable of a substantial arc of motion that augments hand and finger function.

Geometry of the Articulating Surfaces

The global geometry of the carpal bones has been quantified for grasp and active isometric contraction of the elbow flexors [Schuind et al., 1992]. During grasping there is a significant proximal migration of the radius of 0.9 mm, apparent shortening of the capitate, a decrease in the carpal height ratio, and an increase in the lunate uncovering index (Table 3.15). There is also a trend toward increase of the distal radioulnar joint with grasping. The addition of elbow flexion with concomitant grasping did not significantly change the global geometry, except for a significant decrease in the forearm interosseous space [Schuind et al., 1992].

Joint Contact

Studies of the normal biomechanics of the proximal wrist joint have determined that the scaphoid and lunate bones have separate, distinct areas of contact on the distal radius/triangular fibrocartilage complex surface [Viegas et al., 1987] so that the contact areas were localized and accounted for a relatively small fraction of the joint surface, regardless of wrist position (average of 20.6%). The contact areas shift from a more volar location to a more dorsal location as the wrist moves from flexion to extension. Overall, the scaphoid contact area is 1.47 times greater than that of the lunate. The scapho-lunate contact area ratio generally increases as the wrist position is changed from radial to ulnar deviation and/or from flexion to extension. Palmer and Werner [1984] also studied pressures in the proximal wrist joint and found that there are three distinct areas of contact: the ulno-lunate, radio-lunate, and radio-scaphoid. They determined that the peak articular pressure in the ulno-lunate fossa is 1.4 N/mm^2, in the radio-ulnate fossa is 3.0 N/mm^2, and in the radio-scaphoid fossa is 3.3 N/mm^2. Viegas et al. [1989] found a nonlinear relationship between increasing load and the joint contact area (Fig. 3.27). In general, the distribution of load between the scaphoid and lunate was consistent with all loads tested, with 60% of the total contact area involving the scaphoid and 40% involving the lunate. Loads greater than 46 lbs were found to not significantly increase the overall contact area. The overall contact area, even at the highest loads tested, was not more than 40% of the available joint surface.

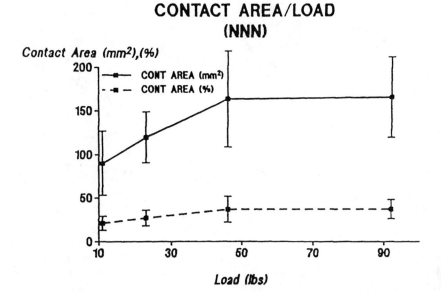

FIGURE 3.27 The nonlinear relation between the contact area and the load at the proximal wrist joint. The contact area was normalized as a percentage of the available joint surface. The load of 11, 23, 46, and 92 lbs. was applied at the position of neutral pronation/supination, neutral radioulnar deviation, and neutral flexion/extension. (*Source:* Viegas SF, Patterson RM, Peterson PD, Roefs J, Tencer A, Choi S. 1989. The effects of various load paths and different loads on the load transfer characteristics of the wrist. *J Hand Surg* 14A(3):458. With permission.)

Horii et al. [1990] calculated the total amount of force born by each joint with the intact wrist in the neutral position in the coronal plane and subjected to a total load of 143 N (Table 3.16). They found that

TABLE 3.16 Force Transmission at the Intercarpal Joints

Joint	Force (N)
Radio-ulno-carpal	
Ulno-triquetral	12 ± 3
Ulno-lunate	23 ± 8
Radio-lunate	52 ± 8
Radio-scaphoid	74 ± 13
Midcarpal	
Triquetral-hamate	36 ± 6
Luno-capitate	51 ± 6
Scapho-capitate	32 ± 4
Scapho-trapezial	51 ± 8

Note: A total of 143 N axial force applied across the wrist.
Source: Horii E, Garcia-Elias M, An KN, Bishop AT, Cooney WP, Linscheid RL, Chao EY. 1990. Effect on force transmission across the carpus in procedures used to treat Kienböck's disease. *J Bone Joint Surg* 15A(3):393.

22% of the total force in the radio-ulno-carpal joint is dissipated through the ulna 14% through the ulno-lunate joint, and 18% through the ulno-triquetral joint) and 78% through the radius (46% through the scaphoid fossa and 32% through the lunate fossa). At the midcarpal joint, the scapho-trapezial joint transmits 31% of the total applied force, the scapho-capitate joint transmits 19%, the luno-capitate joint transmits 29%, and the triquetral-hamate joints transmits 21% of the load.

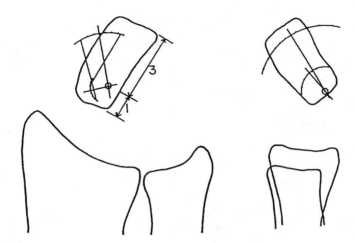

FIGURE 3.28 The location of the center of rotation during ulnar deviation (left) and extension (right), determined graphically using two metal markers embedded in the capitate. Note that during radial-ulnar deviation the center lies at a point in the capitate situated distal to the proximal end of this bone by a distance equivalent to approximately one-quarter of its total longitudinal length. During flexion-extension, the center of rotation is close to the proximal cortex of the capitate. (*Source:* Youm Y, McMurty RY, Flatt AE, Gillespie TE. 1978. Kinematics of the wrist: an experimental study of radioulnar deviation and flexion/extension. *J Bone Joint Surg* 60A(4):423. With permission.)

A limited amount of studies have been done to determine the contact areas in the midcarpal joint. Viegas et al. [1990] have found four general areas of contact: the scapho-trapezial-trapezoid (STT), the scapho-capitate (SC), the capito-lunate (CL), and the triquetral-hamate (TH). The high pressure contact area accounted for only 8% of the available joint surface with a load of 32 lbs and increased to a maximum of only 15% with a load of 118 lbs. The total contact area, expressed as a percentage of the total available joint area for each fossa was: STT = 1.3%, SC = 1.8%, CL = 3.1%, and TH = 1.8%.

The correlation between the pressure loading in the wrist and the progress of degenerative osteoarthritis associated with pathological conditions of the forearm was studied in a cadaveric model [Sato, 1995]. Malunion after distal radius fracture, tear of triangular fibrocartilage, and scapholunate dissociation were all responsible for the alteration of the articulating pressure across the wrist joint. Residual articular incongruity of the distal radius following intra-articular fracture has been correlated with early osteoarthritis. In an *in vitro* model, step-offs of the distal radius articular incongruity were created. Mean contact stress was significantly greater than the anatomically reduced case at only 3 mm of step-off [Anderson et al., 1996].

Axes of Rotation

The complexity of joint motion at the wrist makes it difficult to calculate the instant center of motion. However, the trajectories of the hand during radioulnar deviation and flexion/extension, when they occur in a fixed plane, are circular, and the rotation in each plane takes place about a fixed axis. These axes are located within the head of the capitate and are not altered by the position of the hand in the plane of rotation [Youm et al., 1978]. During radioulnar deviation, the instant center of rotation lies at a point in the capitate situated distal to the proximal end of this bone by a distance equivalent to approximately one-quarter of its total length (Fig. 3.28). During flexion/extension, the instant center is close to the proximal cortex of the capitate, which is somewhat more proximal than the location for the instant center of radioulnar deviation.

Normal carpal kinematics were studied in 22 cadaver specimens using a biplanar radiography method. The kinematics of the trapezium, capitate, hamate, scaphoid, lunate, and triquetrum were determined

TABLE 3.17 Individual Carpal Rotation Relative to the Radius (Degrees) (Sagittal Plane Motion of the Wrist)

	Axis of Rotation											
	X (+) Pronation; (−) Supination				*Y* (+) Flexion; (−) Extension				*Z* (+) Ulnar Deviation; (−) Radial Deviation			
Wrist Motion[a] Carpal Bone	N-E60	N-E30	N-F30	N-F60	N-E60	N-E30	N-F60	N-E60	N-E60	N-E30	N-F30	N-F60
Trapezium (N=13)	−0.9	−1.3	0.9	−1.4	−59.4	−29.3	28.7	54.2	1.2	0.3	−0.4	2.5
S.D.	2.8	2.2	2.6	2.7	2.3	1	1.8	3	4	2.7	1.3	2.8
Capitate (N=22)	0.9	−1	1.3	−1.6	60.3	−30.2	21.5	63.5	0	0	0.6	3.2
S.D.	2.7	1.8	2.5	3.5	2.5	1.1	1.2	2.8	2	1.4	1.6	3.6
Hamate (N=9)	0.4	−1	1.3	−0.3	−59.5	−29	28.8	62.6	2.1	0.7	0.1	1.8
S.D.	3.4	1.7	2.5	2.4	1.4	0.8	10.2	3.6	4.4	1.8	1.2	4.1
Scaphoid (N=22)	−2.5	−0.7	1.6	2	−52.3	−26	20.6	39.7	4.5	0.8	2.1	7.8
S.D.	3.4	2.6	2.2	3.1	3	3.2	2.8	4.3	3.7	2.1	2.2	4.5
Lunate (N=22)	1.2	0.5	0.3	−2.2	−29.7	−15.4	11.5	23	4.3	0.9	3.3	11.1
S.D.	2.8	1.8	1.7	2.8	6.6	3.9	3.9	5.9	2.6	1.5	1.9	3.4
Triquetrum (N=22)	−3.5	−2.5	2.5	−0.7	−39.3	−20.1	15.5	30.6	0	−0.3	2.4	9.8
S.D.	3.5	2	2.2	3.7	4.8	2.7	3.8	5.1	2.8	1.4	2.6	4.3

[a] N-E60: neutral to 60° of extension; N-E30: neutral to 30° of extension; N-F30: neutral to 30° of flexion; N-F60: neutral to 60° of flexion. S.D. = standard deviation.

Source: Kobayashi M, Berger RA, Nagy L, et al. 1997. Normal kinematics of carpal bones: a three-dimensional analysis of carpal bone motion relative to the radius. *J Biomech* 30(8):787.

during wrist rotation in the sagittal and coronal plane [Kobagashi et al., 1997]. The results were expressed using the concept of the screw displacement axis and covered to describe the magnitude of rotation about and translation along three orthogonal axes. The orientation of these axes is expressed relative to the radius during sagittal plane motion of the wrist (Table 3.17). The scaphoid exhibited the greatest magnitude of rotation and the lunate displayed the least rotation. The proximal carpal bones exhibited some ulnar deviation in 60° of wrist flexion. During coronal plane motion (Table 3.18), the magnitude of radial-ulnar deviation of the distal carpal bones was mutually similar and generally of a greater magnitude than that of the proximal carpal bones. The proximal carpal bones experienced some flexion during radial deviation of the wrist and extension during ulnar deviation of the wrist.

3.7 Hand

The hand is an extremely mobile organ that is capable of conforming to a large variety of object shapes and coordinating an infinite variety of movements in relation to each of its components. The mobility of this structure is possible through the unique arrangement of the bones in relation to one another, the articular contours, and the actions of an intricate system of muscles. Theoretical and empirical evidence suggest that limb joint surface morphology is mechanically related to joint mobility, stability, and strength [Hamrick, 1996].

Geometry of the Articulating Surfaces

Three-dimensional geometric models of the articular surfaces of the hand have been constructed. The sagittal contours of the metacarpal head and proximal phalanx grossly resemble the arc of a circle [Tamai et al., 1988]. The radius of curvature of a circle fitted to the entire proximal phalanx surface ranges from 11–13 mm, almost twice as much as that of the metacarpal head, which ranges from 6–7 mm (Table 3.19). The local centers of curvature along the sagittal contour of the metacarpal heads are not fixed. The locus of the center of curvature for the subchondral bony contour approximates the locus of the center for the acute curve of an ellipse (Fig. 3.29). However, the locus of center of curvature for the articular cartilage contour approximates the locus of the obtuse curve of an ellipse.

TABLE 3.18   Individual Carpal Rotation to the Radius (Degrees) (Coronal Plane Motion of the Wrist)

Wrist Motion[a] Carpal Bone	X (+) Pronation; (−) Supination			Y (+) Flexion; (−) Extension			Z (+) Ulnar Deviation; (−) Radial Deviation		
	N-RD15	N-UD15	N-UD30	N-RD15	N-UD15	N-UD30	N-RD15	N-UD15	N-UD30
Trapezium (N=13)	−4.8	9.1	16.3	0	4.9	9.9	−14.3	16.4	32.5
S.D.	2.4	3.6	3.6	1.5	1.3	2.1	2.3	2.8	2.6
Capitate (N=22)	−3.9	6.8	11.8	1.3	2.7	6.5	−14.6	15.9	30.7
S.D.	2.6	2.6	2.5	1.5	1.1	1.7	2.1	1.4	1.7
Hamate (N=9)	−4.8	6.3	10.6	1.1	3.5	6.6	−15.5	15.4	30.2
S.D.	1.8	2.4	3.1	3	3.2	4.1	2.4	2.6	3.6
Scaphoid (N=22)	0.8	2.2	6.6	8.5	−12.5	−17.1	−4.2	4.3	13.6
S.D.	1.8	2.4	3.1	3	3.2	4.1	2.4	2.6	3.6
Lunate (N=22)	−1.2	1.4	3.9	7	−13.9	−22.5	−1.7	5.7	15
S.D.	1.6	0	3.3	3.1	4.3	6.9	1.7	2.8	4.3
Triquetrum (N=22)	−1.1	−1	0.8	4.1	−10.5	−17.3	−5.1	7.7	18.4
S.D.	1.4	2.6	4	3	3.8	6	2.4	2.2	4

[a] N-RD15: neutral to 15° of radial deviation; N-UD30: neutral to 30° of ulnar deviation; N-UD15: neutral to 15° of ulnar deviation.
S.D. = standard deviation.

Source: Kobayashi M, Berger RA, Nagy L, et al. 1997. Normal kinematics of carpal bones: a three-dimensional analysis of carpal bone motion relative to the radius. *J Biomech* 30(8):787.

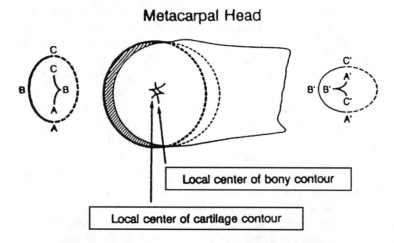

Metacarpal Head

Local center of bony contour

Local center of cartilage contour

FIGURE 3.29   The loci of the local centers of curvature for subchondral bony contour of the metacarpal head approximates the loci of the center for the acute curve of an ellipse. The loci of the local center of curvature for articular cartilage contour of the metacarpal head approximates the loci of the bony center of the obtuse curve of an ellipse. (*Source:* Tamai K, Ryu J, An KN, Linscheid RL, Cooney WP, Chao EYS. 1988. In Three-dimensional geometric analysis of the metacarpophalangeal joint. *J Hand Surg* 13A(4):521. Reprinted with permission of Churchill Livingstone.)

The surface geometry of the thumb carpometacarpal (CMC) joint has also been quantified [Athesian et al., 1992]. The surface area of the CMC joint is significantly greater for males than for females (Table 3.20). The minimum, maximum, and mean square curvature of these joints is reported in Table 3.20. The curvature of the surface is denoted by κ and the radius of curvature is $\rho = 1/\kappa$. The curvature is negative when the surface is concave and positive when the surface is convex.

TABLE 3.19 Radius of Curvature of the Middle Sections of the Metacarpal Head and Proximal Phalanx Base

	Radius (mm)	
	Bony Contour	Cartilage Contour
MCH Index	6.42 ± 1.23	6.91 ± 1.03
Long	6.44 ± 1.08	6.66 ± 1.18
PPB Index	13.01 ± 4.09	12.07 ± 3.29
Long	11.46 ± 2.30	11.02 ± 2.48

Source: Tamai K, Ryu J, An KN, Linscheid RL, Cooney WP, Chao EYS. 1988. Three-dimensional geometric analysis of the metacarpophalangeal joint. *J Hand Surg* 13A(4):521.

Joint Contact

The size and location of joint contact areas of the metacarpophalangeal (MCP) joint changes as a function of the joint flexion angle (Fig. 3.30) The radioulnar width of the contact area becomes narrow in the neutral position and expands in both the hyperextended and fully flexed positions [An and Cooney, 1991]. In the neutral position, the contact area occurs in the center of the phalangeal base, this area being slightly larger on the ulnar than on the radial side.

The contact areas of the thumb carpometacarpal joint under the functional position of lateral key pinch and in the extremes of range of motion were studied using a stereophotogrammetric technique [Ateshian et al., 1995]. The lateral pinch position produced contact predominately on the central, volar, and volar–ulnar regions of the trapezium and the metacarpals (Fig. 3.31). Pelligrini et al. [1993] noted that the palmar compartment of the trapeziometacarpal joint was the primary contact area during flexion adduction of the thumb in lateral pinch. Detachment of the palmar beak ligament resulted in dorsal translation of the contact area producing a pattern similar to that of cartilage degeneration seen in the osteoarthritic joint.

Axes of Rotation

Rolling and sliding actions of articulating surfaces exist during finger joint motion. The geometric shapes of the articular surfaces of the metacarpal head and proximal phalanx, as well as the insertion location of the collateral ligaments, significantly govern the articulating kinematics, and the center of rotation is not fixed but rather moves as a function of the angle of flexion [Pagowski and Piekarski, 1977]. The instant centers of rotation are within 3 mm of the center of the metacarpal head [Walker and Erhman, 1975]. Recently the axis of rotation of the MCP joint has been evaluated *in vivo* by Fioretti [1994]. The instantaneous helical axis of the MCP joint tends to be more palmar and tends to be displaced distally as flexion increases (Fig. 3.32).

The axes of rotation of the CMC joint have been described as being fixed [Hollister et al., 1992], but others believe that a polycentric center of rotation exists [Imaeda et al., 1994]. Hollister et al. [1992] found that axes of the CMC joint are fixed and are not perpendicular to each other, or to the bones, and do not intersect. The flexion/extension axis is located in the trapezium, and the abduction/adduction axis is on the first metacarpal. In contrast, Imaeda et al. [1994] found that there was no single center of rotation, but rather the instantaneous motion occurred reciprocally between centers of rotations within the trapezium and the metacarpal base of the normal thumb. In flexion/extension, the axis of rotation was located within the trapezium, but for abduction/adduction the center of rotation was located distally to the trapezium and within the base of the first metacarpal. The average instantaneous center of circumduction was at approximately the center of the trapezial joint surface (Table 3.21).

The axes of rotation of the thumb interphalangeal and metacarpophalangeal joint were located using a mechanical device [Hollister et al., 1995]. The physiologic motion of the thumb joints occur about these axes (Fig. 3.33 and Table 3.22). The interphalangeal joint axis is parallel to the flexion crease of

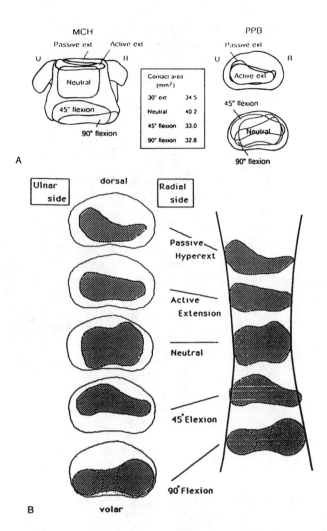

FIGURE 3.30 (A) Contact area of the MCP joint in five joint positions. (B) End on view of the contact area on each of the proximal phalanx bases. The radioulnar width of the contact area becomes narrow in the neutral position and expands in both the hyperextended and fully flexed positions. (*Source:* An KN, Cooney WP. 1991. Biomechanics, Section II. The hand and wrist. In BF Morrey (Ed.), *Joint Replacement Arthroplasty,* pp. 137–146, New York, Churchill Livingstone. By permission of the Mayo Foundation.)

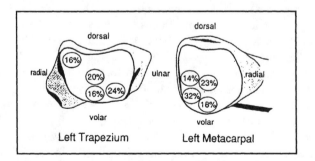

FIGURE 3.31 Summary of the contact areas for all specimens, in lateral pinch with a 25-N load. All results from the right hand are transposed onto the schema of a carpometacarpal joint from the left thumb. (*Source:* Ateshian, GA, Ark JW, Rosenwasser MP, et al. 1995. Contact areas on the thumb carpometacarpal joint. *J Orthop Res* 13:450.)

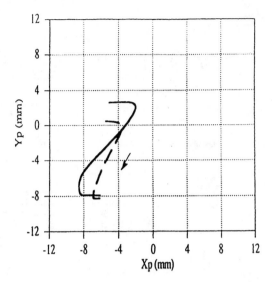

FIGURE 3.32 Intersections of the instantaneous helical angles with the metacarpal sagittal plane. They are relative to one subject tested twice in different days. The origin of the graph is coincident with the calibrated center of the metacarpal head. The arrow indicates the direction of flexion. (*Source:* Fioretti S. 1994. Three-dimensional *in-vivo* kinematic analysis of finger movement. In F Schuind et al. (Eds.), *Advances in the Biomechanics of the Hand and Wrist*, pp. 363–375, New York, Plenum Press. With permission.)

TABLE 3.20   Curvature of Carpometacarpal Joint Articular Surfaces

	n	Area (cm^2)	$\bar{\kappa}_{min}$ (m^{-1})	$\bar{\kappa}_{max}$ (m^{-1})	$\bar{\kappa}_{rms}$ (m^{-1})
Trapezium					
Female	8	1.05 ± 0.21	−61 ± 22	190 ± 36	165 ± 32
Male	5	1.63 ± 0.18	−87 ± 17	114 ± 19	118 ± 6
Total	13	1.27 ± 0.35	−71 ± 24	161 ± 48	147 ± 34
Female vs. male		$p \leq 0.01$	$p \leq 0.05$	$p \leq 0.01$	$p \leq 0.01$
Metacarpal					
Female	8	1.22 ± 0.36	−49 ± 10	175 ± 25	154 ± 20
Male	5	1.74 ± 0.21	−37 ± 11	131 ± 17	116 ± 8
Total	13	1.42 ± 0.40	−44 ± 12	158 ± 31	140 ± 25
Female vs. male		$p \leq 0.01$	$p \leq 0.05$	$p \leq 0.01$	$p \leq 0.01$

Note: Radius of curvature: $\rho = 1/k$

Source: Athesian JA, Rosenwasser MP, Mow VC. 1992. Curvature characteristics and congruence of the thumb carpometacarpal joint: differences between female and male joints. *J Biomech* 25(6):591.

the joint and is not perpendicular to the phalanx. The metacarpophalangeal joint has two fixed axes: a fixed flexion–extension axis just distal and volar to the epicondyles, and an abduction–adduction axis related to the proximal phalanx passing between the sesamoids. Neither axis is perpendicular to the phalanges.

3.8   Summary

It is important to understand the biomechanics of joint-articulating surface motion. The specific characteristics of the joint will determine the musculoskeletal function of that joint. The unique geometry

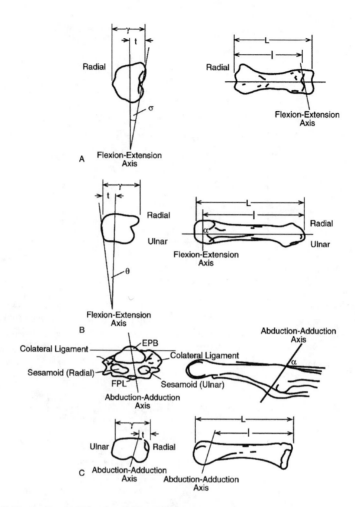

FIGURE 3.33   (A) The angles and length and breadth measurements defining the axis of rotation of the interphalangeal joint of the right thumb. (t/T = ratio of anatomic plane diameter; l/L = ratio of length). (B) The angles and
length and breadth measurements of the metacarpophalangeal flexion-extension axis' position in the metacarpal.
(C) The angles and length and breadth measurements that locate the metacarpophalangeal abduction–adduction
axis. The measurements are made in the metacarpal when the metacapophalangeal joint is at neutral flexion extension.
The measurements are made relative to the metacarpal because the axis passes through this bone, not the proximal
phalanx with which it moves. This method of recording the abduction–adduction measurements allows the measurements of the axes to each other at a neutral position to be made. The metacarpophalangeal abduction–adduction
axis passes through the volar plate of the proximal phalanx. (*Source:* Hollister A, Giurintano DJ, Buford WL, et al.
1995. The axes of rotation of the thumb interphalangeal and metacarpophalangeal joints. *Clin Orthop Rel Res* 320:188.)

of the joint surfaces and the surrounding capsule ligamentous constraints will guide the unique characteristics of the articulating surface motion. The range of joint motion, the stability of the joint, and the
ultimate functional strength of the joint will depend on these specific characteristics. A congruent joint
usually has a relatively limited range of motion but a high degree of stability, whereas a less congruent
joint will have a relatively larger range of motion but less degree of stability. The characteristics of the
joint-articulating surface will determine the pattern of joint contact and the axes of rotation. These
characteristics will regulate the stresses on the joint surface which will influence the degree of degeneration of articular cartilage in an anatomic joint and the amount of wear of an artificial joint.

TABLE 3.21 Location of Center of Rotation
of Trapeziometacarpal Joint

	Mean ± SD (mm)
Circumduction	
x	0.1 ± 1.3
y	−0.6 ± 1.3
z	−0.5 ± 1.4
Flexion/Extension (in x-y plane)	
x	
Centroid	−4.2 ± 1.0
Radius	2.0 ± 0.5
y	
Centroid	−0.4 ± 0.9
Radius	1.6 ± 0.5
Abduction/Adduction (in x-z plane)	
x	
Centroid	6.7 ± 1.7
Radius	4.6 ± 3.1
z	
Centroid	−0.2 ± 0.7
Radius	1.7 ± 0.5

Note: The coordinate system is defined with the x axis corresponding to internal/external rotation, the y axis corresponding to abduction/adduction, and the z axis corresponding to flexion/extension. The x axis is positive in the distal direction, the y axis is positive in the dorsal direction for the left hand and in the palmar direction for the right hand, and the z axis is positive in the radial direction. The origin of the coordinate system was at the intersection of a line connecting the radial and ulnar prominences and a line connecting the volar and dorsal tubercles.

Source: Imaeda T, Niebur G, Cooney WP, Linscheid RL, An KN. 1994. Kinematics of the normal trapeziometacarpal joint. *J Orthop Res* 12:197.

TABLE 3.22   Measurement of Axis Location and Values
for Axis Position in the Bone[a]

Interphalangeal joint flexion-extension axis (Fig. 3.33A)	
t/T	44 ± 17%
l/L	90 ± 5%
Θ	5 ± 2°
β	83 ± 4°
Metacarpophalangeal joint flexion-extension axis (Fig. 3.33B)	
t/T	57 ± 17%
l/L	87 ± 5%
α	101 ± 6°
β	5 ± 2°
Metacarpophalangeal joint abduction-adduction axis (Fig. 3.33C)	
t/T	45 ± 8%
l/L	83 ± 13%
α	80 ± 9°
β	74 ± 8°
M	

[a] The angle of the abduction-adduction axis with respect to the flexion-extension axis is 84.8° ± 12.2°. The location and angulation of the K-wires of the axes with respect to the bones were measured (Θ, α, β) directly with a goniometer. The positions of the pins in the bones were measured (T, L) with a Vernier caliper.

Source: Hollister A, Giurintano DJ, Buford WL, et al. 1995. The axes of rotation of the thumb interphalangeal and metacarpophalangeal joints. *Clin Orthop Rel Res* 320:188.

Acknowledgment

The authors thank Barbara Iverson-Literski for her careful preparation of the manuscript.

References

Ahmed AM, Burke DL, Hyder A. 1987. Force analysis of the patellar mechanism. *J Orthop Res* 5:1, 69.

Allard P. Duhaime M, Labelle H et al. 1987. Spatial reconstruction technique and kinematic modeling of the ankle. *IEEE Eng Med Biol* 6:31.

An KN, Cooney WP. 1991. Biomechanics, Section II. The hand and wrist. In BF Morrey (Ed.), *Joint Replacement Arthroplasty*, pp. 137–146, New York, Churchill Livingstone.

Anderson DD, Bell AL, Gaffney MB et al. 1996. Contact stress distributions in malreduced intra-articular distal radius fractures. *J Orthop Trauma* 10:331.

Andriacchi TP, Stanwyck TS, Galante JO. 1986. Knee biomechanics in total knee replacement. *J Arthroplasty* 1(3):211.

Ateshian GA, Ark JW, Rosenwasser MD et al. 1995. Contact areas in the thumb carpometacarpal joint. *J Orthop Res* 13:450.

Athesian JA, Rosenwasser MP, Mow VC. 1992. Curvature characteristics and congruence of the thumb carpometacarpal joint: differences between female and male joints. *J Biomech* 25(6):591.

Barnett CH, Napier JR. 1952. The axis of rotation at the ankle joint in man. Its influence upon the form of the talus and the mobility of the fibula. *J Anat* 86:1.

Basmajian JV, Bazant FJ. 1959. Factors preventing downward dislocation of the adducted shoulder joint. An electromyographic and morphological study. *J Bone Joint Surg* 41A:1182.

Boileau P, Walch G. 1997. The three-dimensional geometry of the proximal humerus. *J Bone Joint Surg* 79B:857.

D'Ambrosia RD, Shoji H, Van Meter J. 1976. Rotational axis of the ankle joint: comparison of normal and pathological states. *Surg Forum* 27:507.

Davy DT, Kotzar DM, Brown RH et al. 1989. Telemetric force measurements across the hip after total arthroplasty. *J Bone Joint Surg* 70A(1):45.

Doody SG, Freedman L, Waterland JC. 1970. Shoulder movements during abduction in the scapular plane. *Arch Phys Med Rehabil* 51:595.

Draganich LF, Andriacchi TP, Andersson GBJ. 1987. Interaction between intrinsic knee mechanics and the knee extensor mechanism. *J Orthop Res* 5:539.

Driscoll HL, Christensen JC, Tencer AF. 1994. Contact characteristics of the ankle joint. *J Am Podiatr Med Assoc* 84(10):491.

Elftman H. 1945. The orientation of the joints of the lower extremity. *Bull Hosp Joint Dis* 6:139.

Engsberg JR. 1987. A biomechanical analysis of the talocalcaneal joint *in vitro*. *J Biomech* 20:429.

Farahmand F, Senavongse W, Amis AA. 1998. Quantitative study of the quadriceps muscles and trochlear groove geometry related to instability of the patellofemoral joint. *J Orthop Res* 16(1):136.

Fioretti S. 1994. Three-dimensional *in-vivo* kinematic analysis of finger movement. In F Schuind et al. (Eds.), *Advances in the Biomechanics of the Hand and Wrist*, pp. 363–375, New York, Plenum Press.

Freedman L, Munro RR. 1966. Abduction of the arm in the scapular plane: scapular and glenohumeral movements. A roentgenographic study. *J Bone Joint Surg* 48A:1503.

Goel VK, Singh D, Bijlani V. 1982. Contact areas in human elbow joints. *J Biomech Eng* 104:169–175.

Goodfellow J, Hungerford DS, Zindel M. 1976. Patellofemoral joint mechanics and pathology. *J Bone Joint Surg* 58B(3):287.

Hamrick MW. 1996. Articular size and curvature as detriments of carpal joint mobility and stability in strepsirhine primates. *J Morphol* 230:113.

Hartford JM, Gorczyca JT, McNamara JL et al. 1985. Tibiotalar contact area. *Clin Orthop* 320, 82.

Hicks JH. 1953. The mechanics of the foot. The joints. *J Anat* 87:345–357.

Hollister A, Buford WL, Myers LM et al. 1992. The axes of rotation of the thumb carpometacarpal joint. *J Orthop Res* 10:454.

Hollister A, Guirintano DJ, Bulford WL et al. 1995. The axes of rotation of the thumb interphalangeal and metacarpophalangeal joints. *Clin Orthop* 320:188.

Horii E, Garcia-Elias M, An KN et al. 1990. Effect of force transmission across the carpus in procedures used to treat Kienböck's disease. *J Bone Joint Surg* 15A(3):393.

Huberti HH, Hayes WC. 1984. Patellofemoral contact pressures: the influence of Q-angle and tendofemoral contact. *J Bone Joint Surg* 66A(5):715.

Iannotti JP, Gabriel JP, Schneck SL et al. 1992. The normal glenohumeral relationships: an anatomical study of 140 shoulders. *J Bone Joint Surg* 74A(4):491.

Imaeda T, Niebur G, Cooney WP et al. 1994. Kinematics of the normal trapeziometacarpal joint. *J Orthop Res* 12:197.

Inman VT, Saunders JB deCM, Abbott LC. 1944. Observations on the function of the shoulder joint. *J Bone Joint Surg* 26A:1.

Inman VT. 1976. *The Joints of the Ankle*, Baltimore, Williams & Wilkins.

Inman VT, Mann RA. 1979. Biomechanics of the foot and ankle. In VT Inman (Ed.), *DeVrie's Surgery of the Foot*, St. Louis, Mosby.

Iseki F, Tomatsu T. 1976. The biomechanics of the knee joint with special reference to the contact area. *Keio J Med* 25:37.

Isman RE, Inman VT. 1969. Anthropometric studies of the human foot and ankle. *Pros Res* 10–11:97.

Jobe CM, Iannotti JP. 1995. Limits imposed on glenohumeral motion by joint geometry. *J Shoulder Elbow Surg* 4:281.

Kapandji IA. 1987. *The Physiology of the Joints*, Vol 2, *Lower Limb*, Edinburgh, Churchill Livingstone.

Karduna AR, Williams GR, Williams JI, et al. 1996. Kinematics of the glenohumeral joint: influences of muscle forces, ligamentous constraints, and articular geometry. *J Orthop Res* 14:986.

Kent BE. 1971. Functional anatomy of the shoulder complex. A review. *Phys Ther* 51:867.

Kimizuka M, Kurosawa H, Fukubayashi T. 1980. Load-bearing pattern of the ankle joint. Contact area and pressure distribution. *Arch Orthop Trauma Surg* 96:45–49.

Kinzel GL, Hall AL, Hillberry BM. 1972. Measurement of the total motion between two body segments: Part I. Analytic development. *J Biomech* 5:93.

Kirby KA. 1947. Methods for determination of positional variations in the subtalar and transverse tarsal joints. *Anat Rec* 80:397.

Kobayashi M, Berger RA, Nagy L et al. 1997. Normal kinematics of carpal bones: a three-dimensional analysis of carpal bone motion relative to the radius. *J Biomech* 30:787.

Kurosawa H, Walker PS, Abe S et al. 1985. Geometry and motion of the knee for implant and orthotic design. *J Biomech* 18(7):487.

Lewis JL, Lew WD. 1978. A method for locating an optimal "fixed" axis of rotation for the human knee joint. *J Biomech Eng* 100:187.

Libotte M, Klein P, Colpaert H et al. 1982. Contribution à l'étude biomécanique de la pince malléolaire. *Rev Chir Orthop* 68:299.

Lippitts B, Vanderhooft JE, Harris SL et al. 1993. Glenohumeral stability from concavity-compression: a quantitative analysis. *J Shoulder Elbow Surg* 2:27.

Macko VW, Matthews LS, Zwirkoski P et al. 1991. The joint contract area of the ankle: the contribution of the posterior malleoli. *J Bone Joint Surg* 73A(3):347.

Manter JT. 1941. Movements of the subtalar and transverse tarsal joints. *Anat Rec* 80:397–402.

Maquet PG, Vandberg AJ, Simonet JC. 1975. Femorotibial weight bearing areas: experimental determination. *J Bone Joint Surg* 57A(6):766–771.

McPherson EJ, Friedman RJ, An YH et al. 1997. Anthropometric study of normal glenohumeral relationships. *J Shoulder Elbow Surg* 6:105.

Merz B, Eckstein F, Hillebrand S et al: 1997. Mechanical implication of humero-ulnar incongruity-finite element analysis and experiment. *J Biomech* 30:713.

Miyanaga Y, Fukubayashi T, Kurosawa H. 1984. Contact study of the hip joint: load deformation pattern, contact area, and contact pressure. *Arch Orth Trauma Surg* 103:13–17.

Morrey BF, An KN. 1990. Biomechanics of the shoulder. In CA Rockwood and FA Matsen (Eds.), *The Shoulder*, pp. 208–245, Philadelphia, Saunders.

Morrey BF, Chao EYS. 1976. Passive motion of the elbow joint: a biomechanical analysis. *J Bone Joint Surg* 58A(4):501.

Nahass BE, Madson MM, Walker PS. 1991. Motion of the knee after condylar resurfacing—an *in vivo* study. *J Biomech* 24(12):1107.

Paar O, Rieck B, Bernett P. 1983. Experimentelle Untersuchungen über belastungsabhängige Drukund Kontaktflächenverläufe an den Fussgelenken. *Unfallheilkunde* 85:531.

Pagowski S, Piekarski K. 1977. Biomechanics of metacarpophalangeal joint. *J Biomech* 10:205.

Palmer AK, Werner FW. 1984. Biomechanics of the distal radio-ulnar joint. *Clin Orthop* 187:26.

Parlasca R, Shoji H, D'Ambrosia RD. 1979. Effects of ligamentous injury on ankle and subtalar joints. A kinematic study. *Clin Orthop* 140:266.

Pellegrini VS, Olcott VW, Hollenberg C. 1993. Contact patterns in the trapeziometacarpal joint: the role of the palmar beak ligament. *J Hand Surg* 18A:238.

Pereira DS, Koval KJ, Resnick RB et al. 1996. Tibiotalar contact area and pressure distribution: the effect of mortise widening and syndesmosis fixation. *Foot Ankle* 17(5):269.

Poppen NK, Walker PS. 1976. Normal and abnormal motion of the shoulder. *J Bone Joint Surg* 58A:195.

Ramsey PL, Hamilton W. 1976. Changes in tibiotalar area of contact caused by lateral talar shift. *J Bone Joint Surg* 58A:356.

Rastegar J, Miller N, Barmada R. 1980. An apparatus for measuring the load-displacement and load-dependent kinematic characteristics of articulating joints—application to the human ankle. *J Biomech Eng* 102:208.

Root ML, Weed JH, Sgarlato TE, Bluth DR. 1966. Axis of motion of the subtalar joint. *J Am Podiatry Assoc* 56:149.

Saha AK. 1971. Dynamic stability of the glenohumeral joint. *Acta Orthop Scand* 42:491.

Sammarco GJ, Burstein AJ, Frankel VH. 1973. Biomechanics of the ankle: a kinematic study. *Orthop Clin North Am* 4:75–96.

Sato S. 1995. Load transmission through the wrist joint: a biomechanical study comparing the normal and pathological wrist. *Nippon Seikeigeka Gakkai Zasshi-J Jpn Orthop Assoc* 69:470.

Schuind FA, Linscheid RL, An KN et al. 1992. Changes in wrist and forearm configuration with grasp and isometric contraction of elbow flexors. *J Hand Surg* 17A:698.

Shephard E. 1951. Tarsal movements. *J Bone Joint Surg* 33B:258.

Shiba R, Sorbie C, Siu DW et al. 1988. Geometry of the humeral-ulnar joint. *J Orthop Res* 6:897.

Singh AK, Starkweather KD, Hollister AM et al. 1992. Kinematics of the ankle: a hinge axis model. *Foot Ankle* 13(8):439.

Soslowsky LJ, Flatow EL, Bigliani LU et al. 1992. Quantitation of *in situ* contact areas at the glenohumeral joint: a biomechanical study. *J Orthop Res* 10(4):524.

Spoor CW, Veldpaus FE. 1980. Rigid body motion calculated from spatial coordinates of markers. *J Biomech* 13:391.

Stormont TJ, An KA, Morrey BF et al. 1985. Elbow joint contact study: comparison of techniques. *J Biomech* 18(5):329.

Tamai K, Ryu J, An KN et al. 1988. Three-dimensional geometric analysis of the metacarpophalangeal joint. *J Hand Surg* 13A(4):521.

Tanaka S, An KN, Morrey BF. 1998. Kinematics and laxity of ulnohumeral joint under valgus-varus stress. *J Musculoskeletal Res* 2:45.

Tönnis D. 1987. *Congenital Dysplasia and Dislocation of the Hip and Shoulder in Adults*, pp. 1–12, Berlin, Springer-Verlag.

Van Eijden TMGJ, Kouwenhoven E, Verburg J et al. 1986. A mathematical model of the patellofemoral joint. *J Biomech* 19(3):219.

Van Langelaan EJ. 1983. A kinematical analysis of the tarsal joints. An x-ray photogrammetric study. Acta *Orthop Scand* [Suppl] 204:211.

Viegas SF, Tencer AF, Cantrell J et al. 1987. Load transfer characteristics of the wrist: Part I. The normal joint. *J Hand Surg* 12A(6):971.

Viegas SF, Patterson RM, Peterson PD et al. 1989. The effects of various load paths and different loads on the load transfer characteristics of the wrist. *J Hand Surg* 14A(3):458.

Viegas SF, Patterson RM, Todd P et al. October 7, 1990. Load transfer characteristics of the midcarpal joint. Presented at Wrist Biomechanics Symposium, Wrist Biomechanics Workshop, Mayo Clinic, Rochester, MN.

Von Lanz D, Wauchsmuth W. 1938. Das Hüftgelenk, *Praktische Anatomie* I Bd, *Teil 4: Bein und Statik*, pp. 138–175, Berlin, Springer.

Wagner UA, Sangeorzan BJ, Harrington RM et al. 1992. Contact characteristics of the subtalar joint: load distribution between the anterior and posterior facets. *J Orthop Res* 10:535.

Walker PS, Erhman MJ. 1975. Laboratory evaluation of a metaplastic type of metacarpophalangeal joint prosthesis. *Clin Orthop* 112:349.

Wang C-L, Cheng C-K, Chen C-W, et al. 1994. Contact areas and pressure distributions in the subtalar joint. *J Biomech* 28(3):269.

Woltring HJ, Huiskes R, deLange A, Veldpaus FE. 1985. Finite centroid and helical axis estimation from noisy landmark measurements in the study of human joint kinematics. *J Biomech* 18:379.

Yoshioka Y, Siu D, Cooke TDV. 1987. The anatomy and functional axes of the femur. *J Bone Joint Surg* 69A(6):873.

Yoshioka Y, Siu D, Scudamore RA et al. 1989. Tibial anatomy in functional axes. *J Orthop Res* 7:132.

Youm Y, McMurty RY, Flatt AE et al. 1978. Kinematics of the wrist: an experimental study of radioulnar deviation and flexion/extension. *J Bone Joint Surg* 60A(4):423.

Zuckerman JD, Matsen FA. 1989. Biomechanics of the elbow. In M Nordine and VH Frankel (Eds.), *Basic Biomechanics of the Musculoskeletal System*, pp. 249–260, Philadelphia, Lea & Febiger.

4

Joint Lubrication

Michael J. Furey
*Mechanical Engineering Department
and Center for Biomedical
Engineering, Virginia Polytechnic
Institute and State University*

The Fabric of the Joints in the Human Body is a subject so much the more entertaining, as it must strike every one that considers it attentively with an Idea of fine Mechanical Composition. Wherever the Motion of one Bone upon another is requisite, there we find an excellent Apparatus for rendering that Motion safe and free: We see, for Instance, the Extremity of one Bone molded into an orbicular Cavity, to receive the Head of another, in order to afford it an extensive Play. Both are covered with a smooth elastic Crust, to prevent mutual Abrasion; connected with strong Ligaments, to prevent Dislocation; and inclosed in a Bag that contains a proper Fluid Deposited there, for lubricating the Two contiguous Surfaces. So much in general.

The above is the opening paragraph of the surgeon Sir William Hunter's classic paper "Of the Structure and Diseases of Articulating Cartilages," which he read to a meeting of the Royal Society, June 2, 1743 [1]. Since then, a great deal of research has been carried out on the subject of synovial joint lubrication. However, the mechanisms involved are still unknown.

4.1 Introduction

The purpose of this chapter is twofold: (1) to introduce the reader to the subject of tribology—the study of friction, wear, and lubrication; and (2) to extend this to the topic of *biotribology*, which includes the lubrication of natural synovial joints. It is not meant to be an exhaustive review of joint lubrication theories; space does not permit this. Instead, major concepts or principles will be discussed not only in the light of what is known about synovial joint lubrication but perhaps more importantly what is not known. Several references are given for those who wish to learn more about the topic. It is clear that synovial joints are by far the most complex and sophisticated tribological systems that exist. We shall see that although numerous theories have been put forth to attempt to explain joint lubrication, the mechanisms involved are still far from being understood. And when one begins to examine possible connections between tribology and degenerative joint disease or osteoarthritis, the picture is even more complex and controversial. Finally, this article does not treat the (1) tribological behavior of artificial joints or partial joint replacements, (2) the possible use of elastic or poroplastic materials as artificial cartilage, and (3) new developments in cartilage repair using transplanted chondrocytes. These are separate topics, which would require detailed discussion and additional space.

4.2   Tribology

The word *tribology*, derived from the Greek "to rub," covers all frictional processes between solid bodies moving relative to one another that are in contact [2]. Thus tribology may be defined as the study of friction, wear, and lubrication.

Tribological processes are involved whenever one solid slides or rolls against another, as in bearings, cams, gears, piston rings and cylinders, machining and metalworking, grinding, rock drilling, sliding electrical contacts, frictional welding, brakes, the striking of a match, music from a cello, articulation of human synovial joints (e.g., hip joints), machinery, and in numerous less obvious processes (e.g., walking, holding, stopping, writing, and the use of fasteners such as nails, screws, and bolts).

Tribology is a multidisciplinary subject involving at least the areas of materials science, solid and surface mechanics, surface science and chemistry, rheology, engineering, mathematics, and even biology and biochemistry. Although tribology is still an emerging science, interest in the phenomena of friction, wear, and lubrication is an ancient one. Unlike thermodynamics, there are no generally accepted laws in tribology. But there are some important basic principles needed to understand any study of lubrication and wear and even more so in a study of biotribology or biological lubrication phenomena. These basic principles follow.

Friction

Much of the early work in tribology was in the area of friction—possibly because frictional effects are more readily demonstrated and measured. Generally, early theories of friction dealt with dry or unlubricated systems. The problem was often treated strictly from a mechanical viewpoint, with little or no regard for the environment, surface films, or chemistry.

In the first place, *friction may be defined as the tangential resistance that is offered to the sliding of one solid body over another.* Friction is the result of many factors and cannot be treated as something as singular as density or even viscosity. Postulated sources of friction have included (1) the lifting of one asperity over another (increase in potential energy), (2) the interlocking of asperities followed by shear, (3) interlocking followed by plastic deformation or plowing, (4) adhesion followed by shear, (5) elastic hysteresis and waves of deformation, (6) adhesion or interlocking followed by tensile failure, (7) intermolecular attraction, (8) electrostatic effects, and (9) viscous drag. The coefficient of friction, indicated in the literature by μ or f, is defined as the ratio F/W where F = friction force and W = the normal load. It is emphasized that friction is a force and not a property of a solid material or lubricant.

Wear and Surface Damage

One definition of wear in a tribological sense is that it is the *progressive loss of substance from the operating surface of a body as a result of relative motion at the surface.* In comparison with friction, very little theoretical work has been done on the extremely important area of wear and surface damage. This is not too surprising in view of the complexity of wear and how little is known of the mechanisms by which it can occur. Variations in wear can be, and often are, enormous compared with variations in friction. For example, practically all the coefficients of sliding friction for diverse dry or lubricated systems fall within a relatively narrow range of 0.1 to 1. In some cases (e.g., certain regimes of hydrodynamic or "boundary" lubrication), the coefficient of friction may be <0.1 and as low as 0.001. In other cases (e.g., very clean unlubricated metals in vacuum), friction coefficients may exceed one. Reduction of friction by a factor of two through changes in design, materials, or lubricant would be a reasonable, although not always attainable, goal. On the other hand, it is not uncommon for wear rates to vary by a factor of 100, 1000, or even more.

For systems consisting of common materials (e.g., metals, polymers, ceramics), there are at least four main mechanisms by which wear and surface damage can occur between solids in relative motion: (1) abrasive wear, (2) adhesive wear, (3) fatigue wear, and (4) chemical or corrosive wear. A fifth, fretting wear and fretting corrosion, combines elements of more than one mechanism. For complex biological materials such as articular cartilage, most likely other mechanisms are involved.

Again, wear is the removal of material. The idea that friction causes wear and therefore, low friction means low wear, is a common mistake. Brief descriptions of five types of wear; abrasive, adhesive, fatigue, chemical or corrosive, and fretting—may be found in Furey [2] as well as in other references in this chapter. Next, it may be useful to consider some of the major concepts of lubrication.

4.3 Lubrication

Lubrication is a process of reducing friction and/or wear (or other forms of surface damage) between relatively moving surfaces by the application of a solid, liquid, or gaseous substance (i.e., a lubricant). Since friction and wear do not necessarily correlate with each other, the use of the word *and* in place of *and/or* in the above definition is a common mistake to be avoided. The primary function of a lubricant is to reduce friction or wear or both between moving surfaces in contact with each other.

Examples of lubricants are wide and varied. They include automotive engine oils, wheel bearing greases, transmission fluids, electrical contact lubricants, rolling oils, cutting fluids, preservative oils, gear oils, jet fuels, instrument oils, turbine oils, textile lubricants, machine oils, jet engine lubricants, air, water, molten glass, liquid metals, oxide films, talcum powder, graphite, molybdenum disulfide, waxes, soaps, polymers, and the synovial fluid in human joints.

A few general principles of lubrication may be mentioned here.

1. The lubricant must be present at the place where it can function.
2. Almost any substance under carefully selected or special conditions can be shown to reduce friction or wear in a particular test, but that does not mean these substances are lubricants.
3. Friction and wear do not necessarily go together. This is an extremely important principle which applies to nonlubricated (dry) as well as lubricated systems. It is particularly true under conditions of "boundary lubrication," to be discussed later. An additive may reduce friction and increase wear, reduce wear and increase friction, reduce both, or increase both. Although the reasons are not fully understood, this is an experimental observation. Thus, friction and wear should be thought of as separate phenomena—an important point when we discuss theories of synovial joint lubrication.
4. The effective or active lubricating film in a particular system may or may not consist of the original or bulk lubricant phase.

In a broad sense, it may be considered that the main function of a lubricant is to keep the surfaces apart so that interaction (e.g., adhesion, plowing, and shear) between the solids cannot occur; thus, friction and wear can be reduced or controlled.

The following regimes or types of lubrication may be considered in the order of increasing severity or decreasing lubricant film thickness (Fig. 4.1):

1. Hydrodynamic lubrication
2. Elastohydrodynamic lubrication
3. Transition from hydrodynamic and elastohydrodynamic lubrication to boundary lubrication
4. Boundary lubrication

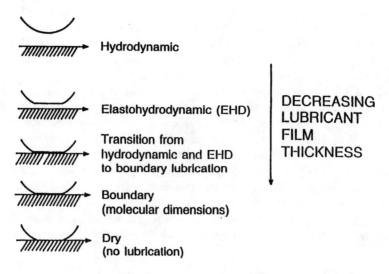

FIGURE 4.1   Regimes of lubrication.

A fifth regime, sometimes referred to as *dry* or *unlubricated*, may also be considered as an extreme or limit. In addition, there is another form of lubrication that does not require relative movement of the bodies either parallel or perpendicular to the surface, i.e., as in externally pressurized hydrostatic or aerostatic bearings.

Hydrodynamic Lubrication Theories

In hydrodynamic lubrication, the load is supported by the pressure developed due to relative motion and the geometry of the system. In the regime of hydrodynamic or fluid film lubrication, there is no contact between the solids. The film thickness is governed by the bulk physical properties of the lubricants, the most important being viscosity; friction arises purely from shearing of viscous lubricant.

Contributions to our knowledge of hydrodynamic lubrication, with special focus on journal bearings, have been made by numerous investigators including Reynolds. The classic Reynolds treatment considered the equilibrium of a fluid element and the pressure and shear forces on this element. In this treatment, eight assumptions were made (e.g., surface curvature is large compared to lubricant film thickness, fluid is Newtonian, flow is laminar, viscosity is constant through film thickness). Velocity distributions due to relative motion and pressure buildup were developed and added together. The solution of the basic Reynolds equation for a particular bearing configuration results in a pressure distribution throughout the film as a function of viscosity, film shape, and velocity.

The total load W and frictional (viscous) drag F can be calculated from this information. For rotating disks with parallel axes, the "simple" Reynolds equation yields:

$$\frac{h_o}{R} = 4.9\left(\frac{\eta U}{W}\right) \tag{4.1}$$

where h_o is the minimum lubricant film thickness, η is the absolute viscosity, U is the average velocity $(U_1 + U_2)/2$, W is the applied normal load per unit width of disk, and R is the reduced radius of curvature $(1/R = 1/R_1 + 1/R_2)$.

The dimensionless term $(\eta U/W)$ is sometimes referred to as the hydrodynamic factor. It can be seen that doubling either the viscosity or velocity doubles the film thickness, and that doubling the applied load halves the film thickness. This regime of lubrication is sometimes referred to as the *rigid isoviscous* or *classical Martin condition*, since the solid bodies are assumed to be perfectly rigid (non-deformable), and the fluid is assumed to have a constant viscosity.

At high loads with systems such as gears, ball bearings, and other high-contact-stress geometries, two additional factors have been considered in further developments of the hydrodynamic theory of lubrication. One of these is that the surfaces deform elastically; this leads to a localized change in geometry more favorable to lubrication. The second is that the lubricant becomes more viscous under the high pressure existing in the contact zone, according to relationships such as:

$$\eta/\eta_o = \exp \alpha \left(p - p_o\right) \tag{4.2}$$

where η is the viscosity at pressure p, η_o is the viscosity at atmospheric pressure p_o, and α is the pressure-viscosity coefficient (e.g., in Pa^{-1}). In this concept, the lubricant pressures existing in the contact zone approximate those of dry contact Hertzian stress. This is the regime of elastohydrodynamic lubrication, sometimes abbreviated as EHL or EHD. It may also be described as the elastic-viscous type or mode of lubrication, since elastic deformation exists and the fluid viscosity is considerably greater due to the pressure effect.

The comparable Dowson–Higginson expression for minimum film thickness between cylinders or disks in contact with parallel axes is:

$$\frac{h_o}{R} = 2.6 \left(\frac{\eta U}{W}\right)^{0.7} \left(\frac{\alpha W}{R}\right)^{0.54} \left(\frac{W}{RE'}\right)^{0.03} \tag{4.3}$$

The term E' represents the reduced modulus of elasticity:

$$\frac{1}{E'} = \frac{\left(1 - v_1^2\right)}{E_1} + \frac{\left(1 - v_2^2\right)}{E_2} \tag{4.4}$$

where E is the modulus, v is Poisson's ratio, and the subscripts *1* and *2* refer to the two solids in contact. All the other terms are the same as previously stated. In addition to the hydrodynamic factor ($\eta U/W$), a pressure-viscosity factor ($\alpha W/R$) and an elastic deformation factor (W/RE') can be considered. Thus, properties of both the lubricant and the solids as materials are included. In examining the elastohydrodynamic film thickness equations, it can be seen that the velocity U is an important factor ($h_o \propto U^{0.7}$) but the load W is rather unimportant ($h_o \propto W^{-0.13}$).

Experimental confirmation of the elastohydrodynamic lubrication theory has been obtained in certain selected systems using electrical capacitance, x-ray transmission, and optical interference techniques to determine film thickness and shape under dynamic conditions. Research is continuing in this area, including studies on micro-EHL or asperity lubrication mechanisms, since surfaces are never perfectly smooth. These studies may lead to a better understanding of not only lubricant film formation in high-contact-stress systems but lubricant film failure as well.

Two other possible types of hydrodynamic lubrication, rigid-viscous and elastic-isoviscous, complete the matrix of four, considering the two factors of elastic deformation and pressure-viscosity effects. In addition, *squeeze film* lubrication can occur when surfaces approach one another. For more information on hydrodynamic and elastohydrodynamic lubrication, see Cameron [3] and Dowson and Higginson [4].

Transition from Hydrodynamic to Boundary Lubrication

Although prevention of contact is probably the most important function of a lubricant, there is still much to be learned about the transition from hydrodynamic and elastohydrodynamic lubrication to boundary lubrication. This is the region in which lubrication goes from the desirable hydrodynamic condition of no contact to the less acceptable "boundary" condition, where increased contact usually leads to higher friction and wear. This regime is sometimes referred to as a condition of *mixed lubrication*.

Several examples of experimental approaches to thin-film lubrication have been reported [3]. It is important in examining these techniques to make the distinction between methods that are used to determine lubricant film thickness under hydrodynamic or elastohydrodynamic conditions (e.g., optical interference, electrical capacitance, or x-ray transmission), and methods that are used to determine the occurrence or frequency of contact. As we will see later, most experimental studies of synovial joint lubrication have focused on friction measurements, using the information to determine the lubrication regime involved; this approach can be misleading.

Boundary Lubrication

Although there is no generally accepted definition of boundary lubrication, it is often described as a condition of lubrication in which the friction and wear between two surfaces in relative motion are determined by the surface properties of the solids and the chemical nature of the lubricant rather than its viscosity. An example of the difficulty in defining boundary lubrication can be seen if the term *bulk viscosity* is used in place of viscosity in the preceding sentence—another frequent form. This opens the door to the inclusion of elastohydrodynamic effects which depend in part on the influence of pressure on viscosity. Increased friction under these circumstances could be attributed to increased viscous drag rather than solid-solid contact. According to another common definition, boundary lubrication occurs or exists when the surfaces of the bearing solids are separated by films of molecular thickness. That may be true, but it ignores the possibility that "boundary" layer surface films may indeed be very thick (i.e., 10, 20, or 100 molecular layers). The difficulty is that boundary lubrication is complex.

Although a considerable amount of research has been done on this topic, an understanding of the basic mechanisms and processes involved is by no means complete. Therefore, definitions of boundary lubrication tend to be nonoperational. This is an extremely important regime of lubrication because it involves more extensive solid–solid contact and interaction as well as generally greater friction, wear, and surface damage. In many practical systems, the occurrence of the boundary lubrication regime is unavoidable or at least quite common. The condition can be brought about by high loads, low relative sliding speeds (including zero for stop-and-go, motion reversal, or reciprocating elements), and low lubricant viscosity—factors that are important in the transition from hydrodynamic to boundary lubrication.

The most important factor in boundary lubrication is the chemistry of the tribological system—the contacting solids and total environment including lubricants. More particularly, the surface chemistry and interactions occurring with and on the solid surfaces are important. This includes factors such as physisorption, chemisorption, intermolecular forces, surface chemical reactions, and the nature, structure, and properties of thin films on solid surfaces. It also includes many other effects brought on by the process of moving one solid over another, such as: (1) changes in topography and the area of contact, (2) high surface temperatures, (3) the generation of fresh reactive metal surfaces by the removal of oxide and other layers, (4) catalysis, (5) the generation of electrical charges, and (6) the emission of charged particles such as electrons.

In examining the action of boundary lubricant compounds in reducing friction or wear or both between solids in sliding contact, it may be helpful to consider at least the following five modes of film formation on or protection of surfaces: (1) physisorption, (2) chemisorption, (3) chemical reactions with the solid surface, (4) chemical reactions on the solid surface, and (5) mere interposition of a solid or other material. These modes of surface protection are discussed in more detail in Furey [2].

The beneficial and harmful effects of minor changes in chemistry of the environment (e.g., the lubricant) are often enormous in comparison with hydrodynamic and elastohydrodynamic effects. Thus, the surface and chemical properties of the solid materials used in tribological applications become especially important. One might expect that this would also be the case in biological (e.g., human joint) lubrication where biochemistry is very likely an important factor.

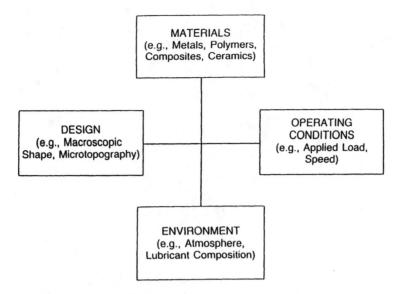

FIGURE 4.2   In any tribological system, friction, wear, and surface damage depend on four interrelated factors.

General Comments on Tribological Processes

It is important to recognize that friction and wear depend upon four major factors, i.e., materials, design, operating conditions, and total environment (Fig. 4.2). This four-block figure may be useful as a guide in thinking about synovial joint lubrication either from a theoretical or experimental viewpoint—the topic discussed in the next section.

Readers are cautioned against the use of various terms in tribology which are either vaguely defined or not defined at all. These would include such terms as "lubricating ability," "lubricity," and even "boundary lubrication." For example, do "boundary lubricating properties" refer to effects on friction or effects on wear and damage? It makes a difference. It is emphasized once again that friction and wear are different phenomena. Low friction does not necessarily mean low wear. We will see several examples of this common error in the discussion of joint lubrication research.

4.4   Synovial Joints

Examples of natural synovial or movable joints include the human hip, knee, elbow, ankle, finger, and shoulder. A simplified representation of a synovial joint is shown in Fig. 4.3. The bones are covered by a thin layer of articular cartilage bathed in synovial fluid confined by synovial membrane. Synovial joints are truly remarkable systems—providing the basis of movement by allowing bones to articulate on one another with minimal friction and wear. Unfortunately, various joint diseases occur even among the young—causing pain, loss of freedom of movement, or instability.

Synovial joints are complex, sophisticated systems not yet fully understood. The loads are surprisingly high and the relative motion is complex. Articular cartilage has the deceptive appearance of simplicity and uniformity. But it is an extremely complex material with unusual properties. Basically, it consists of water (approximately 75%) enmeshed in a network of collagen fibers and proteoglycans with high molecular weight. In a way, cartilage could be considered as one of Nature's composite materials. Articular cartilage also has no blood supply, no nerves, and very few cells (chondrocytes).

The other major component of an articular joint is *synovial fluid*, named by Paracelsus after "synovia" (egg-white). It is essentially a dialysate of blood plasma with added hyaluronic acid. Synovial fluid contains complex proteins, polysaccharides, and other compounds. Its chief constituent is water (approximately 85%). Synovial fluid functions as a joint lubricant, nutrient for cartilage, and carrier for waste products.

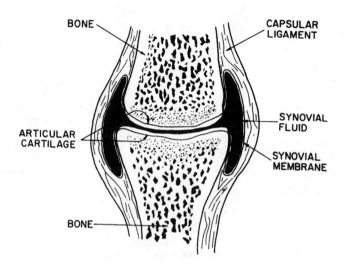

FIGURE 4.3   Representation of a synovial joint.

For more information on the biochemistry, structure, and properties of articular cartilage, Freeman [5], Sokoloff [6], Stockwell [7] and articles referenced in these works are suggested.

4.5   Theories on the Lubrication of Natural and Normal Synovial Joints

As stated, the word *tribology* means the study of friction, wear, and lubrication. Therefore, *biotribology* may be thought of as the study of biological lubrication processes, e.g., as in synovial joints. A surprisingly large number of concepts and theories of synovial joint lubrication have been proposed [8–10] (as shown in Table 4.1). And even if similar ideas are grouped together, there are still well over a dozen fundamentally different theories. These have included a wide range of lubrication concepts, e.g., hydrodynamic, hydrostatic, elasto-hydrodynamic, squeeze-film, "boundary," mixed-regime, "weeping," osmotic, synovial mucin gel, "boosted," lipid, electrostatic, porous layers, and special forms of boundary lubrication (e.g., "lubricating glycoproteins," structuring of boundary water "surface-active" phospholipids). This chapter will not review these numerous theories, but excellent reviews on the lubrication of synovial joints have been written by McCutchen [11], Swanson [12], and Higginsworth and Unsworth [13]. The book edited by Dumbleton is also recommended [14]. In addition, theses by Droogendijk [15] and Burkhardt [16] contain extensive and detailed reviews of theories of joint lubrication.

McCutchen was the first to propose an entirely new concept of lubrication, "weeping lubrication", applied to synovial joint action [17,18]. He considered unique and special properties of cartilage and how this could affect flow and lubrication. The work of Mow et al. continued along a more complex and sophisticated approach in which a biomechanical model is proposed for the study of the dynamic interaction between synovial fluid and articular cartilage [19,20]. These ideas are combined in the more recent work of Ateshian [21] which uses a framework of the biphasic theory of articular cartilage to model interstitial fluid pressurization. Several additional studies have also been made of effects of porosity and compliance, including the behavior of elastic layers, in producing hydrodynamic and squeeze-film lubrication. A good review in this area was given by Unsworth who discussed both human and artificial joints [22].

The following general observations are offered on the theories of synovial joint lubrication that have been proposed:

1. Most of the theories are strictly mechanical or rheological—involving such factors as deformation, pressure, and fluid flow.
2. There is a preoccupation with *friction*, which of course is very low for articular cartilage systems.
3. None of the theories considers *wear*—which is neither the same as friction nor related to it.
4. The detailed structure, biochemistry, complexity, and living nature of the total articular cartilage–synovial fluid system are generally ignored.

These are only general impressions. And although mechanical/rheological concepts seem dominant (with a focus on friction), wear and biochemistry are not completely ignored. For example, Simon [23] abraded articular cartilage from human patellae and canine femoral heads with a stainless steel rotary file, measuring the depth of penetration with time and the amount of wear debris generated. Cartilage wear was also studied experimentally by Bloebaum and Wilson [24], Radin and Paul [25], and Lipshitz, Etheredge, and Glimcher [26–28]. The latter researchers carried out several *in vitro* studies of wear of articular cartilage using bovine cartilage plugs or specimens in sliding contact against stainless steel plates. They developed a means of measuring cartilage wear by determining the hydroxyproline content of both the lubricant and solid wear debris. Using this system and technique, effects of variables such as time, applied load, and chemical modification of articular cartilage on wear and profile changes were determined. This work is of particular importance in that they addressed the question of *cartilage wear and damage* rather than friction, recognizing that wear and friction are different phenomena.

Special note is also made of two researchers, Swann and Sokoloff, who considered biochemistry as an important factor in synovial joint lubrication. Swann et al. very carefully isolated fractions of bovine synovial fluid using sequential sedimentation techniques and gel permeation chromatography. They found a high molecular weight glycoprotein to be the major constituent in the articular lubrication fraction from bovine synovial fluid and called this LGP-I (from lubricating glycoprotein). This was based on friction measurements using cartilage in sliding contact against a glass disc. An excellent summary of this work with additional references is presented in a chapter by Swann in *The Joints and Synovial Fluid: I* [6].

Sokoloff et al. examined the "boundary lubricating ability" of several synovial fluids using a latex-glass test system and cartilage specimens obtained at necropsy from knees [29]. Measurements were made of friction. The research was extended to other *in vitro* friction tests using cartilage obtained from the nasal septum of cows and widely differing artificial surfaces [30]. As a result of this work, a new model of boundary lubrication by synovial fluid was proposed—the structuring of boundary water. The postulate involves adsorption of one part of a glycoprotein on a surface followed by the formation of hydration shells around the polar portions of the adsorbed glycoprotein; the net result is a thin layer of viscous "structured" water at the surface. This work is of particular interest in that it involves not only a specific and more detailed mechanism of boundary lubrication in synovial joints but also takes into account the possible importance of water in this system.

In more recent research by Jay, an interaction between hyaluronic acid and a "purified synovial lubricating factor" (PSLF) was observed, suggesting a possible synergistic action in the boundary lubrication of synovial joints [31]. The definition of "lubricating ability" was based on friction measurements made with a latex-covered stainless steel stud in oscillating contact against polished glass.

The above summary of major synovial joint lubrication theories is taken from Furey and Burkhardt [10] and Jay [31], as well as the thesis by Burkhardt [33].

Two more recent studies are of interest since cartilage wear was considered although not as a part of a theory of joint lubrication. Stachowiak et al. [34] investigated the friction and wear characteristics of adult rat femur cartilage against a stainless steel plate using an environmental scanning microscope (ESM) to examine damaged cartilage. One finding was evidence of a load limit to lubrication of cartilage, beyond which high friction and damage occurred. Another study, by Hayes et al. [35] on the influence of crystals on cartilage wear, is particularly interesting not only in the findings reported (e.g., certain crystals can increase cartilage wear), but also in the full description of the biochemical techniques used.

TABLE 4.1 Examples of Proposed Mechanisms and Studies of Synovial Joint Lubrication

Mechanism	Author	Date
1. Hydrodynamic	MacConnail	1932
2. Boundary	Jones	1934
3. Hydrodynamic	Jones	1936
4. Boundary	Charnley	1959
5. Weeping	McCutchen	1959
6. Floating	Barnett and Cobbold	1962
7. Elastohydrodynamic (EHL)	Tanner	1966
	Dowson	1967
8. Thixotropic/elastic fluid	Dintenfass	1963
9. Osmotic (boundary)	McCutchen	1966
10. Squeeze-film	Fein	1966
	Higginson et al.	1974
11. Synovial gel	Maroudas	1967
12. Thin-film	Faber et al.	1967
13. Combinations of hydrostatic, boundary, and EHL	Linn	1968
14. Boosted	Walker et al.	1968
15. Lipid	Little et al.	1969
16. Weeping + boundary	McCutchen and Wilkins	1969
	McCutchen	1969
17. Boundary	Caygill and West	1969
18. Fat (or mucin)	Freeman et al.	1970
19. Electrostatic	Roberts	1971
20. Boundary + fluid squeeze-film	Radin and Paul	1972
21. Mixed	Unsworth et al.	1974
22. Imbibe/exudate composite model	Ling	1974
23. Complex biomechanical model	Mow et al.	1974
	Mansour and Mow	1977
24. Two porous layer model	Dinnar	1974
25. Boundary	Reimann et al.	1975
26. Squeeze-film + fluid film + boundary	Unsworth, Dowson et al.	1975
27. Compliant bearing model	Rybicki	1977
28. Lubricating glycoproteins	Swann et al.	1977
29. Structuring of boundary water	Sokoloff et al.	1979
30. Surface flow	Kenyon	1980
31. Lubricin	Swann et al.	1985
32. Micro-EHL	Dowson and Jin	1986
33. Lubricating factor	Jay	1992
34. Lipidic component	LaBerge et al.	1993
35. Constitutive modeling of cartilage	Lai et al.	1993
36. Asperity model	Yao et al.	1993
37. Bingham fluid	Tandon et al.	1994
38. Filtration/gel/squeeze film	Hlavacek et al.	1995
39. Surface-active phospholipid	Schwarz and Hills	1998
40. Interstitial fluid pressurization	Ateshian et al.	1998

A special note should be made concerning the doctoral thesis by Lawrence Malcom in 1976 [36]. This is an excellent study of cartilage friction and deformation, in which a device resembling a rotary plate rheometer was used to investigate the effects of static and dynamic loading on the frictional behavior of bovine cartilage. The contact geometry consisted of a circular cylindrical annulus in contact with a concave hemispherical section. It was found that dynamically loaded specimens in bovine synovial fluid yielded the more efficient lubrication based on friction measurements. The Malcom study is thorough and excellent in its attention to detail (e.g., specimen preparation) in examining the influence of type of loading and time effects on cartilage friction. It does not, however, consider cartilage wear and damage except in a very preliminary way. And it does not consider the influence of fluid biochemistry on cartilage friction, wear, and damage. In short, the Malcom work represents a superb piece of systematic research along the lines of mechanical, dynamic, rheological, and viscoelastic behavior—one important dimension of synovial joint lubrication.

4.6 *In Vitro* Cartilage Wear Studies

Over the past 15 years, studies aimed at exploring possible connections between tribology and mechanisms of synovial joint lubrication and degeneration (e.g., osteoarthritis) have been conducted by the author and his graduate and undergraduate students in the Department of Mechanical Engineering at Virginia Polytechnic Institute and State University. The basic approach used involved *in vitro* tribological experiments using bovine articular cartilage, with an emphasis on the effects of fluid composition and biochemistry on cartilage wear and damage. This research is an outgrowth of earlier work carried out during a sabbatical study in the Laboratory for the Study of Skeletal Disorders, The Children's Hospital Medical Center, Harvard Medical School in Boston. In that study, bovine cartilage test specimens were loaded against a polished steel plate and subjected to reciprocating sliding for several hours in the presence of a fluid (e.g., bovine synovial fluid or a buffered saline reference fluid containing biochemical constituents kindly provided by Dr. David Swann). Cartilage wear was determined by sampling the test fluid and determining the concentration of 4-hydroxyproline—a constituent of collagen. The results of that earlier study have been reported and summarized elsewhere [37–40]. Figure 4.4 shows the average hydroxyproline contents of wear debris obtained from these *in vitro* experiments. These numbers are related to the cartilage wear which occurred. However, since the total quantities of collected fluids varied somewhat, the values shown in the bar graph should not be taken as exact or precise measures of fluid effects on cartilage wear.

The main conclusions of that study were as follows:

1. Normal bovine synovial fluid is very effective in reducing cartilage wear under these *in vitro* conditions as compared to the buffered saline reference fluid.
2. There is no significant difference in wear between the saline reference and distilled water.
3. The addition of hyaluronic acid to the reference fluid significantly reduces wear, but its effect depends on the source.
4. Under these tests conditions, Swann's LGP-I (lubricating glycoprotein-I), known to be extremely effective in reducing friction in cartilage-on-glass tests, does not reduce cartilage wear.

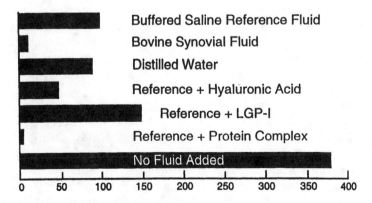

FIGURE 4.4 Relative cartilage wear based on hydroxyproline content of debris (*in vitro* tests with cartilage on stainless steel).

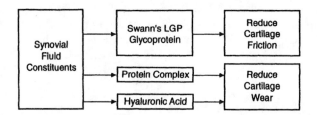

FIGURE 4.5   Friction and wear are different phenomena.

5. However, a protein complex isolated by Swann is extremely effective in reducing wear—producing results similar to those obtained with synovial fluid. The detailed structure of this constituent is complex and has not yet been fully determined.
6. Last, the lack of an added fluid in these experiments leads to extremely high wear and damage of the articular cartilage.

In discussing the possible significance of these findings from a tribological point of view, it may be helpful first of all to emphasize once again that friction and wear are different phenomena. Furthermore, as suggested by Fig. 4.5, certain constituents of synovial fluid (e.g., Swann's lubricating glycoprotein) may act to reduce *friction* in synovial joints while other constituents (e.g., Swann's protein complex or hyaluronic acid) may act to reduce cartilage *wear*. Therefore, it is necessary to distinguish between biochemical anti-friction and anti-wear compounds present in synovial fluid.

In more recent years, this study has been greatly enhanced by the participation of interested faculty and students from the Virginia-Maryland College of Veterinary Medicine and Department of Biochemistry and Animal Science at Virginia Tech. One major hypothesis tested is a continuation of previous work showing that the detailed biochemistry of the fluid-cartilage system has a pronounced and possibly controlling influence on cartilage wear. A consequence of the above hypothesis is that a lack or deficiency of certain biochemical constituents in the synovial joint may be one factor contributing to the initiation and progression of cartilage damage, wear, and possibly osteoarthritis. A related but somewhat different hypothesis concerns synovial fluid constituents which may act to increase the wear and further damage of articular cartilage under tribological contact.

To carry out continued research on biotribology, a new device for studies of cartilage deformation, wear, damage, and friction under conditions of tribological contact was designed by Burkhardt [33] and later modified, constructed, and instrumented. A simplified sketch is shown in Fig. 4.6. The key features of this test device are shown in Table 4.2. The apparatus is designed to accommodate cartilage-on-cartilage specimens. Motion of the lower specimen is controlled by a computer-driven $X–Y$ table, allowing simple oscillating motion or complex motion patterns. An octagonal strain ring with two full semiconductor bridges is used to measure the normal load as well as the tangential load (friction). An LVDT, not shown in the figure, is used to measure cartilage deformation and linear wear during a test. However, hydroxyproline analysis of the wear debris and washings is used for the actual determination of total cartilage wear on a mass basis.

In one study by Schroeder [41], two types of experiments were carried out, i.e., cartilage-on-stainless steel and cartilage-on-cartilage, at applied loads up to 70 N—yielding an average pressure of 2.2 MPa in the contact area. Reciprocating motion (40 cps) was used. The fluids tested included: (1) a buffered saline solution, (2) saline plus hyaluronic acid, and (3) bovine synovial fluid. In cartilage-on-stainless steel tests, scanning electron microscopy, and histological staining showed distinct effects of the lubricants on surface and subsurface damage. Tests with the buffered saline fluid resulted in the most damage, with large wear tracks visible on the surface of the cartilage plug, as well as subsurface voids and cracks. When hyaluronic acid, a constituent of the natural synovial joint lubricant, was added to the saline reference fluid, less severe damage was observed. Little or no cartilage damage was evident in tests in which the natural synovial joint fluid was used as the lubricant.

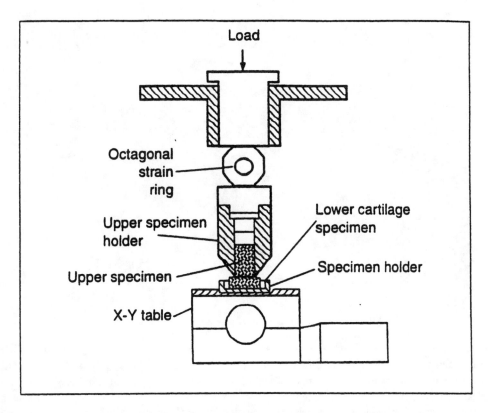

FIGURE 4.6 Device for *in vitro* cartilage-on-cartilage wear studies.

TABLE 4.2 Key Features of Test Device Designed for Cartilage Wear Studies [33]

Contact system	Cartilage-on-cartilage
Contact geometry	Flat-on-flat, convex-on-flat, irregular-on-irregular
Cartilage type	Articular, any source (e.g., bovine)
Specimen size	Upper specimen, 4 to 6 mm diameter, lower specimen, ca. 15 to 25 mm diameter
Applied load	50–660 N
Average pressure	0.44–4.4 MPa
Type of motion	Linear, oscillating; circular, constant velocity; more complex patterns
Sliding velocity	0 to 20 mm/s
Fluid temperature	Ambient (20°C) or controlled humidity
Environment	Ambient or controlled humidity
Measurements	Normal load, cartilage deformation, friction; cartilage wear and damage, biochemical analysis of cartilage specimens, synovial fluid, and wear debris; sub-surface changes

These results were confirmed in a later study by Owellen [42] in which hydroxyproline analysis was used to determine cartilage wear. It was found that increasing the applied load from 20 to 65 N increased cartilage wear by eightfold for the saline solution and approximately threefold for synovial fluid. Furthermore, the coefficient of friction increased from an initial low value of 0.01 to 0.02 to a much higher value, e.g., 0.20 to 0.30 and higher, during a normal test which lasted 3 hours; the greatest change occurred during the first 20 minutes. Another interesting result was that a thin film of transferred or altered material was observed on the stainless steel disks—being most pronounced with the buffered saline lubricant and not observed with synovial fluid. Examination of the film with Fourier transfer infrared microspectrometry shows distinctive bio-organic spectra which differ from that of the original bovine cartilage. We believe this to be an important finding since it suggests a possible biotribochemical effect [43].

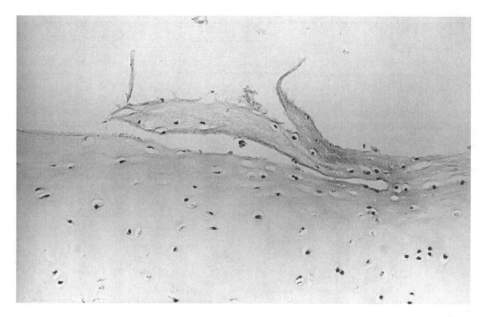

FIGURE 4.7   Cartilage damage produced by sliding contact.

In another phase of this research, the emphasis is on the cartilage-on-cartilage system and the influence of potentially beneficial as well as harmful constituents of synovial fluid on wear and damage. In cartilage-on-cartilage tests, the most severe wear and damage occurred during tests with buffered saline as the lubricant. The damage was less severe than in the stainless steel tests, but some visible wear tracks were detectable with scanning electron microscopy. Histological sectioning and staining of both the upper and lower cartilage samples show evidence of elongated lacunae and coalesced voids that could lead to wear by delamination. An example is shown in Fig. 4.7 (original magnification of 500× on 35-mm slide). The proteoglycan content of the subsurface cartilage under the region of contact was also reduced. When synovial fluid was used as the lubricant, no visible wear or damage was detected [44]. These results demonstrate that even in *in vitro* tests with bovine articular cartilage, the nature of the fluid environment can have a dramatic affect on the severity of wear and subsurface damage.

In a more recent study carried out by Berrien in the biotribology program at Virginia Tech, a different approach was taken to examine the role of joint lubrication in joint disease, particularly osteoarthritis. A degradative biological enzyme, collagenase-3, suspected of playing a role in a cartilage degeneration was used to create a physiologically adverse biochemical fluid environment. Tribological tests were performed with the same device and procedures described previously. The stainless steel disk was replaced with a 1-in. diameter plug of bovine cartilage to create a cartilage sliding on cartilage configuration more closely related to the *in vivo* condition. Normal load was increased to 78.6 N and synovial fluid and buffered saline were used as lubricants. Prior to testing, cartilage plugs were exposed to a fluid medium containing three concentrations of collagenase-3 for 24 h. The major discovery of this work was that exposure to the collagenase-3 enzyme had a substantial adverse effect on cartilage wear *in vitro*, increasing average wear values by three and one-half times those of the unexposed cases. Figure 4.8 shows an example of the effect of enzyme treatment when bovine synovial fluid was used as the lubricant. Scanning electron microscopy showed disruption of the superficial layer and collagen matrix with exposure to collagenase-3, where unexposed cartilage showed none. Histological sections showed a substantial loss of the superficial layer of cartilage and a distinct and abnormal loss of proteoglycans in the middle layer of collagenase-treated cartilage. Unexposed cartilage showed only minor disruption of the superficial layer [45].

This study indicates that some of the biochemical constituents that gain access to the joint space, during normal and pathological functions, can have a significant adverse effect on the wear and damage

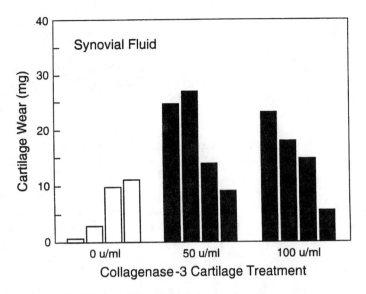

FIGURE 4.8   Effect of collagenase-3 on cartilage wear.

of the articular cartilage. Future studies will include determination of additional constituents that have harmful effects on cartilage wear and damage. This research, using bovine articular cartilage in *in vitro* sliding contact tests, raises a number of interesting questions:

1. Has Nature designed a special biochemical compound that has as its function the protection of articular cartilage?
2. What is the mechanism (or mechanisms) by which biochemical constituents of synovial fluid can act to reduce wear of articular cartilage?
3. Could a lack of this biochemical constituent lead to increased cartilage wear and damage?
4. Does articular cartilage from osteoarthritic patients have reduced wear resistance?
5. Do any of the findings on the importance of synovial fluid biochemistry on cartilage wear in our *in vitro* studies apply to living or *in vitro* systems as well?
6. How does collagenase-3 treatment of cartilage lead to increased wear and does this finding have any significance in the *in vivo* situation? This question is addressed in the next section.

4.7   Biotribology and Arthritis: Are There Connections?

Arthritis is an umbrella term for more than 100 rheumatic diseases affecting joints and connective tissue. The two most common forms are osteoarthritis (OA) and rheumatoid arthritis (RA). Osteoarthritis— also referred to as *osteoarthrosis or degenerative joint disease*—is the most common form of arthritis. It is sometimes simplistically described as the "wear and tear" form of arthritis. The causes and progression of degenerative joint disease are still not understood. Rheumatoid arthritis is a chronic and often progressive disease of the synovial membrane leading to release of enzymes which attack, erode, and destroy articular cartilage. It is an inflammatory response involving the immune system and is more prevalent in females. Rheumatoid arthritis is extremely complex. Its causes are still unknown.

Sokoloff defines degenerative joint disease as "an extremely common, noninflammatory, progressive disorder of movable joints, particularly weight-bearing joints, characterized pathologically by deterioration of articular cartilage and by formation of new bone in the sub-chondral areas and at the margins of the joint" [46]. As mentioned, osteoarthritis or osteoarthrosis is sometimes referred to as the "wear and tear" form of arthritis, but wear itself is rarely a simple process even in well-defined systems.

It has been noted by the author that tribological terms occasionally appear in hypotheses which describe the etiology of osteoarthritis (e.g., "reduced wear resistance of cartilage" or "poor lubricity of synovial fluid"). It has also been noted that there is a general absence of hypotheses connecting normal synovial joint *lubrication* (or lack thereof) and synovial joint *degeneration*. Perhaps it is natural (and unhelpful) for a tribologist to imagine such a connection and that, for example, cartilage wear under certain circumstances might be due to or influenced by a lack of proper "boundary lubrication" by the synovial fluid. In this regard, it may be of interest to quote Swanson [12] who said in 1979 that "there exists at present no experimental evidence which certainly shows that a failure of lubrication is or is not a causative factor in the first stages of cartilage degeneration." A statement made by Professor Glimcher [52] may also be appropriate here. Glimcher fully recognized the fundamental difference between friction and wear as well as the difference between joint lubrication (one area of study) and joint degeneration (another area of study). Glimcher said that wearing or abrading cartilage with a steel file is not osteoarthritis, and neither is digesting cartilage in a test tube with an enzyme. But both forms of cartilage deterioration can occur in a living joint and in a way which is still not understood. It is interesting that essentially none of the many synovial joint lubrication theories consider enzymatic degradation of cartilage as a factor whereas practically all the models of the etiology of degenerative joint disease include this as an important factor.

It was stated earlier that there are at least two main areas to consider, i.e., (1) mechanisms of synovial joint lubrication and (2) the etiology of synovial joint degeneration (e.g., as in osteoarthrosis). Both areas are extremely complex. And the key questions as to what actually happens in each have yet to be answered (and perhaps asked). It may therefore be presumptuous of the present author to suggest possible connections between two areas which in themselves are still not fully understood.

Tribological processes in a movable joint involve not only the contacting surfaces (articular cartilage), but the surrounding medium (synovial fluid) as well. Each of these depends on the synthesis and transport of necessary biochemical constituents to the contact region or interface. As a result of relative motion (sliding, rubbing, rolling, and impact) between the joint elements, friction and/or wear can occur.

It has already been shown and discussed—at least in *in vitro* tests with articular cartilage—that compounds which reduce friction do not necessarily reduce wear; the latter was suggested as being more important [10]. It may be helpful first of all to emphasize once again that friction and wear are different phenomena. Furthermore, certain constituents of synovial fluid (e.g., Swann's lubricating glycoprotein) may act to reduce *friction* in synovial joints while other constituents (e.g., Swann's protein complex or hyaluronic acid) may act to reduce cartilage *wear*.

A significant increase in joint friction could lead to a slight increase in local temperatures or possibly to reduce mobility. But the effects of cartilage wear would be expected to be more serious. When cartilage wear occurs, a very special material is lost and the body is not capable of regenerating cartilage of the same quality nor at the desired rate. Thus, there are at least two major tribological dimensions involved—one concerning the nature of the synovial fluid and the other having to do with the properties of articular cartilage itself. Changes in *either* the synovial fluid or cartilage could conceivably lead to increased wear or damage (or friction) as shown in Fig. 4.9.

A simplified model or illustration of possible connections between osteoarthritis and tribology is offered in Fig. 4.10 taken from Furey [53]. Its purpose is to stimulate discussion. There are other pathways to the disease, pathways which may include genetic factors.

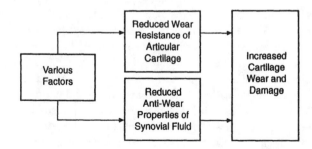

FIGURE 4.9   Two tribological aspects of synovial joint lubrication.

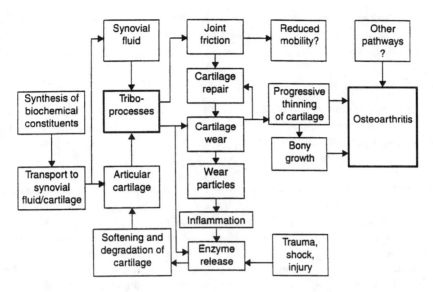

FIGURE 4.10   Osteoarthritis–tribology connections?

In some cases, the body makes an unsuccessful attempt at repair, and bone growth may occur at the periphery of contact. As suggested by Fig. 4.10, this process and the generation of wear particles could lead to joint inflammation and the release of enzymes which further soften and degrade the articular cartilage. This softer, degraded cartilage does not possess the wear resistance of the original. It has been shown previously that treatment of cartilage with collagenase-3 increases wear significantly, thus supporting the idea of enzyme release as a factor in osteoarthritis. Thus, there exists a feedback process in which the occurrence of cartilage wear can lead to even more damage. Degradative enzymes can also be released by trauma, shock, or injury to the joint. Ultimately, as the cartilage is progressively thinned and bony growth occurs, a condition of osteoarthritis or degenerative joint disease may exist. There are other pathways to the disease, pathways which may include genetic factors. It is not argued that arthritis is a tribological problem. However, the inclusion of tribological processes in one set of pathways to osteoarthrosis would not seem strange or unusual.

A specific example of a different tribological dimension to the problem of synovial joint lubrication (i.e., third-body abrasion), was shown by the work of Hayes et al. [54]. In an excellent study of the effect of crystals on the wear of articular cartilage, they carried out *in vitro* tests using cylindrical cartilage subchondral bone plugs obtained from equine fetlock joints in sliding contact against a stainless steel plate. They examined the effects of three types of crystals (orthorhombic calcium pyrophosphate tetrahydrate, monoclinic calcium pyrophosphate dehydrate, and calcium hydroxyapatite) on wear using a Ringer's solution as the carrier fluid. Concentration of cartilage wear debris in the fluid was determined by analyzing for inorganic sulphate derived from the proteoglycans present. Several interesting findings were made, one of them being that the presence of the crystals roughly doubled cartilage wear. This is an important contribution which should be read by anyone seriously contemplating research on the tribology of articular cartilage. The careful attention to detail and potential problems, as well as the precise description of the biochemical procedures and diverse experimental techniques used, set a high standard.

4.8   Recapitulation and Final Comments

It is obvious from the unusually large number of theories of synovial joint lubrication proposed, that very little is known about the subject. Synovial joints are undoubtedly the most sophisticated and complex tribological systems that exist or will ever exist. It will require a great deal more research—possibly very different approaches—before we even begin to understand the processes involved.

Some general comments and specific suggestions are offered—not for the purpose of criticizing any particular study but hopefully to provide ideas which may be helpful in further research as well as in the re-interpretation of some past research.

Terms and Definitions

First of all, as mentioned earlier in this chapter, part of the problem has to do with the use and misuse of various terms in tribology—the study of friction, wear, and lubrication. A glance at any number of the published papers on synovial joint lubrication will reveal such terms and phrases as "lubricating ability," "lubricity," "lubricating properties," "lubricating component," and many others, all undefined. We also see terms like "boundary lubricant," "lubricating glycoprotein," or "lubricin." There is nothing inherently wrong with this but one should remember that lubrication is a process of reducing friction and/or wear between rubbing surfaces. Saying that a fluid is a "good" lubricant does not distinguish between friction and wear. And assuming that friction and wear are correlated and go together is the first pitfall in any tribological study. It cannot be overemphasized that friction and wear are different, though sometimes related, phenomena. Low friction does not mean low wear. The terms and phrases used are therefore extremely important. For example, in a brief and early review article by Wright and Dowson [55], it was stated that "Digestion of hyaluronate does not alter the boundary lubrication," referring to the work of Radin, Swann, and Weisser [56]. In another article, McCutchen re-states this conclusion in another way, saying "... the lubricating ability did not reside in the hyaluronic acid" and later asks the question "Why do the glycoprotein molecules (of Swann) lubricate?" [57]. These statements are based on effects of various constituents of friction, not wear. The work of the present author showed that in tests with bovine articular cartilage, Swann's lubricating glycoprotein (LGP-I) which was effective in reducing friction did not reduce cartilage wear. However, hyaluronic acid—shown earlier not to be responsible for friction reduction—did reduce cartilage wear. Thus, it is important to make the distinction between friction reduction and wear reduction. It is suggested that operational definitions be used in place of vague "lubricating ability," etc. terms in future papers on the subject.

Experimental Contact Systems

Second, some comments are made on the experimental approaches that have been reported in the literature on synovial joint lubrication mechanisms. Sliding contact combinations in *in vitro* studies have consisted of (1) cartilage-on-cartilage, (2) cartilage-on-some other surface (e.g., stainless steel, glass), and (3) solids other than cartilage sliding against each other in X-on-X or X-on-Y combinations.

The cartilage-on-cartilage combination is of course the most realistic and yet most complex contact system. But variations in shape or macroscopic geometry, microtopography, and the nature of contact present problems in carrying out well-controlled experiments. There is also the added problem of acquiring suitable specimens which are large enough and reasonably uniform.

The next combination—cartilage-on-another material—allows for better control of contact, with the more elastic, deformable cartilage loaded against a well-defined hard surface (e.g., a polished, flat solid made of glass or stainless steel). This contact configuration can provide useful tribological information on effects of changes in biochemical environment (e.g., fluids), on friction, wear, and sub-surface damage. It also could parallel the situation in a partial joint replacement in which healthy cartilage is in contact with a metal alloy.

The third combination, which appears in some of the literature on synovial joint lubrication, does not involve any articular cartilage at all. For example, Jay made friction measurements using a latex-covered stainless steel stud in oscillating contact against polished glass [31]. Williams et al., in a study of a lipid component of synovial fluid, used reciprocating contact of borosilicate glass-on-glass [58]. And in a recent paper on the action of a surface-active phospholipid as the "lubricating component of lubricin," Schwarz and Hills carried out friction measurements using two optically flat quartz plates in sliding contact [59]. In another study, a standard four-ball machine using alloy steel balls was used to examine the "lubricating ability" of synovial fluid constituents. Such tests, in the absence of cartilage, are easiest to control and carry out. However, they are not relevant to the study of synovial joint lubrication. With a glass sphere sliding against a glass flat, almost anything will reduce friction—including a wide variety of chemicals, biochemicals, semi-solids, and fluids. This has little if anything to do with the lubrication of synovial joints.

Fluids and Materials Used as Lubricants in *in Vitro* Biotribology Studies

Fluids used as lubricants in synovial joint lubrication studies have consisted of (1) "normal" synovial fluid (e.g., bovine), (2) buffered saline solution containing synovial fluid constituents (e.g., hyaluronic acid), and (3) various aqueous solutions of surface active compounds neither derived from nor present in synovial fluid. In addition, a few studies used synovial fluids from patients suffering from either osteoarthritis or rheumatoid arthritis.

The general comment made here is that the use of synovial fluids—whether derived from human or animal sources and whether "healthy" or "abnormal"—is important in *in vitro* studies of synovial joint lubrication. The documented behavior of synovial fluid in producing low friction and wear with articular cartilage sets a reference standard and demonstrates that useful information can indeed come from *in vitro* tests.

Studies that are based on adding synovial fluid constituents to a reference fluid (e.g., a buffered saline solution) can also be useful in attempting to identify which biochemical compound or compounds are responsible for reductions in frictions or wear. But if significant interactions between compounds exist, then such an approach may require an extensive program of tests. It should also be mentioned that in the view of the present author, the use of a pure undissolved constituent of synovial fluid, either derived or synthetic, in a sliding contact test is not only irrelevant but may be misleading. An example would be the use of a pure lipid (e.g., phospholipid) at the interface rather than in the concentration and solution form in which this compound would normally exist in synovial fluid. This is basic in any study of lubrication and particularly in the case of boundary lubrication where major effects on wear or friction can be brought on by minor, seemingly trivial, changes in chemistry.

The Preoccupation with Rheology and Friction

The synovial joint as a system—the articular cartilage and underlying bone structure as well as the synovial fluid as important elements—is extremely complex and far from being understood. It is noted that there is a proliferation of mathematical modeling papers stressing rheology and the mechanics of deformation, flow, and fluid pressures developed in the cartilage model. One recent example is the paper "The Role of Interstitial Fluid Pressurization and Surface Properties on the Boundary Friction of Articular Cartilage" by Ateshian et al. [21]. This study, a genuine contribution, grew out of the early work by Mow and connects also with the "weeping lubrication" model of McCutchen. Both McCutchen and Mow have made significant contributions to our understanding of synovial joint lubrication, although each approach is predominantly rheological and friction oriented with little regard for biochemistry and wear. This is not to say that rheology is unimportant. It could well be that, as suggested by Ateshian, the mechanism of interstitial fluid pressurization that leads to low friction in cartilage could also lead to low wear rates [60].

The Probable Existence of Various Lubrication Regimes

In an article by Wright and Dowson, it is suggested that a variety of types of lubrication operate in human synovial joints at different parts of a walking cycle: "At heel-strike a squeeze-film situation may develop, leading to elastohydrodynamic lubrication and possibly both squeeze-film and boundary lubrication, while hydrodynamic lubrication may operate during the free-swing phase of walking" [55].

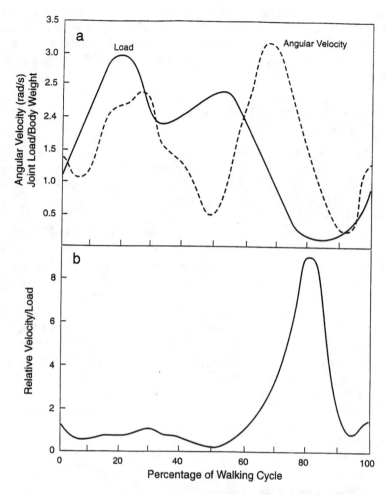

FIGURE 4.11 (a) Hip joint forces and angular velocities at different parts of the walking cycle (after Graham and Walker [61]). (b) Calculated ratio of velocity to force for the hip joint (from Figure 4.11a).

In a simplified approach to examining the various regimes of lubrication that could exist in a human joint, it may be useful to look at Fig. 4.11a which shows the variation in force (load) and velocity for a human hip joint at different parts of the walking cycle (taken from Graham and Walker [61]). As discussed earlier in this chapter, theories of hydrodynamic and elastohydrodynamic lubrication all include the hydrodynamic factor ($\eta U/W$) as the key variable, where η = fluid viscosity, U = the relative sliding velocity, and W = the normal load. High values of ($\eta U/W$) lead to thicker hydrodynamic films—a more desirable condition if one wants to keep surfaces apart. It can be seen from Fig. 4.11a that there is considerable variation in load and velocity, with peaks and valleys occurring at different parts of the cycle. Note also that in this example, the loads can be quite high (e.g., up to three times body weight). The maximum load occurs at 20% of the walking cycle illustrated in Fig. 4.11a, with a secondary maximum occurring at a little over the 50% point. The maximum angular velocity occurs at approximately 67% of the cycle. If one now creates a new curve of relative velocity/load or (U/W) from Fig. 4.11a, the result obtained is shown in Fig. 4.11b. We see now a very different and somewhat simplified picture. There is a clear and distinct maximum in the ratio of velocity to load (U/W) at 80% of walking cycle, favoring the formation of a hydrodynamic film of maximum thickness. However, for most of the cycle (e.g., from 0 to 60%), the velocity/load ratio is significantly lower, thus favoring a condition of minimum film thickness and "boundary lubrication." However, we also know that synovial fluid is non-Newtonian;

at higher rates of shear, its viscosity decreases sharply, approaching that of water. The shear rate is equal to the relative velocity divided by fluid film thickness (U/h) and is expressed in s^{-1}. This means that at the regions of low (U/W) ratios or thinner hydrodynamic films, the viscosity term in ($\eta U/W$) is even lower, thus pushing the minima to lower values favoring a condition of boundary lubrication. This is only a simplified view and does not consider those periods in which the relative sliding velocity is zero at motion reversal and where squeeze-film lubrication may come into play. A good example of the complexity of load and velocity variation in a human knee joint—including several zero-velocity periods—may be found in the chapter by Higginson and Unsworth [62] citing the work of Seedhom et al., which deals with biomechanics in the design of a total knee replacement [63].

The major point made here is that (1) there are parts of a walking cycle that would be expected to approach a condition of minimum fluid film thickness and boundary lubrication, and (2) it is during these parts of the cycle that cartilage wear and damage resulting from contact is more likely to occur. Thus, approaches to reducing cartilage wear in a synovial joint could be broken down into two categories (i.e., promoting thicker hydrodynamic films and providing special forms of "boundary lubrication").

Recent Developments

Recent developments in addressing some of the problems that involve cartilage damage and existing joint replacements include (1) progress in promoting cartilage repair [64], (2) possible use of artificial cartilage materials (e.g., synthetic hydrogels) [65,66], and (3) the development and application of more compliant joint replacement materials to promote a more favorable formation of an elastohydrodynamic film [67]. Although these are not strictly "lubricant-oriented" developments, they do and will involve important tribological aspects discussed in this chapter. For example, if new cartilage growth can be promoted by transplanting healthy chondrocytes to a platform in a damaged region of a synovial joint, how long will this cartilage last? If a hydrogel is used as an artificial cartilage, how long will it last? And if softer, elastomeric materials are used as partial joint replacements or coatings, how long will they last? These are questions of wear, not friction. And although the early fundamental studies of hydrogels as artificial cartilage measured only friction, and often only after a few moments of sliding, we know from recent work that even for hydrogels, low friction does not mean low wear [68].

4.9   Conclusions

The following main conclusions relating to the tribological behavior of natural, "normal" synovial joints are presented:

1. An unusually large number of theories and studies of joint lubrication have been proposed over the years. All of the theories focus on friction, none address wear, many do not involve experimental studies with cartilage, and very few consider the complexity and detailed biochemistry of the synovial-fluid articular-cartilage system.

2. It was shown by *in vitro* tests with bovine articular cartilage that the detailed biochemistry of synovial fluid has a significant effect on cartilage wear and damage. "Normal" bovine synovial fluid was found to provide excellent protection against wear. Various biochemical constituents isolated from bovine synovial fluid by Dr. David Swann, of the Shriners Burn Institutes in Boston, showed varying effects on cartilage wear when added back to a buffered saline reference fluid. This research demonstrates once again the importance of distinguishing between friction and wear.

3. In a collaborative study of biotribology involving researchers and students in Mechanical Engineering, the Virginia-Maryland College of Veterinary Medicine, and Biochemistry, *in vitro* tribological tests using bovine articular cartilage demonstrated among other things that (1) normal synovial fluid provides better protection than a buffered saline solution in a cartilage-on-cartilage system; (2) tribological contact in cartilage systems can cause subsurface damage, delamination, changes in proteoglycan content, and in chemistry via a "biotribochemical" process not

understood; and (3) pre-treatment of articular cartilage with the enzyme collagenase-3—suspected as a factor in osteoarthritis—significantly increases cartilage wear.

4. It is suggested that these results could change significantly the way mechanisms of synovial joint lubrication are examined. Effects of biochemistry of the system on wear of articular cartilage are likely to be important; such effects may not be related to physical/rheological models of joint lubrication.

5. It is also suggested that connections between tribology/normal synovial joint lubrication and degenerative joint disease are not only possible but likely; however, such connections are undoubtedly complex. It is *not* argued that osteoarthritis is a tribological problem or that it is necessarily the result of a tribological deficiency. Ultimately, a better understanding of how normal synovial joints function from a tribological point of view could conceivably lead to advances in the prevention and treatment of osteoarthritis.

6. Several problems exist that make it difficult to understand and interpret many of the published works on synovial joint lubrication. One example is the widespread use of non-operational and vague terms such as "lubricating activity," "lubricating factor," "boundary lubricating ability," and similar undefined terms which not only fail to distinguish between friction (which is usually measured) and cartilage wear (which is rarely measured), but tend to lump these phenomena together—a common error. Another problem is that a significant number of the published experimental studies of biotribology do not involve cartilage at all—relying on the use of glass-on-glass, rubber-on-glass, and even steel-on-steel. Such approaches may be a reflection of the incorrect view that "lubricating activity" is a property of a fluid and can be measured independently. Some suggestions are offered.

7. Last, the topic of synovial joint lubrication is far from being understood. It is a complex subject involving at least biophysics, biomechanics, biochemistry, and tribology. For a physical scientist or engineer, carrying out research in this area is a humbling experience.

Acknowledgments

The author wishes to acknowledge the support of the Edward H. Lane, G. Harold, and Leila Y. Mathers Foundations for their support during his sabbatical study at The Children's Hospital Medical Center. He also wishes to thank Dr. David Swann, for his invaluable help in providing the test fluids and carrying out the biochemical analyses, as well as Ms. Karen Hodgens, for conducting the early scanning electron microscopy studies of worn cartilage specimens.

The author is also indebted to the following researchers for their encouraging and stimulating discussions of this topic over the years and for teaching a tribologist something of the complexity of synovial joints, articular cartilage, and arthritis: Drs. Leon Sokoloff, Charles McCutchen, Melvin Glimcher, David Swann, Henry Mankin, Clement Sledge, Helen Muir, Paul Dieppe, and Heikki Helminen, as well as his colleagues at Virginia Tech—Hugo Veit, E. T. Kornegay, and E. M. Gregory.

Last, the author expresses his appreciation for and recognition of the valuable contributions made by students interested in biotribology over the years. These include graduate students Bettina Burkhardt, Michael Owellen, Matt Schroeder, Mark Freeman, and especially La Shaun Berrien, who contributed to this chapter, as well as the following summer undergraduate research students: Jean Yates, Elaine Ashby, Anne Newell, T. J. Hayes, Bethany Revak, Carolina Reyes, Amy Diegelman, and Heather Hughes.

References

1. Hunter, W. Of the structure and diseases of articulating cartilages, *Philosophical Transactions*, 42, 514–521, pp. 1742–1743.
2. Furey, M. J. Tribology, *Encyclopedia of Materials Science and Engineering*, Pergamon Press, Oxford, 1986, pp. 5145–5158.
3. Cameron, A. *The Principles of Lubrication*, Longmans Green & Co., London, 1966.

4. Dowson, D. and Higginson, G. R. *Elastohydrodynamic Lubrication*, SI ed., Pergamon Press, Oxford, 1977.

5. Freeman, M. A. R. *Adult Articular Cartilage*, 2nd ed., Pitman Medical Publishing Co., Tunbridge Wells, Kent, England, 1979.

6. Sokoloff, L., Ed. *The Joints and Synovial Fluid*, Vol. I, Academic Press, New York, 1978.

7. Stockwell, R. A. *Biology of Cartilage Cells*, Cambridge University Press, Cambridge, 1979.

8. Furey, M. J. Biochemical aspects of synovial joint lubrication and cartilage wear, *European Society of Osteoarthrology Symposium on Joint Destruction in Arthritis and Osteoarthritis*, Noordwijkerhout, the Netherlands, May 24–27, 1992.

9. Furey, M. J. Biotribology: cartilage lubrication and wear, *6th International Congress on Tribology*, *EUROTRIB '93*, Budapest, Hungary, August 30–September 2, 1993.

10. Furey, M. J. and Burkhardt, B. M. Biotribology: friction, wear, and lubrication of natural synovial joints, *Lubrication Science*, pp. 255–271, May 1997.

11. McCutchen, C. W. Lubrication of joints. *The Joints and Synovial Fluid*, Vol. I, Academic Press, New York, 1978, pp. 437–483.

12. Swanson, S. A. V. Friction, wear and lubrication. In *Adult Articular Cartilage*, 2nd ed., M.A.R. Freeman, Ed., Pitman Medical Publishing, Tunbridge Wells, Kent, England, 1979, pp. 415–460.

13. Higginson, G. R. and Unsworth, T. The lubrication of natural joints. In *Tribology of Natural and Artificial Joints* J. H. Dumbleton, Ed., Elsevier Scientific, Amsterdam, 1981, pp. 47–73.

14. Dumbleton, J. H. *Tribology of Natural and Artificial Joints*, Elsevier Scientific, Amsterdam, 1981.

15. Droogendijk, L. On the Lubrication of Synovial Joints, Ph.D. thesis, Twente University of Technology, the Netherlands, 1984.

16. Burkhardt, B. M. Development and Design of a Test Device for Cartilage Wear Studies, M.S. thesis, Mechanical Engineering, Virginia Polytechnic Institute and State University, Blacksburg, VA, December 1988.

17. McCutchen, C. W. Mechanisms of animal joints: sponge-hydrostatic and weeping bearings, *Nature (London)*, 184, pp. 1284–1285, 1959.

18. McCutchen, C. W. The frictional properties of animal joints, *Wear*, 5(1), pp. 1–17, 1962.

19. Torzilli, P. A. and Mow, V. C. On the fundamental fluid transport mechanisms through normal and pathological articular cartilage during friction. 1. The formulation. *Journal of Biomechanics*, 9, pp. 541–552, 1976.

20. Mansour, J. M. and Mow, V. C. On the natural lubrication of synovial joints: normal and degenerated, *Journal of Lubrication Technology*, pp. 163–173, April 1977.

21. Ateshian, G. A., Wang, H., and Lai, W. M. The role of interstitial fluid pressurization and surface porosities on the boundary friction of articular cartilage, *ASMS Journal of Biomedical Engineering*, 120, pp. 241–251, 1998.

22. Unsworth, A. Tribology of human and artificial joints. Proc. I. Mech. E., Part II, *Journal of Engineering in Medicine*, 205, 1991.

23. Simon, W. H. Wear properties of articular cartilage. *In vitro*, Section on Rheumatic Diseases, Laboratory of Experimental Pathology, National Institute of Arthritis and Metabolic Diseases, National Institutes of Health, February 1971.

24. Bloebaum, R. D. and Wilson, A. S. The morphology of the surface of articular cartilage in adult rats, *Journal Anatomy*, 131, pp. 333–346, 1980.

25. Radin, E. L. and Paul, I. L. Response of joints to impact loading. I. *In vitro* wear tests, *Arthritis and Rheumatism*, 14(3), May–June 1971.

26. Lipshitz, H. and Glimcher, M. J. A technique for the preparation of plugs of articular cartilage and subchondral bone, *Journal of Biomechanics*, 7, pp. 293–304, 1974.

27. Lipshitz, H. and Etheredge, III, R. *In vitro* wear of articular cartilage, *Journal of Bone Joint Surgery*, 57-A, pp. 527–534, June 1975.

28. Lipshitz, H. and Glimcher, M. J. *In vitro* studies of wear of articular cartilage, II. Characteristics of the wear of articular cartilage when worn against stainless steel plates having characterized surfaces, *Wear*, 52, pp. 297–337, 1979.

29. Sokoloff, L. Davis, W. H., and Lee, S. L. Boundary lubricating ability of synovial fluid in degenerative joint disease, *Arthritis and Rheumatism*, 21, pp. 754–760, 1978.

30. Sokoloff, L., Davis, W. H., and Lee, S. L. A proposed model of boundary lubrication by synovial fluid: structuring of boundary water, *Journal of Biomechanical Engineering*, 101, pp. 185–192, 1979.

31. Jay, D. J. Characterization of bovine synovial fluid lubricating factor, I. Chemical surface activity and lubrication properties, *Connective Tissue Research*, 28, pp. 71–88, 1992.

32. Furey, M. J. Joint lubrication. In *The Biomedical Engineering Handbook*, J. D. Bronzino, Ed., CRC Press, Boca Raton, FL, 1995, pp. 333–351.

33. Burkhardt, B. M. Development and Design of a Test Device for Cartilage Wear Studies, M.S. thesis, Mechanical Engineering, Virginia Polytechnic Institute and State University, Blacksburg, VA, December 1988.

34. Stachowiak, G. W., Batchelor, A. W., and Griffiths, L. J. Friction and wear changes in synovial joints, *Wear*, 171, pp. 135–142, 1994.

35. Hayes, A., Harris, B., Dieppe, P. A., and Clift, S. E. Wear of articular cartilage: the effect of crystals, *Proceedings of the Institution of Mechanical Engineers*, pp. 41–58, 1993.

36. Malcolm, L. L. An Experimental Investigation of the Frictional and Deformational Responses of Articular Cartilage Interfaces to Static and Dynamic Loading, Ph.D. thesis, University of California, San Diego, 1976.

37. Furey, M. J. Biotribology: an *in vitro* study of the effects of synovial fluid constituents on cartilage wear. *Proceedings 15th Symposium of the European Society of Osteoarthrology*, Kuopio, Finland, June 25–27, 1986, abstract in *Scandanavian Journal of Rheumatology, Supplement.*

38. Furey, M. J. The influence of synovial fluid constituents on cartilage wear: a scanning electron microscope study. *Conference on Joint Destruction, 15th Symposium of the European Society of Osteoarthrology*, Sochi, U.S.S.R., September 28–October 3, 1987.

39. Furey, M. J. Biochemical aspects of synovial joint lubrication and cartilage wear. *European Society of Osteoarthrology Symposium on Joint Destruction in Arthritis and Osteoarthritis*, Noordwigkerhout, the Netherlands, May 24–27, 1992.

40. Furey, M. J. Biotribology: cartilage lubrication and wear. *Proceedings, 6th International Congress on Tribology EUROTRIB; '93*, Vol. 2, pp. 464–470, Budapest, Hungary, August 30–September 2, 1993.

41. Schroeder, M. O. Biotribology: Articular Cartilage Friction, Wear, and Lubrication, M.S. thesis, Mechanical Engineering, Virginia Polytechnic Institute and State University, Blacksburg, VA, July 1995.

42. Owellen, M. C. Biotribology: The Effect of Lubricant and Load on Articular Cartilage Wear and Friction, M.S. thesis, Mechanical Engineering, Virginia Polytechnic Institute and State University, Blacksburg, VA, July 1997.

43. Furey, M. J., Schroeder, M. O., Hughes, H. L., Owellen, M. C., Berrien, L. S., Veit, H., Gregory, E. M., and Kornegay, E. T. Observations of subsurface damage and cartilage degradation in *in vitro* tribological tests using bovine articular cartilage, *21st Symposium of the European Society for Osteoarthrology*, Vol. 15, Ghent, Belgium, September 1996, 5, 3.2.

44. Furey, M. J., Schroeder, M. O., Hughes, H. L., Owellen, M. C., Berrien, L. S., Veit, H., Gregory, E. M., and Kornegay, E. T. *Biotribology, Synovial Joint Lubrication and Osteoarthritis*, Paper in Session W5 on Biotribology, *World Tribology Congress*, London, September 8–12, 1997.

45. Berrien, L. S., Furey, M. J., Veit, H. P. and Gregory, E. M. The effect of collagenase-3 on the *in vitro* wear of bovine articular cartilage, paper, Biotribology Session, *Fifth International Tribology Conference*, Brisbane, Australia, December 6–9, 1998.

46. Sokoloff, L. *The Biology of Degenerative Joint Disease*, University of Chicago Press, Chicago, IL, 1969.

47. Kelley, W. N., Harris, Jr., E. D., Ruddy, S, and Sledge, C. B. *Textbook of Rheumatology*, W. B. Saunders, Philadelphia, 1981.

48. Moskowitz, R. W., Howell, D. S., Goldberg, V. M., and Mankin, H. J. *Osteoarthritis: Diagnosis and Management*, W. B. Saunders, Philadelphia, 1984.

49. Verbruggen, G. and Veyes, E. M. Degenerative joints: test tubes, tissues, models, and man. *Proceedings First Conference on Degenerative Joint Diseases*, Excerpta Medica, Amsterdam, 1982.

50. Gastpar, H. Biology of the articular cartilage in health and disease. *Proc. Second Munich Symposium on Biology of Connective Tissue,* Munich, July 23–24, 1979; F. K. Schattauer Verlag, Stuttgart, 1980.

51. Dieppe, P. and Calvert, P. *Crystals and Joint Disease,* Chapman & Hall, London, 1983.

52. Discussions with M. J. Glimcher, The Children's Hospital Medical Center, Boston, MA, Fall 1983.

53. Furey, M. J. Exploring possible connections between tribology and osteoarthritis, *Lubrication Science,* p. 273, May 1997.

54. Hayes, A., Harris, B., Dieppe, P. A., and Clift, S. E. Wear of cartilage: the effect of crystals, *Proceedings of the Institution of Mechanical Engineers,* pp. 41–58, 1993.

55. Wright, V. and Dowson, D. Lubrication and cartilage, *Journal of Anatomy,* 121, pp. 107–118, 1976.

56. Radin, E. L., Swann, D. A., and Weisser, P. A. Separation of a hyaluronate-free lubricating fraction from synovial fluid, *Nature,* 228, pp. 377–378, 1970.

57. McCutchen, C. W. Joint lubrication, *Bulletin of the Hospital for Joint Diseases Orthopaedic Institute,* 43(2), pp. 118–129, 1983.

58. Williams, III, P. F., Powell, G. L., and LaBerge, M. Sliding friction analysis of phosphatidylcholine as a boundary lubricant for articular cartilage, *Proceedings of the Institution of Mechanical Engineers,* 207, pp. 41–166, 1993.

59. Schwarz, I. M. and Hills, B. A. Surface-active phospholipid as the lubricating component of lubrician, *British Journal of Rheumatology,* 37, pp. 21–26, 1998.

60. Private communication, letter to Michael J. Furey from Gerard A. Ateshian, July 1998.

61. Graham, J. D. and Walker, T. W. Motion in the hip: the relationship of split line patterns to surface velocities, a paper in *Perspectives in Biomedical Engineering,* R. M. Kenedi, Ed., University Park Press, Baltimore, MD, 1973, pp. 161–164.

62. Higginson, G. R. and Unsworth, T. The lubrication of natural joints. In *Tribology by Natural and Artificial Joints,* J.H. Dumbleton, Ed., Elsevier Scientific, Amsterdam, 1981, pp. 47–73.

63. Seedhom, B. B., Longton, E. B., Dowson, D., and Wright, V. Biomechanics background in the design of total replacement knee prosthesis. *Acta Orthopaedica Belgica,* Tome 39, Fasc 1, pp. 164–180, 1973.

64. Brittberg, M. *Cartilage Repair,* 2nd ed., a collection of five articles on cartilaginous tissue engineering with an emphasis on chondrocyte transplantation. Institute of Surgical Sciences and Department of Clinical Chemistry and Institute of Laboratory Medicine, Göteborg University, Sweden, 1996.

65. Corkhill, P. H., Trevett, A. S., and Tighe, B. J. The potential of hydrogels as synthetic articular cartilage, *Proceedings of the Institution of Mechanical Engineers,* 204, pp. 147–155, 1990.

66. Caravia, L., Dowson, D., Fisher, J., Corkhill, P. H., and Tighe, B. J. A comparison of friction in hydrogel and polyurethane materials for cushion form joints, *Journal of Materials Science: Materials in Medicine,* 4, pp. 515–520, 1993.

67. Caravia, L., Dowson, D., Fisher, J., Corkhill, P.H., and Tighe, B. J. Friction of hydrogel and polyurethane elastic layers when sliding against each other under a mixed lubrication regime, *Wear,* 181–183, pp. 236–240, 1995.

68. Freeman, M. E., Furey, M. J., Love, B. J., and Hampton, J. M. Friction, wear, and lubrication of hydrogels as synthetic articular cartilage, paper, Biotribology Session, *Fifth International Tribology Conference, AUSTRIB '98,* Brisbane, Australia, December 6–9 1998.

Further Information

For more information on synovial joints and arthritis, the following references are suggested: *The Biology of Degenerative Joint Disease* [46], *Adult Articular Cartilage* [5], *The Joints and Synovial Fluid,* Vol. I [6], *Textbook of Rheumatology* [47], *Osteoarthritis: Diagnosis and Management* [48], "Degenerative Joints: Test Tubes, Tissues, Models, and Man" [49], "Biology of the Articular Cartilage in Health and Disease" [50], and *Crystals and Joint Disease* [51].

5

Musculoskeletal Soft Tissue Mechanics

Richard L. Lieber
*University of California and
Veterans Administration
Medical Centers*

Thomas J. Burkholder
The Georgia Institute of Technology

5.1 Introduction

Skeletal muscles are length- and velocity-dependent force generators [Zajac, 1989]. Numerous extrinsic and intrinsic factors determine the magnitude of muscle force generated, maximum muscle contractile velocity, and the sensitivity of force generation to length and velocity. In this chapter, we briefly describe and summarize experimental values reported for these factors. Such information is necessary to generate accurate and physiologically relevant musculoskeletal models. Since muscles transmit force to bones via tendons, a summary of tendon biomechanical properties is also presented.

5.2 Fundamentals of Soft Tissue Biomechanics

Muscle Architecture

Muscles are organized arrays of multinucleated myofibers. Their mechanical properties depend on both the intrinsic properties of those fibers and their extrinsic arrangement, or architecture. Muscle architecture is typically described in terms of muscle length, mass, myofiber length, pennation angle (the angle between the line of action and the myofiber long axis), and physiological cross-sectional area (PCSA). PCSA is an approximation of the total cross-sectional area of all muscle fibers, projected along the muscle's line of action; it is calculated as:

$$\text{PCSA}\left(\text{mm}^2\right) = \frac{M\left(\text{g}\right) \cdot \cos\theta}{\rho\left(\text{g}/\text{mm}^3\right) \cdot L_f\left(\text{mm}\right)}$$

where M = muscle mass, ρ = muscle density (1.056 g/cm³ in fresh tissue), θ = surface pennation angle, and L_f = myofiber length. This formulation provides a good estimate of experimentally measured isometric muscle force output [Powell et al., 1984].

TABLE 5.1 Skeletal Muscle Specific Tension

Species	Muscle Type	Preparation	Specific Tension (kPa)	Ref.
Rat	SO	Single fiber	134	Fitts et al. [1991]
Human	Slow	Single fiber	133	Fitts et al. [1991]
Rat	FOG	Single fiber	108	Fitts et al. [1991]
Rat	FG	Single fiber	108	Fitts et al. [1991]
Human	Fast	Single fiber	166	Fitts et al. [1991]
Cat	1	Motor unit	59	Dum et al. [1982]
Cat	S	Motor unit	172	Bodine et al. [1987]
Cat	2A	Motor unit	284	Dum et al. [1982]
Cat	FR	Motor unit	211	Bodine et al. [1987]
Cat	2B + 2AB	Motor unit	343	Dum et al. [1982]
Cat	FF/FI	Motor unit	249	Bodine et al. [1987]
Human	Elbow flexors	Whole muscle	230–420	Edgerton et al. [1990]
Human	Ankle plantarflexors	Whole muscle	45–250	Fukunaga et al. [1996]
Rat	Tibialis anterior	Whole muscle	272	Wells [1965]
Rat	Soleus	Whole muscle	319	Wells [1965]
Guinea pig	Hindlimb muscles	Whole muscle	225	Powell et al. [1984]
Guinea pig	Soleus	Whole muscle	154	Powell et al. [1984]

Maximum Muscle Stress

Maximum active stress (often termed specific tension) varies somewhat among fiber types and species (Table 5.1) around a typical value of 250 kPa. Specific tension can be determined in any system in which it is possible to measure force and estimate the area of contractile material. In practice, area measurements may be difficult to make, giving rise to the large variability of reported values. Given muscle PCSA, the maximum force produced by a muscle can be predicted by multiplying this PCSA by specific tension (Table 5.1). Specific tension can also be calculated for isolated muscle fibers or motor units in which estimates of cross-sectional area have been made.

Maximum Muscle Contraction Velocity

Muscle maximum contraction velocity is primarily dependent on the type and number of sarcomeres in series along the muscle fiber length [Gans, 1982]. This number has been experimentally determined for a number of skeletal muscles (Table 5.2). Maximum contraction velocity of a given muscle can thus be calculated based on a knowledge of the number of serial sarcomeres along the muscle length

TABLE 5.2 Skeletal Muscle Dynamic Properties

Species	Muscle Type	Preparation	V_{max}[a]	a/Po	b/V_{max}	Ref.
Rat	SO	Single fiber	1.49 L/s			Fitts et al. [1991]
Human	Slow	Single fiber	0.86 L/s			Fitts et al. [1991]
Rat	FOG	Single fiber	4.91 L/s			Fitts et al. [1991]
Rat	FG	Single fiber	8.05 L/s			Fitts et al. [1991]
Human	Fast	Single fiber	4.85 L/s			Fitts et al. [1991]
Mouse	Soleus	Whole muscle	31.7 µm/s			Close [1972]
Rat	Soleus	Whole muscle	18.2 µm/s			Close [1972]
Rat	Soleus	Whole muscle	5.4 cm/s	0.214	0.23	Wells [1965]
Cat	Soleus	Whole muscle	13 µm/s			Close [1972]
Mouse	EDL	Whole muscle	60.5 µm/s			Close [1972]
Rat	EDL	Whole muscle	42.7 µm/s			Close [1972]
Cat	EDL	Whole muscle	31 µm/s			Close [1972]
Rat	TA	Whole muscle	14.4 cm/s	0.356	0.38	Wells [1965]

[a] L/s, fiber or sarcomere lengths per second; µm/s, sarcomere velocity; cm/s, whole muscle velocity.

TABLE 5.3 Architectural Properties of the Human Arm and Forearm[a,b]

Muscle	Muscle Mass (g)	Muscle Length (mm)	Fiber Length (mm)	Pennation Angle (°)	Cross-Sectional Area (cm²)	FL/ML Ratio
BR ($n = 8$)	16.6 ± 2.8	175 ± 8.3	121 ± 8.3	2.4 ± 0.6	1.33 ± 0.22	0.69 ± 0.062
PT ($n = 8$)	15.9 ± 1.7	130 ± 4.7	36.4 ± 1.3	9.6 ± 0.8	4.13 ± 0.52	0.28 ± 0.012
PQ ($n = 8$)	5.21 ± 1.0	39.3 ± 2.3	23.3 ± 2.0	9.9 ± 0.3	2.07 ± 0.33	0.58 ± 0.021
EDC I ($n = 8$)	3.05 ± 0.45	114 ± 3.4	56.9 ± 3.6	3.1 ± 0.5	0.52 ± 0.08	0.49 ± 0.024
EDC M ($n = 5$)	6.13 ± 1.2	112 ± 4.7	58.8 ± 3.5	3.2 ± 1.0	1.02 ± 0.20	0.50 ± 0.014
EDC R ($n = 7$)	4.70 ± 0.75	125 ± 10.7	51.2 ± 1.8	3.2 ± 0.54	0.86 ± 0.13	0.42 ± 0.023
EDC S ($n = 6$)	2.23 ± 0.32	121 ± 8.0	52.9 ± 5.2	2.4 ± 0.7	0.40 ± 0.06	0.43 ± 0.029
EDQ ($n = 7$)	3.81 ± 0.70	152 ± 9.2	55.3 ± 3.7	2.6 ± 0.6	0.64 ± 0.10	0.36 ± 0.012
EIP ($n = 6$)	2.86 ± 0.61	105 ± 6.6	48.4 ± 2.3	6.3 ± 0.8	0.56 ± 0.11	0.46 ± 0.023
EPL ($n = 7$)	4.54 ± 0.68	138 ± 7.2	43.6 ± 2.6	5.6 ± 1.3	0.98 ± 0.13	0.31 ± 0.020
PL ($n = 6$)	3.78 ± 0.82	134 ± 11.5	52.3 ± 3.1	3.5 ± 1.2	0.69 ± 0.17	0.40 ± 0.032
FDS I(P) ($n = 6$)	6.0 ± 1.1	92.5 ± 8.4	31.6 ± 3.0	5.1 ± 0.2	1.81 ± 0.83	0.34 ± 0.022
FDS I(D) ($n = 9$)	6.6 ± 0.8	119 ± 6.1	37.9 ± 3.0	6.7 ± 0.3	1.63 ± 0.22	0.32 ± 0.013
FDS I(C) ($n = 6$)	12.4 ± 2.1	207 ± 10.7	67.6 ± 2.8	5.7 ± 0.2	1.71 ± 0.28	0.33 ± 0.025
FDS M ($n = 9$)	16.3 ± 2.2	183 ± 11.5	60.8 ± 3.9	6.9 ± 0.7	2.53 ± 0.34	0.34 ± 0.014
FDS R ($n = 9$)	10.2 ± 1.1	155 ± 7.7	60.1 ± 2.7	4.3 ± 0.6	1.61 ± 0.18	0.39 ± 0.023
FDS S ($n = 9$)	1.8 ± 0.3	103 ± 6.3	42.4 ± 2.2	4.9 ± 0.7	0.40 ± 0.05	0.42 ± 0.014
FDP I ($n = 9$)	11.7 ± 1.2	149 ± 3.8	61.4 ± 2.4	7.2 ± 0.7	1.77 ± 0.16	0.41 ± 0.018
FDP M ($n = 9$)	16.3 ± 1.7	200 ± 8.2	68.4 ± 2.7	5.7 ± 0.3	2.23 ± 0.22	0.34 ± 0.011
FDP R ($n = 9$)	11.9 ± 1.4	194 ± 7.0	64.6 ± 2.6	6.8 ± 0.5	1.72 ± 0.18	0.33 ± 0.009
FDP S ($n = 9$)	13.7 ± 1.5	150 ± 4.7	60.7 ± 3.9	7.8 ± 0.9	2.20 ± 0.30	0.40 ± 0.015
FPL ($n = 9$)	10.0 ± 1.1	168 ± 10.0	45.1 ± 2.1	6.9 ± 0.2	2.08 ± 0.22	0.24 ± 0.010

[a] Data from Lieber et al. [1990, 1992].

[b] BR: brachioradialis; EDC I, EDC M, EDC R, and EDC S: extensor digitorum communis to the index, middle, ring and small fingers, respectively; EDQ: extensor digiti quinti; EIP: extensor indicis proprious; EPL: extensor pollicis longus; FDP I, FDP M, FDP R, and FDP S: flexor digitorum profundus muscles; FDS I, FDS M, FDS R, and FDS S: flexor digitorum superficialis muscles; FDS I(P) and FDS I(D): proximal and distal bellies of the FDS I; FDS I(C): the combined properties of the two bellies as if they were a single muscle; FPL: flexor pollicis longus; PQ: pronator quadratus; PL: palmaris longus; PT: pronator teres, FL, fiber length; ML, muscle length.

(Tables 5.3, 5.4, and 5.5) multiplied by the maximum contraction velocity of an individual sarcomere (Table 5.2). Sarcomere shortening-velocity varies widely among species and fiber types (Table 5.2).

Types of Muscle Models

There are three general classes of models for predicting muscle force: (1) Huxley biochemical models, (2) Hill phenomenological models, and (3) constitutive models. Huxley style models attempt to determine muscle force based on rates of attachment and detachment of muscle cross-bridges [Huxley, 1957]. Though this model is an excellent predictor of steady-state muscle force and energetics, it is generally computationally prohibitive to model a whole muscle in this manner. Constitutive models, such as those described by Ma and Zahalak [1991] model muscle behavior by describing populations of cross-bridges. While this represents a novel and potentially powerful approach, this technique has not yet been widely adopted. The vanguard of muscle modeling remains the phenomenological model first described by Hill [1938]. The Hill model describes a muscle as an active contractile component with series and parallel passive elastic components. The contractile component is generally described by its steady-state isometric force-length relation (Fig. 5.1) and isotonic force-velocity relation (Fig. 5.2).

Muscle Force–Length Relationship

Under conditions of constant length, the muscle force generated is proportional to the magnitude of the interaction between the actin and myosin contractile filament arrays. Since myosin filament length in

TABLE 5.4   Architectural Properties of Human Lower Limb[a,b]

Muscle	Muscle Mass (g)	Muscle Length (mm)	Fiber Length (mm)	Pennation Angle (°)	Cross-Sectional Area (cm^2)	FL/ML Ratio
RF ($n = 3$)	84.3 ± 14	316 ± 5.7	66.0 ± 1.5	5.0 ± 0.0	12.7 ± 1.9	0.209 ± 0.002
VL ($n = 3$)	220 ± 56	324 ± 14	65.7 ± 0.88	5.0 ± 0.0	30.6 ± 6.5	0.203 ± 0.007
VM ($n = 3$)	175 ± 41	335 ± 15	70.3 ± 3.3	5.0 ± 0.0	21.1 ± 4.3	0.210 ± 0.005
VI ($n = 3$)	160 ± 59	329 ± 15	68.3 ± 4.8	3.3 ± 1.7	22.3 ± 8.7	0.208 ± 0.007
SM ($n = 3$)	108 ± 13	262 ± 1.5	62.7 ± 4.7	15 ± 2.9	16.9 ± 1.5	0.239 ± 0.017
BF$_l$ ($n = 3$)	128 ± 28	342 ± 14	85.3 ± 5.0	0.0 ± 0.0	12.8 ± 2.8	0.251 ± 0.022
BF$_s$ ($n = 3$)		271 ± 11	139 ± 3.5	23 ± 0.9		0.517 ± 0.032
ST ($n = 2$)	76.9 ± 7.7	317 ± 4	158 ± 2.0	5.0 ± 0.0	5.4 ± 1.0	0.498 ± 0.0
SOL ($n = 2$)	215 ($n = 1$)	310 ± 1.5	19.5 ± 0.5	25 ± 5.0	58.0 ($n = 1$)	0.063 ± 0.002
MG ($n = 3$)	150 ± 14	248 ± 9.9	35.3 ± 2.0	16.7 ± 4.4	32.4 ± 3.1	0.143 ± 0.010
LG ($n = 3$)		217 ± 11	50.7 ± 5.6	8.3 ± 1.7		0.233 ± 0.016
PLT ($n = 3$)	5.30 ± 1.9	85.0 ± 15	39.3 ± 6.7	3.3 ± 1.7	1.2 ± 0.4	0.467 ± 0.031
FHL ($n = 3$)	21.5 ± 3.3	222 ± 5.0	34.0 ± 1.5	10.0 ± 2.9	5.3 ± 0.6	0.154 ± 0.010
FDL ($n = 3$)	16.3 ± 2.8	260 ± 15	27.0 ± 0.58	6.7 ± 1.7	5.1 ± 0.7	0.104 ± 0.004
PL ($n = 3$)	41.5 ± 8.5	286 ± 17	38.7 ± 3.2	10.0 ± 0.0	12.3 ± 2.9	0.136 ± 0.010
PB ($n = 3$)	17.3 ± 2.5	230 ± 13	39.3 ± 3.5	5.0 ± 0.0	5.7 ± 1.0	0.170 ± 0.006
TP ($n = 3$)	53.5 ± 7.3	254 ± 26	24.0 ± 4.0	11.7 ± 1.7	20.8 ± 3	0.095 ± 0.015
TA ($n = 3$)	65.7 ± 10	298 ± 12	77.3 ± 7.8	5.0 ± 0.0	9.9 ± 1.5	0.258 ± 0.015
EDL ($n = 3$)	35.2 ± 3.6	355 ± 13	80.3 ± 8.4	8.3 ± 1.7	5.6 ± 0.6	0.226 ± 0.024
EHL ($n = 3$)	12.9 ± 1.6	273 ± 2.4	87.0 ± 8.0	6.0 ± 1.0	1.8 ± 0.2	0.319 ± 0.030
SAR ($n = 3$)	61.7 ± 14	503 ± 27	455 ± 19	0.0 ± 0.0	1.7 ± 0.3	0.906 ± 0.017
GR ($n = 3$)	35.3 ± 7.4	335 ± 20	277 ± 12	3.3 ± 1.7	1.8 ± 0.3	0.828 ± 0.017
AM ($n = 3$)	229 ± 32	305 ± 12	115 ± 7.9	0.0 ± 0.0	18.2 ± 2.3	0.378 ± 0.013
AL ($n = 3$)	63.5 ± 16	229 ± 12	108 ± 2.0	6.0 ± 1.0	6.8 ± 1.9	0.475 ± 0.023
AB ($n = 3$)	43.8 ± 8.4	156 ± 12	103 ± 6.4	0.0 ± 0.0	4.7 ± 1.0	0.663 ± 0.036
PEC ($n = 3$)	26.4 ± 6.0	123 ± 4.5	104 ± 1.2	0.0 ± 0.0	2.9 ± 0.6	0.851 ± 0.040
POP ($n = 2$)	20.1 ± 2.4	108 ± 7.0	29.0 ± 7.0	0.0 ± 0.0	7.9 ± 1.4	0.265 ± 0.048

[a] Data from Wickiewicz et al. [1983].

[b] AB, adductor brevis; AL, adductor longus; AM, adductor magnus; BF$_l$, biceps femoris, long head; BF$_S$, biceps femoris, short head; EDL, extensor digitorum longus; EHL, extensor hallucis longus; FDL, flexor digitorum longus; GR, gracilis; FHL, flexor hallucis longus; LG, lateral gastrocnemius; MG, medial gastrocnemius; PEC, pectineus; PB, peroneus brevis; PL, peroneus longus; PLT, plantaris; POP, popliteus; RF, rectus femoris; SAR, sartorius; SM, semi-membranosus; SOL, soleus; ST, semitendinsus; TA, tibialis anterior; TP, tibialis posterior; VI, vastus intermedius; VL, vastus lateralis; VM, vastus medialis, FL, fiber length; ML, muscle length.

all species is approximately 1.6 μm and actin filament length varies (Table 5.6), optimal sarcomere length and maximum sarcomere length are calculated using these values. For optimal force generation, each half-myosin filament must completely overlap a half-actin filament, without opposing actin filaments overlapping. No active force is produced at sarcomere spacings shorter than the myosin filament length nor longer than the sum of actin and myosin filament lengths. The range of operating sarcomere lengths varies between muscles, but generally spans the range of optimal length ± 20% (Fig. 5.3).

Muscle Force–Velocity Relationship

Under conditions of constant contractile load the relationship between muscle force and velocity is given by the Hill equation [Hill, 1938]. The behavior of muscle is quite different when it is allowed to shorten (often termed "concentric" contraction) compared to when it is forced to lengthen (often termed "eccentric" contraction) and thus, the shortening force-velocity relation can be described by:

$$\left(P + a\right)v = b\left(P_o - P\right) \tag{5.1}$$

TABLE 5.5   Architectural Properties of Human Foot Muscles[a,b]

Muscle	Muscle Volume (cm³)	Muscle Length (mm)	Fiber Length (mm)	Cross-Sectional Area (cm²)
ABDH	15.2 ± 5.3	115.8 ± 4.9	23.0 ± 5.5	6.68 ± 2.07
ABDM	8.8 ± 4.7	112.8 ± 19.0	23.9 ± 7.4	3.79 ± 1.83
ADHO	9.1 ± 3.1	67.4 ± 4.6	18.6 ± 5.3	4.94 ± 1.36
ADHT	1.1 ± 0.6	24.8 ± 4.2	18.7 ± 5.2	0.62 ± 0.26
DI1	2.7 ± 1.4	51.0 ± 4.9	16.1 ± 4.4	1.70 ± 0.64
DI2	2.5 ± 1.4	49.9 ± 5.1	15.3 ± 4.0	1.68 ± 0.80
DI3	2.5 ± 1.2	44.3 ± 5.6	15.6 ± 5.4	1.64 ± 0.58
DI4	4.2 ± 2.0	61.4 ± 4.5	16.0 ± 4.8	2.72 ± 1.33
EDB2	2.1 ± 1.2	69.8 ± 16.8	28.0 ± 6.5	0.79 ± 0.43
EDB3	1.3 ± 0.7	82.2 ± 20.7	26.4 ± 5.1	0.51 ± 0.30
EDB4	1.0 ± 0.7	70.4 ± 21.1	23.1 ± 3.8	0.44 ± 0.29
EHB	3.6 ± 1.5	65.7 ± 8.5	27.9 ± 5.7	1.34 ± 0.66
FDB2	4.5 ± 2.3	92.9 ± 15.0	25.4 ± 4.5	1.78 ± 0.79
FDB3	3.2 ± 1.5	98.8 ± 18.1	22.8 ± 4.0	1.49 ± 0.71
FDB4	2.6 ± 1.0	103.0 ± 9.2	20.8 ± 4.5	1.26 ± 0.47
FDB5	0.7 ± 0.3	83.2 ± 3.0	18.2 ± 2.2	0.35 ± 0.16
FDMB	3.4 ± 1.7	51.0 ± 5.3	17.7 ± 3.8	2.00 ± 1.02
FHBL	3.4 ± 1.4	65.3 ± 7.1	16.5 ± 3.4	2.12 ± 0.84
FHBM	3.1 ± 1.3	76.0 ± 19.8	17.5 ± 4.8	1.80 ± 0.75
LB2	0.6 ± 0.4	53.9 ± 11.8	22.4 ± 6.5	0.28 ± 0.17
LB3	0.5 ± 0.4	45.2 ± 8.7	22.3 ± 6.7	0.28 ± 0.09
LB4	0.6 ± 0.4	37.3 ± 19.9	21.1 ± 9.3	0.30 ± 0.32
LB5	0.4 ± 0.4	41.0 ± 12.1	16.2 ± 7.0	0.18 ± 0.13
PI1	1.5 ± 0.5	46.2 ± 4.0	13.6 ± 3.7	1.23 ± 0.65
PI2	1.9 ± 0.7	56.6 ± 6.6	13.9 ± 3.5	1.41 ± 0.48
PI3	1.8 ± 0.6	48.8 ± 9.9	14.2 ± 5.9	1.38 ± 0.55
QPL	2.4 ± 1.2	55.3 ± 3.9	23.4 ± 7.1	1.00 ± 0.41
QPM	5.6 ± 3.4	81.3 ± 20.1	27.5 ± 7.0	1.96 ± 0.94

[a] Data from Kura et al. [1997].
[b] ABDH, abductor hallucis; ABDM, abductor digiti minimi; ADHO, adductor hallucis oblique; ADHT, adductor hallucis transverse; DI, dorsal interosseous; EDB, extensor digitorum brevis; EHB, extensor hallucis brevis; FDB, flexor digitorum brevis; FDMB, flexor digiti minimi brevis; FHBL, flexor hallucis brevis lateralis; FHBM flexor hallucis brevis medialis; LB, lumbrical; PI, plantar interosseous; QPL, quadratus plantaris lateralis; QPM, quadratus plantaris medialis.

where a and b are constants derived experimentally (usually about 0.25, Table 5.2), P is muscle force, P_o is maximum tetanic tension, and v is muscle velocity. The lengthening relation can be described by:

$$F = 1.8 - 0.8 \frac{V_{max} + V}{V_{max} - 7.6\,V}.$$

The dynamic parameters (a, b, and V_{max}) vary with species and fiber type (Table 5.2).

Tendon Biomechanics

Tendons, as most biological structures, have highly nonlinear stress–strain properties. This tissue is composed of a network of collagen fibrils in a hydrated matrix of elastin and proteoglycans. At rest, the collagen fibrils are significantly crimped or wavy so that initial loading acts primarily to straighten these fibrils. At higher strains, the straightened collagen fibrils must be lengthened which requires greater stress.

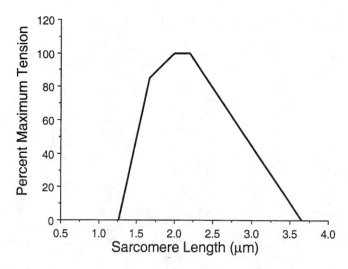

FIGURE 5.1   Sarcomere length-tension curve elucidated for isolated frog skeletal muscle fibers by Gordon, Huxley, and Julian [1966]. This demonstrates the length dependence of skeletal muscle on isometric strength.

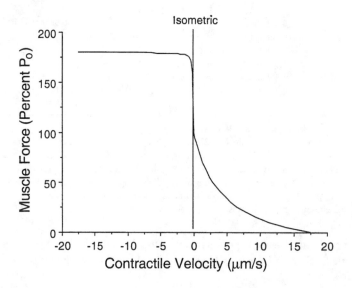

FIGURE 5.2   Force-velocity relationship for isolated skeletal muscle for shortening and lengthening. These data are plots of Eqs. 5.1 and 5.2 normalized for the maximum contraction velocity of the muscle.

As a result, tendons are more compliant at low loads and less compliant at high loads. The highly nonlinear low-load region has been referred to as the "toe" region and occurs up to approximately 3% strain and 5 MPa [Butler et al., 1979; Zajac, 1989]. Typically tendons have linear properties from about 3% strain until ultimate strain which is approximately 10% (Table 5.7). The tangent modulus in this linear region is approximately 1.5 GPa. Ultimate tensile stress reported for tendons is approximately 100 MPa (Table 5.7). Tendons operate physiologically at stresses of approximately 5 to 10 MPa (Table 5.7). Thus, tendons operate with a safety factor of 10 to 20 MPa.

TABLE 5.6   Actin Filament Lengths

Species	1/2 Actin Filament Length (μm)	Ref.
Human	1.18	Lieber et al. [1994]
Cat	1.12	Herzog et al. [1992]
Rat	1.09	Herzog et al. [1992]
Rabbit	1.09	Herzog et al. [1992]
Frog	0.98	Page and Huxley [1963]
Monkey	1.16	Walker and Schrodt [1973]
Human	1.27	Walker and Schrodt [1973]
Hummingbird	1.75	Mathieu-Costello et al. [1992]
Chicken	0.95	Page [1969]
Wild rabbit	1.12	Dimery [1985]
Carp	0.98	Sosnicki et al. [1991]

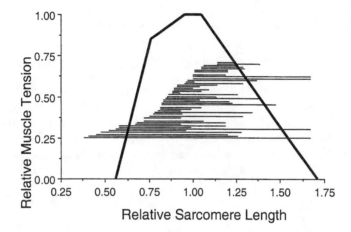

FIGURE 5.3   Sarcomere length excursion range during normal movement summarized from 48 different scientific studies. Since thin filament length varies among species, these lengths were normalized to the thin filament plus the thick filament length. Thick filaments are 1.65 μm long while thin filament length varies from 2.0 to 2.5 μm.

TABLE 5.7   Tendon Biomechanical Properties

Tendon	Ultimate Stress (MPa)	Ultimate Strain (%)	Stress under Normal Loads (MPa)	Strain under Normal Loads (%)	Tangent Modulus (GPa)	Ref.
Swine	40	9				Woo et al. [1980]
Wallaby			15–40		1.56	Bennett et al. [1986]
Porpoise					1.53	Bennett et al. [1986]
Dolphin					1.43	Bennett et al. [1986]
Deer			28–74		1.59	Bennett et al. [1986]
Sheep					1.65	Bennett et al. [1986]
Donkey			22–44		1.25	Bennett et al. [1986]
Human leg			53		1.0–1.2	Bennett et al. [1986]
Cat leg					1.21	Bennett et al. [1986]
Pig tail					0.9	Bennett et al. [1986]
Rat tail					0.8–1.5	Bennett et al. [1986]
Horse				4–10		Ker et al. [1988]
Dog leg			84			Ker et al. [1988]
Camel ankle			18			Ker et al. [1988]
Human limb (various)	60–120					McElhaney et al. [1976]
Human calcaneal	55	9.5				McElhaney et al. [1976]
Human wrist	52–74	11–17	3.2–3.3	1.5–3.5		Loren and Lieber [1994]

References

Butler, D.L., Grood, E.S., Noyes, F.R., and Zernicke, R.F. 1979. Biomechanics of ligaments and tendons. *Exer. Sport Sci. Rev.* 6:125–181.

Gans, C. 1982. Fiber architecture and muscle function. *Exer. Sport Sci. Rev.* 10:160–207.

Hill, A.V. 1938. The heat of shortening and the dynamic constants of muscle. *Proc. R. Soc. Lond. [Biol].* 126:136–195.

Huxley, A.F. 1957. Muscle structure and theories of contraction. *Prog. Biophys. Mol. Biol.* 7:255–318.

Kura, H., Luo, Z., Kitaoka, H.B., and An, K. 1997. Quantitative analysis of the intrinsic muscles of the foot. *Anat. Rec.* 249:143–151.

Lieber, R.L., Fazeli, B.M., and Botte, M.J. 1990. Architecture of selected wrist flexor and extensor muscles. *J. Hand Surg.* 15:244–250.

Lieber, R.L., Jacobson, M.D., Fazeli, B.M., Abrams, R.A., and Botte, M.J. 1992. Architecture of selected muscles of the arm and forearm: anatomy and implications for tendon transfer *J. Hand Surg.* 17:787–798.

Lieber, R.L., Loren, G.J., and Fridén, J. 1994. *In vivo* measurement of human wrist extensor muscle sarcomere length changes. *J. Neurophysiol.* 71:874–881.

Ma, S.P. and Zahalak, G.I. 1991. A distribution-moment model of energetics in skeletal muscle. *J. Biomech.* 24:21–35.

Powell, P.L., Roy, R.R., Kanim, P., Bello, M.A., and Edgerton, V.R. 1984. Predictability of skeletal muscle tension from architectural determinations in guinea pig hindlimbs. *J. Appl. Physiol.* 57:1715–1721.

Wickiewicz, T.L., Roy, R.R., Powell, P.L., and Edgerton, V.R. 1983. Muscle architecture of the human lower limb. *Clin. Orthop. Rel. Res.* 179:275–283.

Zajac, F.E. 1989. Muscle and tendon: Properties, models, scaling and application to biomechanics and motor control. *CRC Crit. Rev. Biomed. Eng.* 17:359–411.

6

Mechanics of the Head/Neck

Albert I. King
Wayne State University

David C. Viano
Wayne State University

Injury is a major societal problem in the United States. Approximately 140,000 fatalities occur each year due to both intentional and unintentional injuries. Two-thirds of these are unintentional, and of these about one-half are attributable to automotive-related injuries. In 1993, the estimated number of automotive-related fatalities dipped under 40,000 for the first time in the last three decades due to a continuing effort by both the industry and government to render vehicles safer in crash situations. However, for people under 40 years of age, automotive crashes, falls, and other unintentional injuries are the highest risks of fatality in the United States in comparison with all other causes.

The principal aim of impact biomechanics is the prevention of injury through environmental modification, such as the provision of an airbag for automotive occupants to protect them during a frontal crash. To achieve this aim effectively, it is necessary that workers in the field have a clear understanding of the *mechanisms of injury,* be able to describe the *mechanical response* of the tissues involved, have some basic information on *human tolerance* to impact, and be in possession of tools that can be used as *human surrogates* to assess a particular injury [Viano et al., 1989]. This chapter deals with the biomechanics of blunt impact injury to the head and neck.

6.1 Mechanisms of Injury

Head Injury Mechanisms

Among the more popular theories of brain injury due to blunt impact are changes in intracranial pressure and the development of shear strains in the brain. Positive pressure increases are found in the brain behind the site of impact on the skull. Rapid acceleration of the head, in-bending of the skull, and the propagation of a compressive pressure wave are proposed as mechanisms for the generation of intracranial compression that causes local contusion of brain tissue. At the contrecoup site, there is an opposite response in the form of a negative-pressure pulse that also causes bruising. It is not clear as to whether the injury is due to the negative pressure itself (tensile loading) or to a cavitation phenomenon similar to that seen on the surfaces of propellers of ships (compression loading). The pressure differential across the brain necessarily results in a pressure gradient that can give rise to shear strains developing within the deep structures of the brain. Furthermore, when the head is impacted, it not only translates but also rotates about the neck. It is postulated that the relative motion of the brain surface with respect to the rough inner surface of the skull results in surface contusions and the tearing of bridging veins between the brain and the dura mater, the principal membrane protecting the brain beneath the skull. Gennarelli [1983] has found that rotational acceleration of the head can cause a diffuse injury to the white matter of the brain in animal models, as evidenced by retraction balls developing along the axons of injured nerves. This injury was described by Strich [1961] as diffuse axonal injury (DAI) that she found in the white matter of autopsied human brains. Other researchers, including Lighthall et al. [1990], have been able to cause the development of DAI in the brain of an animal model (ferrets) by the application of direct impact to the brain without the associated head angular acceleration. Adams et al. [1986] indicated that DAI is the most important factor in severe head injury because it is irreversible and leads to incapacitation and dementia. It is postulated that DAI occurs as a result of the mechanical insult but cannot be detected by staining techniques at autopsy unless the patient survives the injury for at least several hours.

Neck Injury Mechanisms

The neck or the cervical spine is subjected to several forms of unique injuries that are not seen in the thoracolumbar spine. Injuries to the upper cervical spine, particularly at the atlanto-occipital joint, are considered to be more serious and life-threatening than those at the lower level. The atlanto-occipital joint can be dislocated either by an axial torsional load or a shear force applied in the anteroposterior direction, or vice versa. A large compression force can cause the arches of C1 to fracture, breaking it up into two or four sections. The odontoid process of C2 is also a vulnerable area. Extreme flexion of the neck is a common cause of odontoid fractures, and a large percentage of these injuries are related to automotive accidents [Pierce and Barr, 1983]. Fractures through the pars interarticularis of C2, commonly known as "hangman's fractures" in automotive collisions, are the result of a combined axial compression and extension (rearward bending) of the cervical spine. Impact of the forehead and face of unrestrained occupants with the windshield can result in this injury. Garfin and Rothman [1983] discussed this injury in relation to hanging and traced the history of this mode of execution. It was estimated by a British judiciary committee that the energy required to cause a hangman's fracture was 1708 N·m (1260 ft·lb).

In automotive-type accidents, the loading on the neck due to head contact forces is usually a combination of an axial or shear load with bending. Bending loads are almost always present, and the degree of axial of shear force depends on the location and direction of the contact force. For impacts near the crown of the head, compressive forces predominate. If the impact is principally in the transverse plane, there is less compression and more shear. Bending modes are infinite in number because the impact can come from any angle around the head. To limit the scope of the discussion, the following injury modes are considered: tension–flexion, tension–extension, compression–flexion, and compression–extension in the midsagittal plane and lateral bending.

Tension–Flexion Injuries.   Forces resulting from inertial loading of the head–neck system can result in flexion of the cervical spine while it is being subjected to a tensile force. In experimental impacts of restrained subjects undergoing forward deceleration, Thomas and Jessop [1983] reported atlanto-occipital separation and C1–C2 separation occurring in subhuman primates at 120 *g*. Similar injuries in human cadavers were found at 34 to 38 *g* by Cheng et al. [1982], who used a preinflated driver airbag system that restrained the thorax but allowed the head and neck to rotate over the bag.

Tension–Extension Injuries. The most common type of injury due to combined tension and extension of the cervical spine is the "whiplash" syndrome. However, a large majority of such injuries involve the soft tissues of the neck, and the pain is believed to reside in the joint capsules of the articular facets of the cervical vertebrate [Lord et al., 1993]. In severe cases, teardrop fractures of the anterosuperior aspect of the vertebral body can occur. Alternately, separation of the anterior aspect of the disk from the vertebral endplate is known to occur. More severe injuries occur when the chin impacts the instrument panel or when the forehead impacts the windshield. In both cases, the head rotates rearward and applies a tensile and bending load on the neck. In the case of windshield impact by the forehead, hangman's fracture of C2 can occur. Garfin and Rothman [1983] suggested that it is caused by spinal extension combined with compression on the lamina of C2, causing the pars to fracture.

Compression–Flexion Injuries. When a force is applied to the posterosuperior quadrant of the head or when a crown impact is administered while the head is in flexion, the neck is subjected to a combined load of axial compression and forward bending. Anterior wedge fractures of vertebral bodies are commonly seen, but with increased load burst fractures and fracture-dislocations of the facets can result. The latter two conditions are unstable and tend to disrupt or injure the spinal cord, and the extent of the injury depends on the penetration of the vertebral body or its fragments into the spinal canal. Recent experiments by Pintar et al. [1989, 1990] indicate that burst fractures of lower cervical vertebrae can be reproduced in cadaveric specimens by a crown impact to a flexed cervical spine. A study by Nightingale et al. [1993] showed that fracture-dislocations of the cervical spine occur very early in the impact event (within the first 10 ms) and that the subsequent motion of the head or bending of the cervical spine cannot be used as a reliable indicator of the mechanism of injury.

Compression–Extension Injuries. Frontal impacts to the head with the neck in extension will cause compression–extension injuries. These involve the fracture of one or more spinous processes and, possibly, symmetrical lesions of the pedicles, facets, and laminae. If there is a fracture-dislocation, the inferior facet of the upper vertebra is displaced posteriorly and upward and appears to be more horizontal than normal on x-ray.

Injuries Involving Lateral Bending. If the applied force or inertial load on the head has a significant component out of the midsagittal plane, the neck will be subjected to lateral or oblique along with axial and shear loading. The injuries characteristic of lateral bending are lateral wedge fractures of the vertebral body and fractures to the posterior elements on one side of the vertebral column.

Whenever there is lateral or oblique bending, there is the possibility of twisting the neck. The associated torsional loads may be responsible for unilateral facet dislocations or unilateral locked facets [Moffat et al., 1978]. However, the authors postulated that pure torsional loads on the neck are rarely encountered in automotive accidents. It was shown by Wismans and Spenny [1983] that, in a purely lateral impact, the head rotated axially about the cervical axis while it translated laterally and vertically and rotated about an anteroposterior axis. These responses were obtained from lateral impact tests performed by the Naval Biodynamics Laboratory on human subjects who were fully restrained at and below the shoulders.

6.2 Mechanical Response

Mechanical Response of the Head

A large number of cadaveric studies on blunt head impact have been carried out over the past 50 years. The head was impacted by rigid and padded surfaces and by impactors of varying shapes to simulate flat surfaces and knobs encountered in the automotive environment. In general, the impact responses were described in terms of head acceleration or impact force. Both of these responses are dependent on a variety of factors, including the inertial properties of the head and the surface impacted by the head.

TABLE 6.1   Average Male Head Mass

Ref.	No. of Subjects	Average Body Mass (kg)	Average Head Mass (kg)
Walker et al. [1973]	16	67.1	4.49
Hubbard and McLeod [1974]	11		4.54
Reynolds et al. [1975]	6	65.2	3.98
Adjusted per HMRTF	6	76.9	4.69
Beier et al. [1980]	19	74.7	4.32
McConville et al. [1980]	31	77.5	4.55[a]
Robbins [1983]	25	76.7	4.54[b]

[a] Based on adjusted head volume of 95% of the reported head volume (4396 cm^3) and a head specific gravity of 1.097.

[b] Based on an estimated head volume of 4137 cm^3 and a head specific gravity of 1.097.

TABLE 6.2   Average Mass Moments of Inertia[a] of the Male Head (kg·m$^2 \times 10^{-3}$)

Ref.	I_{xx}	I_{yy}	I_{zz}
Walker et al. [1973]		23.3	
Hubbard and McLeod [1974]	17.4	16.4	20.3
Adjusted per HMRTF	22.6	21.3	26.3
Beier et al. [1980] (16 male subjects only)	20.7	22.6	14.9
McConville et al. [1980]	20.4	23.2	15.1
Adjusted by sp. gr. 1.097	22.4	25.5	16.6
Robbins [1983]	20.0	22.2	14.5
Adjusted by sp. gr. 1.097	22.0	24.2	15.9

[a] The mass moments of inertia given are about the *x, y,* or *z* anatomic axes through the center of gravity of the head.

In this section, the inertial properties of the head will be described, and response data for head impact against a flat, rigid surface will be provided. It should be noted that while response data against surfaces with a variety of shapes and stiffnesses are of interest, the only generally applicable and reproducible data are those of impacts to flat, rigid surfaces.

Inertial Properties of the Head.   There are several sources of data on the inertial properties of the human head. Mass data are shown in Table 6.1, while mass moment of inertial data can be found in Table 6.2. The data by Walker et al. [1973] shown in Table 6.1 were analyzed by Hubbard and McLeod [1974], who found that 16 of the 20 heads used by Walker et al. [1973] had dimensions that were close to those of the average male and provided the data for the average body and head mass. Similarly, Hubbard and McLeod [1974] analyzed the head-mass data of Hodgson and Thomas [1971] and Hodgson et al. [1972] for 11 heads, the dimensions of which were consistent with those of a fiftieth percentile male head and found an average value of 4.54 kg, as shown in Table 6.1. The data of Reynolds et al. [1975] were obtained from 6 cadavers that were of low body weight. The Human Mechanical Response Task Force (HMRTF) of the Society of Automotive Engineers (SAE) made an adjustment for the body and head mass, and the results are shown just below the original data of Reynolds et al. [1975] in Table 6.1. The data of Beier et al. [1980] are based on their original study. The studies of McConville et al. [1980] and Robbins [1983] were derived from anthropometric measurement of living subjects. In these studies, head mass was estimated from volumetric measurements and a previously determined value of the specific gravity of the head, 1.097. For the average values of the mass moment of inertia, shown in Table 6.2, the origin of the coordinate axes is at the center of gravity (cg) of the head, and the *x* axis is in the posteroanterior direction, while the *z* axis is in the inferosuperior direction and the *y* axis is in a lateral direction. The adjustments made in Table 6.1 are applicable to Table 6.2 as well.

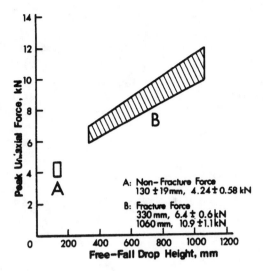

FIGURE 6.1 Impact response of the head in terms of peak force.

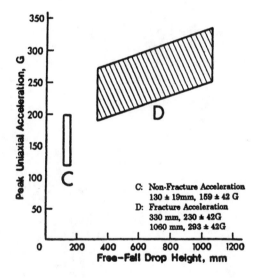

FIGURE 6.2 Impact response of the head in terms of peak acceleration.

Cranial Impact Response.   Impact response of the head against a flat, rigid surface was obtained by Hodgson and Thomas [1973, 1975], who performed a series of drop tests using embalmed cadavers. The responses in terms of peak force and acceleration are shown in Figs. 6.1 and 6.2, respectively, as a function of an equivalent free-fall drop height. Details of adjustments made to the original data are described by Prasad et al. [1985]. It should be noted that the data shown were from the frontal, lateral, and occipital directions and that there was a large scatter in the peak values. For this reason, the data were pooled, and individual data points were not shown.

The difficulty with acceleration measurements in head impact is twofold. The head is not a rigid body, and accelerometers cannot be mounted at the center of gravity of the head. The center of gravity acceleration can be computed if head angular acceleration is measured, but the variation in skull stiffness cannot be corrected for easily. It is recommended that in all future head-impact studies, head angular acceleration be measured. Several methods for measuring this parameter have been proposed. At present, the most reliable method appears to be that proposed by Padgaonkar et al. [1975] using an array of 9 linear accelerometers arranged in a 3-2-2-2 cluster.

The data presented in this section do not refer to the response of the brain during an injury-producing impact. For intact heads, the motion of the brain inside the skull has not been studied exhaustively. There is evidence that relative motion of the brain with respect to the skull occurs [Nusholtz et al., 1984], particularly during angular acceleration of the head. However, this motion does not fully explain injuries seen in the center of the brain and in the brainstem. More research is needed to explore the mechanical response of the brain to both linear and angular acceleration and to relate this response to observed injuries, such as diffuse axonal injury.

Mechanical Response of the Neck

The mechanical response of the cervical spine was studied by Mertz and Patrick [1967, 1971], Patrick and Chou [1976], Schneider et al. [1975], and Ewing et al. [1978]. Mertz et al. [1973] quantified the response in terms of rotation of the head relative to the torso as a function of bending moment at the occipital condyles. Loading corridors were obtained for flexion and extension, as shown in Figs. 6.3 and 6.4. An exacting definition of the impact environments to be used in evaluating dummy necks relative to the loading corridors illustrated in these figures is included in SAE J 1460 [1985]. It should be noted that the primary basis for these curves is volunteer data and that the extension of these corridors to dummy tests in the injury-producing range is somewhat surprising. Static and dynamic lateral response data were provided by Patrick and Chou [1976]. A response envelope for lateral flexion is shown in Fig. 6.5. A limited amount of the voluminous data obtained by Ewing et al. [1978] (six runs) was analyzed by Wismans and Spenny [1983, 1984] for lateral and sagittal flexion. The rotations were represented in three dimensions by a rigid link of fixed length pivoted at T1 at the bottom and within the head at the

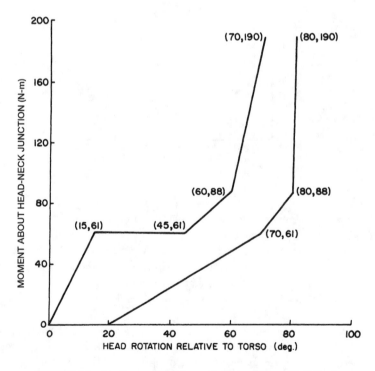

FIGURE 6.3   Loading corridor for neck flexion (forward bending).

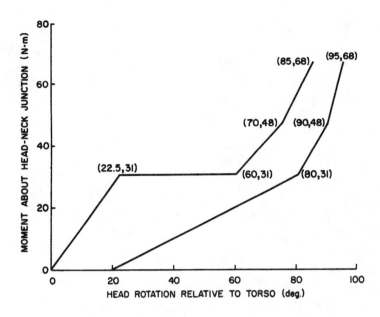

FIGURE 6.4   Loading corridor for neck extension (rearward bending).

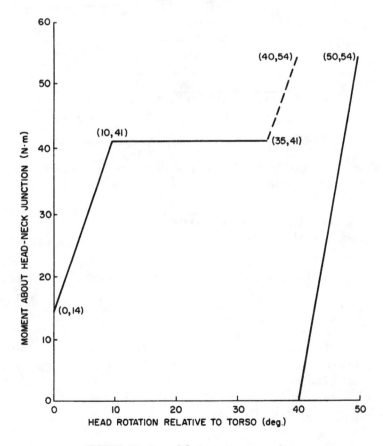

FIGURE 6.5   Lateral flexion response envelope.

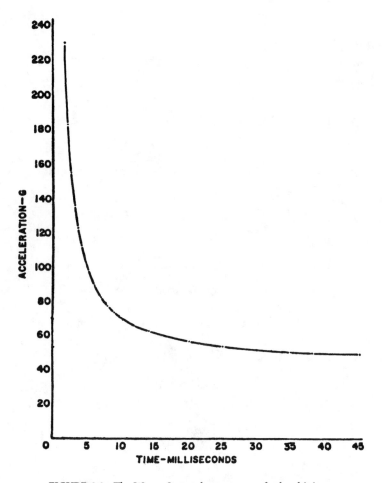

FIGURE 6.6   The Wayne State tolerance curve for head injury.

top. In terms of torque at the occipital condyles and head rotation, the results fell within the corridor for forward and lateral flexion shown in Figs. 6.3 and 6.5, respectively.

6.3   Regional Tolerance of the Head and Neck to Blunt Impact

Regional Tolerance of the Head

The most commonly measured parameter for head injury is acceleration. It is therefore natural to express human tolerance to injury in terms of head acceleration. The first known tolerance criterion is the Wayne State tolerance curve, proposed by Lissner et al. [1960] and subsequently modified by Patrick et al. [1965] by the addition of animal and volunteer data to the original cadaveric data. The modified curve is shown in Fig. 6.6. The head can withstand higher accelerations for shorter durations, and any exposure above the curve is injurious. When this curve is plotted on logarithmic paper, it becomes a straight line with a slope of −2.5. This slope was used as an exponent by Gadd [1961] in his proposed severity index, now known as the *Gadd Severity Index* (GSI):

$$\text{GSI} = \int_0^T a^{2.5}\, dt$$

where a = instantaneous acceleration of the head
T = duration of the pulse

If the integrated value exceeds 1000, a severe injury will result. A modified form of the GSI, now known as the *Head Injury Criterion* (HIC), was proposed by Versace [1970] to identify the most damaging part of the acceleration pulse by finding the maximum value of the following integral:

$$\text{HIC} = \left(t_2 - t_1\right)\left[\left(t_2 - t_1\right)^{-1}\int_{t_1}^{t_2} a(t)\,dt\right]_{\max}^{2.5}$$

where $a(t)$ = resultant instantaneous acceleration of the head
$t_2 - t_1$ = time interval over which HIC is a maximum

A severe but not life-threatening injury would have occurred if the HIC reached or exceeded 1000. Subsequently, Prasad and Mertz [1985] proposed a probabilistic method of assessing head injury and developed the curve shown in Fig. 6.7. At an HIC of 1000, approximately 16% of the population would sustain a severe to fatal injury. It is apparent that this criterion is useful in automotive safety design and in the design of protective equipment for the head, such as football and bicycle helmets. However, there is another school of thought that believes in the injurious potential of angular acceleration in its ability to cause cerebral contusion of the brain surface and rupture of the parasagittal bridging veins between the brain and the dura mater. A proposed limit for angular acceleration is 4500 rad/s², based on a

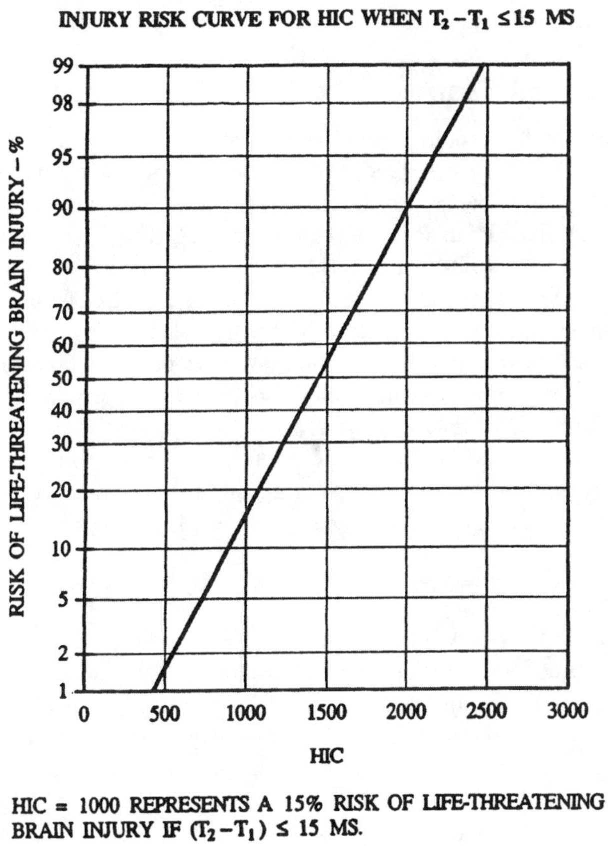

INJURY RISK CURVE FOR HIC WHEN T₂ – T₁ ≤ 15 MS

HIC = 1000 REPRESENTS A 15% RISK OF LIFE-THREATENING
BRAIN INJURY IF (T₂ – T₁) ≤ 15 MS.

FIGURE 6.7 Head injury risk curve based on HIC.

mathematical model developed by Lowenhielm [1974]. This limit has not received universal acceptance. Many other criteria have been proposed, but HIC is the current criterion for Federal Motor Vehicle Safety Standard (FMVSS) 214, and attempts to replace it have so far been unsuccessful.

Regional Tolerance of the Neck

Currently, there are no tolerance values for the neck for the various injury modes. This is not due to a lack of data but rather to the many injury mechanisms and several levels of injury severity, ranging from life-threatening injuries to the spinal cord to minor soft-tissue injuries that cannot be identified on radiographic or magnetic scans. It is likely that a combined criterion of axial load and bending moment about one or more axes will be adopted as a future FMVSS.

6.4 Human Surrogates of the Head and Neck

The Experimental Surrogate

The most effective experimental surrogate for impact biomechanics research is the unembalmed cadaver. This is also true for the head and neck, despite the fact that the cadaver is devoid of muscle tone because the duration of impact is usually too short for the muscles to respond adequately. It is true, however, that muscle pretensioning in the neck may have to be added under certain circumstances. Similarly, for the brain, the cadaveric brain cannot develop DAI, and the mechanical properties of the brain change rapidly after death. If the pathophysiology of the central nervous system is to be studied, the ideal surrogate is an animal brain. Currently, the rat is frequently used as the animal of choice, and there is some work in progress using the minipig.

The Injury-Assessment Tool

The response and tolerance data acquired from cadaveric studies are used to design human-like surrogates, known as *anthropomorphic test devices* (ATD). These surrogates are required to have biofidelity, the ability to simulate human response, but also need to provide physical measurements that are representative of human injury. In addition, they are designed to be repeatable and reproducible. The current frontal impact dummy is the Hybrid III, which is human-like in many of its responses, including that of the head and neck. The head consists of an aluminum headform covered by an appropriately designed vinyl "skin" to yield human-like acceleration responses for frontal and lateral impacts against a flat, rigid surface. Two-dimensional physical models of the brain were proposed by Margulies et al. [1990] using a silicone gel in which preinscribed grid lines would deform under angular acceleration. No injury criterion is associated with this gel model.

The dummy neck was designed to yield responses in flexion and extension that would fit within the corridors shown in Figs. 6.3 and 6.4. The principal function of the dummy neck is to place the head in the approximate position of a human head in the same impact involving a human occupant.

Computer Models

Models of head impact first appeared over 50 years ago [Holbourn, 1943]. Extensive reviews of such models were made by King and Chou [1977] and Hardy et al. [1994]. The use of the finite-element method (FEM) to simulate the various components of the head appears to be the most effective and popular means of modeling brain response. Despite the large numbers of nodes and elements used, the models are still not detailed enough to predict the location of DAI development following a given impact. The research is also hampered by the limited amount of animal DAI data currently available.

A large number of neck and spinal models also have been developed over the past four decades. A recent paper by Kleinberger [1993] provides a brief and incomplete review of these models. However, the method of choice for modeling the response of the neck is the finite-element method, principally because of the complex geometry of the vertebral components and the interaction of several different materials. A fully validated model for impact responses is still not available.

References

Adams JH, Doyle D, Graham DI et al. 1986. Gliding contusions in nonmissile head injury in humans. *Arch Pathol Lab Med* 110:485.

Beier G, Schuller E, Schuck M et al. 1980. Center of gravity and moments of inertia of human head. In *Proceedings of the 5th International Conference on the Biokinetics of Impacts*, pp. 218–228.

Cheng R, Yang KH, Levine RS et al. 1982. Injuries to the cervical spine caused by a distributed frontal load to the chest. In *Proceedings of the 26th Stapp Car Crash Conference*, pp. 1–40.

Ewing CL, Thomas DJ, Lustick L et al. 1978. Effect of initial position on the human head and neck response to +*Y* impact acceleration. In *Proceedings of the 22nd Stapp Car Crash Conference*, pp. 101–138.

Gadd CW. 1961. Criteria for injury potential. In *Impact Acceleration Stress Symposium*, National Research Council Publication No. 977, pp. 141–144. Washington, D.C., National Academy of Sciences.

Garfin SR, Rothman RH. 1983. Traumatic spondylolisthesis of the axis (Hangman's fracture). In RW Baily (Ed.), *The Cervical Spine*, pp. 223–232. Philadelphia, Lippincott.

Gennarelli TA. 1983. Head injuries in man and experimental animals: clinical aspects. *Acta Neurochir Suppl* 32:1.

Hardy WN, Khalil TB, King AI. 1994. Literature review of head injury biomechanics. *Int J Impact Eng* 15:561–586.

Hodgson VR, Mason MW, Thomas LM. 1972. In *Proceedings of the 16th Stapp Car Crash Conference*, pp. 1–13.

Hodgson VR, Thomas LM. 1971. Comparison of head acceleration injury indices in cadaver skull fracture. In *Proceedings of 15th Stapp Car Crash Conference*, pp. 190–206.

Hodgson VR, Thomas LM. 1973. *Breaking Strength of the Human Skull versus Impact Surface Curvature*. Detroit, Wayne State University.

Hodgson VR, Thomas LM. 1975. *Head Impact Response*. Warrendale, PA, Vehicle Research Institute, Society of Automotive Engineers.

Holbourn AHS. 1943. Mechanics of head injury. *Lancet* 2:438.

Hubbard RP, McLeod DG. 1974. Definition and development of a crash dummy head. In *Proceedings of the 18th Stapp Car Crash Conference*, pp. 599–628.

King AI, Chou C. 1977. Mathematical modelling, simulation and experimental testing of biomechanical system crash response. *J Biomech* 9:310.

Kleinberger M. 1993. Application of finite element techniques to the study of cervical spine mechanics. In *Proceedings of the 37th Stapp Car Crash Conference*, pp. 261–272.

Lighthall JW, Goshgarian HG, Pinderski CR. 1990. Characterization of axonal injury produced by controlled cortical impact. *J Neurotrauma* 7(2):65.

Lissner HR, Lebow M, Evans FG. 1960. Experimental studies on the relation between acceleration and intracranial pressure changes in man. *Surg Gynecol Obstet* 111:329.

Lord S, Barnsley L, Bogduk N. 1993. Cervical zygapophyseal joint pain in whiplash. In RW Teasell and AP Shapiro (Eds.), *Cervical Flexion–Extension/Whiplash Injuries*, pp. 355–372. Philadelphia, Hanley & Belfus.

Lowenhielm P. 1975. Mathematical simulation of gliding contusions. *J Biomech* 8:351.

Margulies SS, Thibault LE, Gennarelli TA. 1990. Physical model simulation of brain injury in the primate. *J Biomech* 23:823.

McConville JT, Churchill TD, Kaleps I et al. 1980. *Anthropometric Relationships of Body and Body Segment Moments of Inertia*, AMRL-TR-80-119. Wright-Patterson AFB, OH, Aerospace Medical Research Lab.

Mertz HJ, Patrick LM. 1967. Investigation of the kinematics and kinetics of whiplash. In *Proceedings of the 11th Stapp Car Crash Conference*, pp. 267–317.

Mertz HJ, Patrick LM. 1971. Strength and response of the human neck. In *Proceedings of the 15th Stapp Car Crash Conference*, pp. 207–255.

Mertz HJ, Neathery RF, Culver CC. 1973. Performance requirements and characteristics of mechanical necks. In WF King and HJ Mertz (Eds.), *Human Impact Response Measurement and Simulations*, pp. 263–288. New York, Plenum Press.

Moffat EA, Siegel AW, Huelke DF. 1978. The biomechanics of automotive cervical fractures. In *Proceedings of the 22nd Conference of American Association for Automotive Medicine*, pp. 151–168.

Nightingale RW, McElhaney JH, Best TM et al. 1993. In *Proceedings of the 39th Meeting of the Orthopedic Research Society*, p. 233.

Nuscholtz G, Lux P, Kaiker P, Janicki MA. 1984. Head impact response: skull deformation and angular accelerations. In *Proceedings of the 28th Stapp Car Crash Conference*, pp. 41–74.

Padgaonkar AJ, Krieger KW, King AI. 1975. Measurement of angular acceleration of a rigid body using linear accelerometers. *J Appl Mech* 42:552.

Patrick LM, Lissner HR, Gurdjian ES. 1965. Survival by design: head protection. In *Proceedings of the 7th Stapp Car Crash Conference*, pp. 483–499.

Patrick LM, Chou C. 1976. *Response of the Human Neck in Flexion, Extension, and Lateral Flexion*, Vehicle Research Institute Report No. VRI-7-3. Warrendale, PA, Society of Automotive Engineers.

Pierce DA, Barr JS. 1983. Fractures and dislocations at the base of the skull and upper spine. In RW Baily (Ed.), *The Cervical Spine*, pp. 196–206. Philadelphia, Lippincott.

Pintar FA, Yoganandan N, Sances A Jr et al. 1989. Kinematic and anatomical analysis of the human cervical spinal column under axial loading. In *Proceedings of the 33rd Stapp Car Crash Conference*, pp. 191–214.

Pintar FA, Sances A Jr, Yoganandan N et al. 1990. Biodynamics of the total human cadaveric spine. In *Proceedings of the 34th Stapp Car Crash Conference*, pp. 55–72.

Prasad P, Melvin JW, Huelke DF et al. 1985. Head. In *Review of Biomechanical Impact Response and Injury in the Automotive Environment*: Phase 1 Task B Report: *Advanced Anthropomorphic Test Device Development Program*, DOT Report No. DOT HS 807 042. Ann Arbor, University of Michigan.

Prasad P, Mertz HJ. 1985. *The Position of the United States Delegation to the ISO Working Group 6 on the Use of HIC in the Automotive Environment*, SAE Paper No. 851246. Warrendale, PA, Society of Automotive Engineers.

Reynolds HM, Clauser CE, McConville J et al. 1975. *Mass Distribution Properties of the Male Cadaver*, SAE Paper No. 750424. Warrendale, PA, Society of Automotive Engineers.

Robbins DH. 1983. *Development of Anthropometrically Based Design Specifications for an Advanced Adult Anthropomorphic Dummy Family*, Vol 2: *Anthropometric Specifications for a Midsized Male Dummy*, Report No. UMTRI 83-53-2. Ann Arbor, University of Michigan Transportation Research Institute.

Schneider LW, Foust DR, Bowman BM et al. 1975. Biomechanical properties of the human neck in lateral flexion. In *Proceedings of the 19th Stapp Car Crash Conference*, pp. 455–486.

Society of Automotive Engineers, Human Mechanical Response Task Force. 1985. *Human Mechanical Response Characteristics*, SAE J 1460. Warrendale, PA, Society of Automotive Engineers.

Strich SJ. 1961. Shearing of nerve fibres as a cause of brain damage due to head injury. *Lancet* 2:443.

Thomas DJ, Jessop ME. 1983. Experimental head and neck injury. In CL Ewing et al. (Eds.), *Impact Injury of the Head and Spine*, pp. 177–217. Springfield, IL, Charles C Thomas.

Versace J. 1970. A review of the severity index. In *Proceedings of the 15th Stapp Car Crash Conference*, pp. 771–796.

Viano DC, King AI, Melvin JW, Weber K. 1989. Injury biomechanics research: an essential element in the prevention of trauma. *J Biomech* 21:403.

Walker LB Jr, Harris EH, Pontius UR. 1973. Mass, volume, center of mass, and mass moment of inertia of head and neck of human body. In *Proceedings of the 17th Stapp Car Crash Conference*, pp. 525–537.

Wismans J, Spenny DH. 1983. Performance requirements for mechanical necks in lateral flexion. In *Proceedings of the 27th Stapp Car Crash Conference*, pp. 137–148.

Wismans J, Spenny DH. 1984. Head-neck response in frontal flexion. In *Proceedings of the 28th Stapp Car Crash Conference*, pp. 161–171.

7

Biomechanics of Chest and Abdomen Impact

David C. Viano
Wayne State University

Albert I. King
Wayne State University

Injury is caused by energy transfer to the body by an impacting object. It occurs when the body is struck by a blunt object, such as a vehicle instrument panel or side interior, and sufficient force is concentrated on the chest or abdomen. The risk of injury is influenced by the object's shape, stiffness, point of contact, and orientation. It can be reduced by energy-absorbing padding or crushable materials, which allow the surfaces in contact to deform, extend the duration of impact, and reduce loads. The torso is viscoelastic, so reaction force substantially increases with the speed of body deformation.

The biomechanical response of the body has three components: (1) inertial resistance by acceleration of body masses, (2) elastic resistance by compression of stiff structures and tissues, and (3) viscous resistance by rate-dependent properties of tissue. For low impact speeds, the elastic stiffness protects from crush injuries, whereas, for high rates of body deformation, the inertial and viscous properties determine the force developed and limit deformation. The risk of skeletal and internal organ injury relates to energy stored or absorbed by the elastic and viscous properties. The reaction load is related to these responses and inertial resistance of body masses that combine to resist deformation and prevent injury. When tissues are deformed beyond their recoverable limit, injuries result.

7.1 Chest and Abdomen Injury Mechanisms

The primary mechanism of chest and abdomen injury is compression of the body at high rates of loading. This causes deformation and stretching of internal organs and vessels. When the compression of the torso exceeds the ribcage tolerance, fractures occur and internal organs and vessels can be contused or ruptured. In some chest impacts, however, internal injury occurs without skeletal damage. This can happen during high-speed loading. It is due to the viscous or rate-sensitive nature of human tissue as biomechanical responses differ for low- and high-speed impact.

When organs or vessels are loaded slowly, the input energy is absorbed gradually through deformation, which is resisted by elastic properties and pressure buildup in tissue. When loaded rapidly, reaction force is proportional to the speed of tissue deformation as the viscous properties of the body resist deformation and provide a natural protection from impact. However, there is also a considerable inertial component to the reaction force. In this case, the body develops high internal pressure, and injuries can occur before the ribs deflect much. The ability of an organ or other biologic system to absorb impact energy without failure is called *tolerance.*

If an artery is stretched beyond its tensile strength, the tissue will tear. Organs and vessels can be stretched in different ways, which result in different types of injury. Motion of the heart during chest compression stretches the aorta along its axis from points of tethering in the body. This elongation generally leads to a transverse laceration when the strain limit is exceeded. In contrast, an increase in vascular pressure dilates the vessel and produces biaxial strain that is larger in the transverse than axial direction. If pressure rises beyond the vessel's limit, it will burst. For severe impacts, intraaortic pressure exceeds 500 to 1000 mmHg, which is a significant, nonphysiologic level but is tolerable for short durations. When laceration occurs, the predominant mode of aortic failure is axial, so the combined effects of stretch and internal pressure contribute to injury. Chest impact also compresses the ribcage, causing tensile strain on the outer surface of the ribs. As compression increases, the risk of rib fracture increases. In both cases, the mechanism of injury is tissue deformation.

The abdomen is more vulnerable to injury than the chest because there is little bony structure below the ribcage to protect internal organs in front and lateral impacts. Blunt impact of the upper abdomen can compress and injure the liver and spleen before significant whole-body motion occurs. In the liver, compression increases intrahepatic pressure and generates tensile or shear strains. If the tissue is sufficiently deformed, laceration of the major hepatic vessels can result in hemoperitoneum. Abdominal deformation also causes lobes of the liver to move relative to each other, stretching and shearing the vascular attachment at the hilar region.

Effective occupant restraints, safety systems, and protective equipment not only spread impact energy over the strongest body structures but also reduce contact velocity between the body and the impacted surface or object. The design of protective systems is aided by an understanding of injury mechanisms, quantification of human tolerance levels, and development of numerical relationships between measurable engineering parameters, such as force, acceleration, or deformation, and injury. These relationships are called *injury criteria.*

7.2 Injury Tolerance Criteria

Acceleration Injury

Stapp [1970] conducted rocket-sled experiments on belt-restraint systems and achieved a substantial human tolerance to long-duration, whole-body acceleration. Safety belts protected military personnel exposed to rapid but sustained acceleration. The experiments enabled Eiband [1959] to show in Fig. 7.1

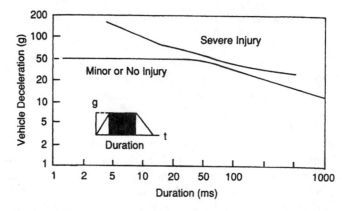

FIGURE 7.1   Whole-body human tolerance to vehicle acceleration based on impact duration. (Redrawn from Eiband [1959] and from Viano [1988]. With permission.)

that the tolerance to whole-body acceleration increased as the exposure duration decreased. This linked human tolerance and acceleration for exposures of 2- to 1000-ms duration. The tolerance data are based on average sled acceleration rather than the acceleration of the volunteer subject, which would be higher due to compliance of the restraint system. Even with this limitation, the data provide useful early guidelines for the development of military and civilian restraint systems.

More recent side impact tests have led to other tolerance formulas for chest injury. Morgan et al. [1986] evaluated rigid, side-wall cadaver tests and developed TTI, a thoracic trauma index, which is the average rib and spine acceleration. TTI limits human tolerance to 85 to 90 g in vehicle crash tests. Better injury assessment was achieved by Cavanaugh [1993] using average spinal acceleration (ASA), which is the average slope of the integral of spinal acceleration. ASA is the rate of momentum transfer during side impact, and a value of 30 g is proposed. In most cases, the torso can withstand 60 to 80 g peak whole-body acceleration by a well-distributed load.

Force Injury

Whole-body tolerance is related to Newton's second law of motion, where acceleration of a rigid mass is proportional to the force acting on it, or $F = ma$. While the human body is not a rigid mass, a well-distributed restraint system allows the torso to respond as though it were fairly rigid when load is applied through the shoulder and pelvis. The greater the acceleration, the greater is the force and risk of injury. For a high-speed frontal crash, a restrained occupant can experience 60 g acceleration. For a body mass of 76 kg, the inertial load is 44.7 kN (10,000 lb) and is tolerable if distributed over strong skeletal elements.

The ability to withstand high acceleration for short durations implies that tolerance is related to momentum transfer, because an equivalent change in velocity can be achieved by increasing the acceleration and decreasing its duration, since $\Delta V = a\Delta t$. The implication for occupant protection systems is that the risk of injury can be decreased if the crash deceleration is extended over a greater period of time. For occupant restraint in 25 ms, a velocity change of 14.7 m/s (32.7 mi/h) occurs with 60 g whole-body acceleration. This duration can be achieved by crushable vehicle structures and occupant restraints.

Prior to the widespread use of safety belts, safety engineers needed information on the tolerance of the chest to design energy-absorbing instrument panels and steering systems. The concept was to limit impact force below human tolerance by crushable materials and structures. Using the highest practical crush force, safety was extended to the greatest severity of vehicle crashes. GM Research and Wayne State University collaborated on the development of the first crash sled that was used to simulate progressively more severe frontal impacts. Embalmed human cadavers were exposed to head, chest, and knee impact on 15-cm-diameter (6-in) load cells until bone fracture was observed on x-ray. Patrick [1965] demonstrated that blunt chest loading of 3.3 kN (740 lb) could be tolerated with minimal risk of serious injury. This is a pressure of 187 kPa. Gadd and Patrick [1968] later found a tolerance of 8.0 kN (1800 lb) if the load was distributed over the shoulders and chest by a properly designed steering wheel and column. Cavanaugh [1993] found that side-impact tolerance is similar to frontal tolerance and that shoulder contact is also an important load path.

Compression Injury

High-speed films of cadaver impacts show that whole-body acceleration does not describe torso impact biomechanics. Tolerance of the chest and abdomen must consider body deformation. Force acting on the body causes two simultaneous responses: (1) compression of the complaint structures of the torso and (2) acceleration of body masses. The previously neglected mechanism of injury was compression, which causes the sternum to displace toward the spine as ribs bends and possibly fracture. Acceleration and force per se are not sufficient indicators of impact tolerance because they cannot discriminate between the two responses. Numerous studies have shown that acceleration is less related to injury than compression.

The importance of chest deformation was confirmed by Kroell [1971, 1974] in blunt thoracic impacts of unembalmed cadavers. Peak spinal acceleration and impact force were poorer injury predictors than

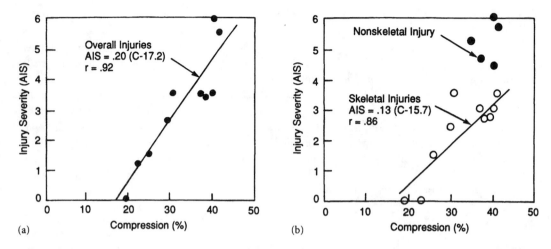

FIGURE 7.2 (a) Injury severity from blunt impact of human cadavers as a function of the maximum chest compression. (From Viano [1988], with permission.) (b) Severity of skeletal injury and incidence of internal organ injury as a function of maximum chest compression for blunt impacts of human cadavers. (From Viano [1988]. With permission.)

the maximum compression of the chest, as measured by the percentage change in the anteroposterior thickness of the body. A relationship between injury risk and compression involves energy stored by elastic deformation of the body. Stored energy E_s by a spring representing the ribcage and soft tissues is related to the displacement integral of force: $E_s = \int F\, dx$. Force in a spring is proportional to deformation: $F = kx$, where k is a spring constant. Stored energy is $E_s = k\int x\, dx = 0.5kx^2$. Over a reasonable range, stored energy is proportional to deformation or compression, so $E_s \approx C$.

Tests with human volunteers showed that compression up to 20% during moderate-duration loading was fully reversible. Cadaver impacts with compression greater than 20% (Fig. 7.2a) an increase in rib fractures and internal organ injury as the compression increased to 40%. The deflection tolerance was originally set at 8.8 cm (3.5 in) for moderate but recoverable injury. This represents 39% compression. However, at this level of compression, multiple rib fractures and serious injury can occur, so a more conservative tolerance of 34% is used to avert the possibility of flail chest (Fig. 7.2b). This reduces the risk of direct loading on the heart, lungs, and internal organs by a loss of the protective function of the ribcage.

Viscous Injury

The velocity of body deformation is an important factor in impact injury. For example, when a fluid-filled organ is compressed slowly, energy can be absorbed by tissue deformation without damage. When loaded rapidly, the organ cannot deform fast enough, and rupture may occur without significant change in shape, even though the load is substantially higher than for the slow-loading condition.

The viscoelastic behavior of soft tissues is important when the velocity of deformation exceeds 3 m/s. For lower speeds, such as in slow crushing loads or for a belt-restrained occupant in a frontal crash, tissue compression is limited by elastic properties resisting skeletal and internal organ injury. For higher speeds of deformation, such as occupant loading by the door in a side impact or for an unrestrained occupant or pedestrian, maximum compression does not adequately address the viscous and inertial properties of the torso, nor the time of greatest injury risk. In these conditions, the tolerance to compression is progressively lower as the speed of deformation increases, and the velocity of deformation becomes a dominant factor in injury.

Insight on a rate-dependent injury mechanism came from over 20 years of research by Clemedson and Jonsson [1979] on high-speed impact and blast-wave exposures. The studies confirmed that tolerable compression inversely varied with the velocity of impact. The concept was studied further in relation to the abdomen by Lau [1981] for frontal impacts in the range of 5 to 20 m/s (10 to 45 mi/h). The liver was the target organ. Using a maximum compression of 16%, the severity of injury increased with the speed of loading, including serious mutilation of the lobes and major vessels in the highest-speed impacts. While the compression was within limits of volunteer loading at low speeds, the exposure produced critical injury at higher speeds. Subsequent tests on other animals and target organs verified an interrelationship between body compression, deformation velocity, and injury.

The previous observations led Viano and Lau [1988] to propose a viscous injury mechanism for soft biologic tissues. The *viscous response VC* is defined as the product of velocity of deformation V and compression C, which is a time-varying function in an impact. The parameter has physical meaning to absorbed energy E_a by a viscous dashpot under impact loading. Absorbed energy is related to the displacement integral of force: $E_a = \int F \, dx$, and force in a dashpot is proportional to the velocity of deformation: $F = cV$, where c is a dashpot parameter. Absorbed energy is $E_a = c \int V \, dx$, or a time integral by substitution: $E_a = c \int V^2 \, dt$. The integrand is composed of two responses, so $E_a = c[\int d(Vx) - \int ax \, dt]$, where a is acceleration across the dashpot. The first term is the viscous response, and the second is an inertial term related to the deceleration of fluid set in motion. Absorbed energy is given by $E_a = c(Vx - \int ax \, dt)$. The viscous response is proportional to absorbed energy, or $E_a \approx VC$, during the rapid phase of impact loading prior to peak compression.

Subsequent tests by Lau and Viano [1988, 1986] verified that serious injury occurred at the time of peak *VC*. For blunt chest impact, peak *VC* occurs in about half the time for maximum compression. Rib fractures also occur progressively with chest compression, as early as 9 to 14 ms—at peak *VC*—in a cadaver impact requiring 30 ms to reach peak compression. Upper abdominal injury by steering wheel contact also relates to viscous loading. Lau [1987] showed that limiting the viscous response by a self-aligning steering wheel reduced the risk of liver injury, as does force limiting an armrest in side impacts. Animal tests also have shown that *VC* is a good predictor of functional injury to the heart and respiratory systems. In these experiments, Stein [1982] found that the severity of cardiac arrhythmia and traumatic apnea was related to *VC*. This situation is important to baseball impact protection of children [Viano et al., 1992] and in the design of bulletproof protective vests [Quatros, 1994].

Figure 7.3 summarizes injury mechanisms associated with impact deformation. For low speeds of deformation, the limiting factor is crush injury from compression C of the body. This occurs at $C = 35$ to 40% depending on the contact area and orientation of loading. For deformation speeds above 3 m/s,

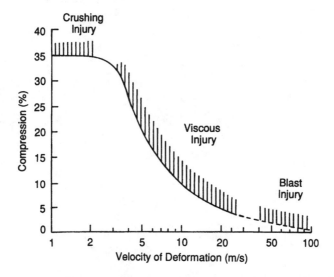

FIGURE 7.3 Biomechanics of chest injury by a crushing injury mechanism limited by tolerable compression at C_{max} = 35%, a viscous injury mechanism limited by the product of velocity and extent of deformation at $VC_{max} = 1.0$ m/s, and a blast injury mechanism for shock-wave loading.

injury is related to peak viscous response of $VC = 1.0$ m/s. In a particular situation, injury can occur by a compression or viscous responses, or both, since these responses occur at different times in an impact. At extreme rates of loading, such as in a blast-wave exposure, injury occurs with less than 10 to 15% compression by high energy transfer to viscous elements of the body.

7.3 Biomechanical Responses during Impact

The reaction force developed by the chest varies with the velocity of deformation and biomechanics is best characterized by a family of force-deflection responses. Figure 7.4 summarizes frontal and lateral chest biomechanics for various impact speeds. The dynamic compliance is related to viscous, inertial, and elastic properties of the body. The initial rise in force is due to inertia as the sternal mass is rapidly accelerated to the impact speed. The plateau force is related to the viscous component, which is rate dependent, and a superimposed elastic stiffness that increases force with chest compression. Unloading provides a hysteresis loop representing the energy absorbed by body deformation.

Melvin [1988] analyzed frontal biomechanics and modeled the force-deflection response as an initial stiffness $A = 0.26 + 0.60(V - 1.3)$ and a plateau force $B = 1.0 + 0.75(V - 3.7)$, where A is in kN/cm, B is in kN, and V is in m/s. The force B reasonably approximates the plateau level for lateral chest and abdominal impacts, but the initial stiffness is considerably lower at $A = 0.12(V - 1.2)$ for side loadings.

A simple, but relevant, lumped-mass model of the chest was developed by Lobdell [1973] and is shown in Fig. 7.5. The impacting mass is m_1, and skin compliance is represented by k_{12}. An energy-absorbing

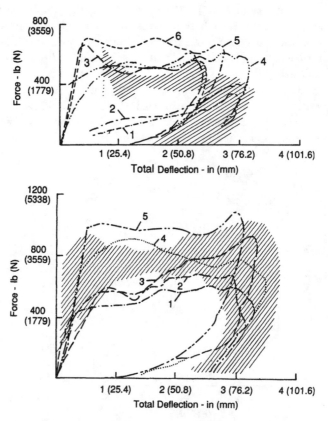

FIGURE 7.4 Frontal and lateral force-deflection response of the human cadaver chest at various speeds of blunt pendulum impact. The initial stiffness is followed by a plateau force until unloading. (From Kroell [1974] and Viano [1989], summarized by Cavanaugh [1993]. With permission.)

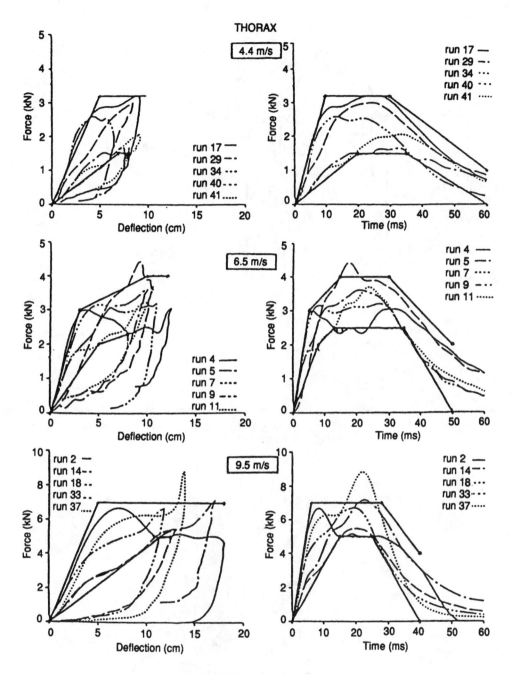

FIGURE 7.4 (continued)

interface was added by Viano [1987] to evaluate protective padding. Chest structure is represented by a parallel Voigt and Maxwell spring-dashpot system that couples the sternal m_2 and spinal m_3 masses. When subjected to a blunt sternal impact, the model follows established force-deflection corridors. The biomechanical model is effective in studying compression and viscous responses. It also simulates military exposure to high-speed nonpenetrating projectiles (Fig. 7.6), even though the loading conditions are quite different from the cadaver database used to develop the model. This mechanical system characterizes the elastic, viscous, and inertial components of the torso.

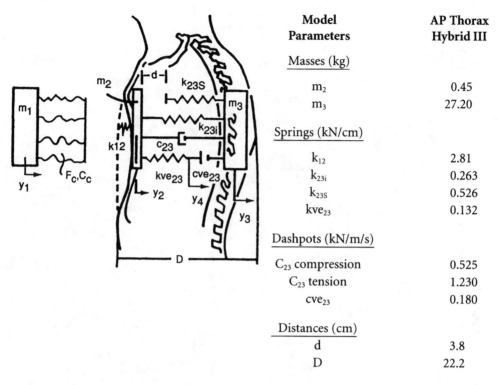

Model Parameters	AP Thorax Hybrid III
Masses (kg)	
m_2	0.45
m_3	27.20
Springs (kN/cm)	
k_{12}	2.81
k_{23i}	0.263
k_{23S}	0.526
kve_{23}	0.132
Dashpots (kN/m/s)	
C_{23} compression	0.525
C_{23} tension	1.230
cve_{23}	0.180
Distances (cm)	
d	3.8
D	22.2

FIGURE 7.5 Lumped-mass model of the human thorax with impacting mass and energy-absorbing material interface. The biomechanical parameters are given for mass, spring, and damping characteristics of the chest in blunt frontal impact. (Modified from Lobdell [1973] by Viano [1987]. With permission.)

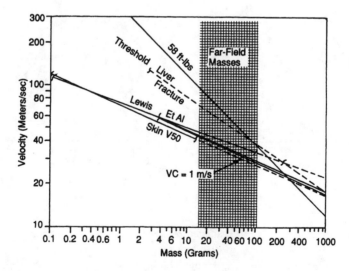

FIGURE 7.6 Tolerance levels for blunt loading as a function of impact mass and velocity. The plot includes information from automotive impact situations and from high-speed military projectile impacts. The Lobdell model is effective over the entire range of impact conditions. (Modified from Quatros [1993]. With permission.)

The Hybrid III dummy reported on by Foster [1977] was the first to demonstrate human-like chest responses typical of the biomechanical data for frontal impacts. Rouhana [1989] developed a frangible abdomen, useful in predicting injury for lap-belt submarining. More recent work by Schneider [1992] led to a new prototype frontal dummy. Lateral impact tests of cadavers against a rigid wall and blunt pendulum led to side-impact dummies such as the Eurosid and Biosid [Mertz, 1993].

7.4 Injury Risk Assessment

Over years of study, tolerances have been established for most responses of the chest and abdomen. Table 7.1 provides tolerance levels from reviews by Cavanaugh [1993] and Rouhana [1993]. While these are single thresholds, they are commonly used to evaluate safety systems. The implication is that for biomechanical responses below tolerance there is no injury, and for responses above tolerance there is injury. An additional factor is biomechanical response scaling for individuals of different size and weight. The commonly accepted procedure involves equal stress and velocity, which enabled Mertz et al. [1989] to predict injury tolerances and biomechanical responses for different-sized adult dummies.

Injury risk assessment is frequently used. It evaluates the probability of injury as a continuous function of a biomechanical response. A logist function relates injury probability p to a biomechanical response x by $p(x) = [1 + \exp(\alpha - \beta x)]^{-1}$, where α and β are parameters derived from statistical analysis of biomechanical data. This function provides a sigmoidal relationship with three distinct regions in Fig. 7.7. For low biomechanical response levels, there is a low probability of injury. Similarly, for very high levels, the risk asymptotes to 100%. The transition region between the two extremes involves risk that is proportional to the biomechanical response. A sigmoidal function is typical of human tolerance because it represents the distribution in weak through strong subjects in a population exposed to impact. Table 7.2 summarizes available parameters for chest and abdominal injury risk assessment.

TABLE 7.1 Human Tolerance for Chest and Abdomen Impact

Criteria	Chest Frontal	Chest Lateral	Abdomen Frontal	Abdomen Lateral	Criteria
Acceleration					Acceleration
3-ms limit	60 *g*				
TTI		85–90 *g*			
ASA		30 *g*			
AIS 4+		45 *g*		39 *g*	AIS 4+
Force					Force
Sternum	3.3 kN				
Chest + shoulder	8.8 kN	10.2 kN			
AIS 3+			2.9 kN	3.1 kN	AIS 3+
AIS 4+		5.5 kN	3.9 kN	6.7 kN	AIS 4+
Pressure					Pressure
			166 kPa		AIS 3+
			216 kPa		AIS 4+
Compression					Compression
Rib fracture	20%				
Ribcage	32%		38%		AIS 3+
Flail chest	40%	38%	48%	44%	AIS 4+
Viscous					Viscous
AIS 3+	1.0 m/s				AIS 3+
AIS 4+	1.3 m/s	1.47 m/s	1.4 m/s	1.98 m/s	AIS 4+

Source: Adapted from Cavanaugh [1993] and Rouhana [1993].

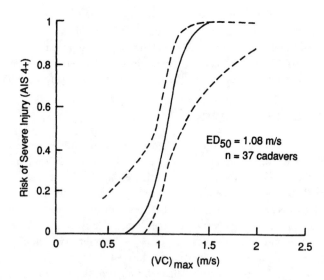

FIGURE 7.7 Typical logist injury-probability function relating the risk of serious injury to the viscous response of the chest. (From Viano [1988]. With permission.)

TABLE 7.2 Injury Probability Functions for Blunt Impact

Body Region	$ED_{25\%}$	α	β	X^2	p	R
			Frontal Impact			
Chest (AIS 4+)						
VC	1.0 m/s	11.42	11.56	25.6	0.000	0.68
C	34%	10.49	0.277	15.9	0.000	0.52
			Lateral Impact			
Chest (AIS 4+)						
VC	1.5 m/s	10.02	6.08	13.7	0.000	0.77
C	38%	31.22	0.79	13.5	0.000	0.76
Abdomen (AIS 4+)						
VC	2.0 m/s	8.64	3.81	6.1	0.013	0.60
C	47%	16.29	0.35	4.6	0.032	0.48
Pelvis (pubic ramus fracture)						
C	27%	84.02	3.07	11.5	0.001	0.91

Source: Adapted from Viano [1989]. With permission.

References

Cavanaugh JM. 1993. The biomechanics of thoracic trauma. In Nahum AM, Melvin JW (Eds.), *Accidental Injury: Biomechanics and Prevention*, pp. 362–391. New York, Springer-Verlag.

Cavanaugh JM, Zhu Y, Huang Y et al. 1993. Injury and response of the thorax in side impact cadaveric tests. In *Proceedings of the 37th Stapp Car Crash Conference*, pp. 199–222, SAE Paper no. 933127. Warrendale, PA, Society of Automotive Engineers.

Eiband AM. 1959. *Human Tolerance to Rapidly Applied Acceleration: A Survey of the Literature*, NASA Memo No. 5-19-59E. Washington, D.C., National Aeronautics and Space Administration.

Foster JK, Kortge JO, Wolanin MJ. 1977. Hybrid III—a biomechanically-based crash test dummy. In *Proceedings of the 21st Stapp Car Crash Conference*, pp. 975–1014, SAE Paper no. 770938. Warrendale, PA, Society of Automotive Engineers.

Gadd CW, Patrick LM. 1968. *Systems versus Laboratory Impact Tests for Estimating Injury Hazards*, SAE Paper no. 680053. Warrendale, PA, Society of Automotive Engineers.

Jonsson A, Clemedson CJ et al. 1979. Dynamic factors influencing the production of lung injury in rabbits subjected to blunt chest wall impact. *Aviat Space Environ Med* 50:325.

King Al. 1984. Regional tolerance to impact acceleration. In SP-622. Warrendale, PA, Society of Automotive Engineers.

Kroell CK, Schneider DC, Nahum AM. 1971. Impact tolerance and response to the human thorax. In *Proceedings of the 15th Stapp Car Crash Conference*, pp. 84–134, SAE Paper no. 710851. Warrendale, PA, Society of Automotive Engineers.

Kroell CK, Schneider DC, Nahum AM. 1974. Impact tolerance and response to the human thorax II. In *Proceedings of the 18th Stapp Car Crash Conference*, pp. 383–457, SAE Paper no. 741187. Warrendale, PA, Society of Automotive Engineers.

Lau IV, Viano DC. 1981. Influence of impact velocity on the severity of nonpenetrating hepatic injury. *Trauma* 21(2):115.

Lau IV, Viano DC. 1986. The viscous criterion—bases and application of an injury severity index for soft tissue. In *Proceedings of the 30th Stapp Car Crash Conference*, pp. 123–142, SAE Paper no. 861882. Warrendale, PA, Society of Automotive Engineers.

Lau IV, Viano DC. 1988. How and when blunt injury occurs: implications to frontal and side impact protection. In *Proceedings of the 32nd Stapp Car Crash Conference*, pp. 81–100, SAE Paper no. 881714. Warrendale, PA, Society of Automotive Engineers.

Lau IV, Horsch JD, Andrzejak D et al. 1987. Biomechanics of liver injury by steering wheel loading. *Trauma* 27:225.

Lobdell TE, Kroell CK, Schneider DC et al. 1973. Impact response of the human thorax. In King WF, Mertz HJ (Eds.), *Human Impact Response Measurement and Simulation*, pp. 201–245. New York, Plenum Press.

Melvin JW, Weber K (Eds). 1988. Review of biomechanical response and injury in the automotive environment, AATD Task B Final Report, DOT-HS-807-224. Washington, D.C., U.S. Department of Transportation, National Highway Traffic Safety Administration.

Melvin JW, King Al, Alem, NM. 1988. AATD system technical characteristics, design concepts, and trauma assessment criteria. AATD Task E-F Final Report, DOT-HS-807-224. Washington, D.C., U.S. Department of Transportation, National Highway Traffic Safety Administration.

Mertz HJ. 1993. Anthropomorphic test devices. In Nahum AM, Melvin JW (Eds.), *Accidental Injury: Biomechanics and Prevention*, pp. 66–84. New York, Springer-Verlag.

Mertz HJ, Gadd CW. 1971. Thoracic tolerance to whole-body deceleration. In *Proceedings of the 15th Stapp Car Crash Conference*, pp. 135–137, SAE Paper no. 710852. Warrendale, PA, Society of Automotive Engineers.

Mertz HJ, Irwin A, Melvin J et al. 1989. *Size, Weight and Biomechanical Impact Response Requirements for Adult Size Small Female and Large Male Dummies*, SAE Paper no. 890756. Warrendale, PA, Society of Automotive Engineers.

Morgan RM, Marcus JH, Eppinger RH. 1986. Side impact—the biofidelity of NHTSA's proposed ATD and efficacy of TTI. In *Proceedings of the 30th Stapp Car Crash Conference*, pp. 27–40, SAE Paper no. 861877. Warrendale, PA, Society of Automotive Engineers.

Patrick LM, Kroell CK, Mertz HJ. 1965. Forces on the human body in simulated crashes. In *Proceedings of the 9th Stapp Car Crash Conference*, pp. 237–260. Warrendale, PA, Society of Automotive Engineers.

Patrick LM, Mertz HJ, Kroell CK. 1967. Cadaver knee, chest, and head impact loads. In *Proceedings of the 11th Stapp Car Crash Conference*, pp. 168–182, SAE Paper no. 670913. Warrendale, PA, Society of Automotive Engineers.

Rouhana SW, Viano D, Jedrzejczak E et al. 1989. Assessing submarining and abdominal injury risk in the Hybrid III family of dummies. In *Proceedings of the 33rd Stapp Car Crash Conference*, pp. 257–279, SAE Paper no. 892440. Warrendale, PA, Society of Automotive Engineers.

Rouhana SW. 1993. Biomechanics of abdominal trauma. In Nahum AM, Melvin JW (Eds.), *Accidental Injury: Biomechanics and Prevention*, pp. 391–428. New York, Springer-Verlag.

Schneider LW, Haffner MP et al. 1992. Development of an advanced ATD thorax for improved injury assessment in frontal crash environments. In *Proceedings of the 36th Stapp Car Crash Conference*, pp. 129–156, SAE Paper no. 922520. Warrendale, PA, Society of Automotive Engineers.

Society of Automotive Engineers. 1986. *Human Tolerance to Impact Conditions as Related to Motor Vehicle Design*, SAE J885. Warrendale, PA, Society of Automotive Engineers.

Stalnaker RL, McElhaney JH, Roberts VL, Trollope ML. 1973. Human torso response to blunt trauma. In King WF, Mertz HJ (Eds.), *Human Impact Response Measurement and Simulation*, pp 181–199. New York, Plenum Press.

Stapp JP. 1970. Voluntary human tolerance levels. In Gurdjian ES, Lange WA, Patrick LM, Thomas LM (Eds.), *Impact Injury and Crash Protection*, pp. 308–349. Springfield, IL, Charles C Thomas.

Stein PD, Sabbah HN, Viano D et al. 1982. Response of the heart to nonpenetrating cardiac trauma. *J Trauma* 22(5):364.

Viano DC. 1987. Evaluation of the benefit of energy-absorbing materials for side impact protection. In *Proceedings of the 31st Stapp Car Crash Conference*, pp. 185–224, SAE Paper no. 872213. Warrendale, PA, Society of Automotive Engineers.

Viano DC. 1988. Cause and control of automotive trauma. *Bull NY Acad Med* 64:376.

Viano DC. 1989. Biomechanical responses and injuries in blunt lateral impact. In *Proceedings of the 33rd Stapp Car Crash Conference*, pp. 113–142, SAE Paper no. 892432. Warrendale, PA, Society of Automotive Engineers.

Viano DC, Lau IV. 1988. A viscous tolerance criterion for soft tissue injury assessment. *J Biomech* 21:387.

Viano DC, King AI, Melvin J et al. 1989. Injury biomechanics research: an essential element in the prevention of trauma. *J Biomech* 22:403.

Viano DC, Andrzejak DV, Polley TZ, King AI. 1992. Mechanism of fatal chest injury by baseball impact: development of an experimental model. *Clin J Sports Med* 2:166.

8
Analysis of Gait

Roy B. Davis
Motion Analysis Laboratory
Shriners Hospitals for Children

Peter A. DeLuca
Gait Analysis Laboratory
Connecticut Children's
Medical Center

Sylvia Õunpuu
Gait Analysis Laboratory
Connecticut Children's
Medical Center

Gait analysis is the quantitative measurement and assessment of human locomotion including both walking and running. A number of different disciplines use gait analysis. Basic scientists seek a better understanding of the mechanisms that normal ambulators use to translate muscular contractions about articulating joints into functional accomplishment, e.g., level walking and stair climbing. Increasingly, researchers endeavor to better appreciate the relationship between the human motor control systems and gait dynamics. With respect to running, athletes and their coaches use gait analysis techniques in a ceaseless quest for meaningful improvements in performance while avoiding injury. Sports equipment manufacturers seek to quantify the perceived advantages of their products relative to a competitor's offering.

In the realm of clinical gait analysis, medical professionals apply an evolving knowledge base in the interpretation of the walking patterns of impaired ambulators for the planning of treatment protocols, e.g., orthotic prescription and surgical intervention. Clinical gait analysis is an evaluation tool that allows the clinician to determine the extent to which an individual's gait has been affected by an already diagnosed disorder [Brand and Crowninshield, 1981]. Examples of clinical pathologies currently served by gait analysis include amputation, cerebral palsy (CP), degenerative joint disease, poliomyelitis, multiple sclerosis, muscular dystrophy, myelodysplasia, rheumatoid arthritis, stroke, and traumatic brain injury.

Generally, gait analysis data collection protocols, measurement precision, and data reduction models have been developed to meet the requirements specific to the research, sport, or clinical setting. For example, gait measurement protocols in a research setting might include an extensive physical examination to characterize the anthropometrics of each subject. This time expenditure may not be possible in a clinical setting. Also, sport assessments generally require higher data acquisition rates because of increased velocity amplitudes relative to walking. The focus of this chapter is on the methods for the assessment of walking patterns of persons with locomotive impairment, i.e., clinical gait analysis. The discussion will include a description of the available measurement technology, the components of data collection and reduction, the type of gait information produced for clinical interpretation, and the strengths and limitations of clinical gait analysis.

8.1 Fundamental Concepts

Clinical Gait Analysis Information

Gait is a cyclic activity for which certain discrete events have been defined as significant. Typically, the *gait cycle* is defined as the period of time from the point of initial contact (also referred to as *foot contact*) of the subject's foot with the ground to next point of initial contact for that same limb. Dividing the gait cycle in stance and swing phases is the point in the cycle where the stance limb leaves the ground, called *toe off* or *foot off*. Gait variables that change over time such as the patient's joint angular displacements are normally presented as a function of the individual's gait cycle for clinical analysis. This is done to facilitate the comparison of different walking trials and the use of a normative data base [Õunpuu et al., 1991]. Data that are currently provided for the clinical interpretation of gait include:

- Static physical examination measures, such as passive joint range of motion, muscle strength and tone, and the presence and degree of bony deformity
- Stride and temporal parameters, such as step length and walking velocity
- Segment and joint angular displacements commonly referred to as *kinematics*
- The forces and torque applied to the subject's foot by the ground, or ground reaction forces
- The reactive joint moments produced about the lower extremity joints by active and passive soft tissue forces as well as the associated mechanical power of the joint moment, collectively referred to as *kinetics*
- Indications of muscle activity during gait, i.e., voltage potentials produced by contracting muscles, known as *dynamic electromyography* (EMG)
- A measure of metabolic energy expenditure, e.g., oxygen consumption, energy cost.
- A videotape of the individual's gait trial for qualitative review and quality control purposes

Data Collection Protocol

The steps involved in the gathering of data for the interpretation of gait pathologies usually include a complete physical examination, biplanar videotaping, and multiple walks of the "instrumented" subject along a walkway that is commonly both level and smooth. The time to complete these steps can range from three to five hours (Table 8.1). Although the standard for analysis is barefoot gait, subjects are tested in other conditions as well, e.g., lower extremity orthoses and crutches. Requirements and constraints associated with clinical gait data gathering include the following:

- The patient should not be intimidated or distracted by the testing environment.
- The measurement equipment and protocols should not alter the subject's gait.
- Patient preparation and testing time must be minimized, and rest (or play) intervals must be included in the process.
- Data collection techniques must be reasonably repeatable.
- Methodology must be sufficiently robust and flexible to allow the evaluation of a variety of pathological gait abnormalities where the dynamic range of motion and anatomy may be significantly different from normal.
- The collected data must be validated before the end of the test period, e.g., raw data fully processed before the patient leaves the facility.

TABLE 8.1 A Typical Gait Data Collection Protocol

Test Component	Estimate Time (min)
Pretest tasks: Test explanation to the adult subject or child and parent, system calibration	10
Videotaping: Brace, barefoot, close-up, standing	15–25
Clinical examination: Range of motion, muscle strength, etc.	30–45
Motion marker placement	15–20
Motion data collection: Subject calibration and multiple walks, per test condition (barefoot and orthosis)	30–60
Electromyography (surface electrodes and fine-wire electrodes)	20–60
Data reduction of all trials	30–90
Data interpretation	20–30
Report dictation, generation, and distribution	120–180

Measurement Approaches and Systems

The purpose of this section is to provide an overview of the several technologies that are available to measure the dynamic gait variables listed above, including stride and temporal parameters, kinematics, kinetics, and dynamic EMG. Methods of data reduction will be described in a section that follows.

Stride and Temporal Parameters

The timing of the gait cycle events of initial contact and toe off must be measured for the computation of the stride and temporal quantities. These measures may be obtained through a wide variety of approaches ranging from the use of simple tools such as a stop watch and tape measure to sophisticated arrays of photoelectric monitors. Foot switches may be applied to the plantar aspect of the subject's foot over the bony prominences of the heel and metatarsal heads in different configurations depending on the information desired. A typical configuration is the placement of a switch on the heel, first (and fifth) metatarsal heads and great toe. In a clinical population, foot switch placement is challenging because of the variability of foot deformities and foot-ground contact patterns. This switch placement difficulty is avoided through the use of either shoe insoles instrumented with one or two large foot switches or entire contact sensitive walkways. Alternatively, video cameras may be employed with video frame counters to determine the timing of initial contact and toe-off events. These gait events may also be measured using either the camera-based motion measurement systems or the force platform technology described below.

Motion Measurement

A number of alternative technologies are available for the measurement of body segment spatial position and orientation. These include the use of electrogoniometry, high-speed photography, accelerometry, and video-based digitizers. These approaches are described below.

Electrogoniometry. A simple electrogoniometer consists of a rotary potentiometer with arms fixed to the shaft and base for attachment to the extremity across the joint of interest. The advantages of multiaxial goniometers (more appropriate for human joint motion measurement) include the capability for real-time display and rapid collection of single-joint information on many subjects. Electrogoniometers are limited to the measurement of relative angles and may be cumbersome in many typical clinical applications such as the simultaneous, bilateral assessment of hip, knee, and ankle motion.

Cinefilm. High-speed photography offers particular advantages in the assessment of activities such as sprinting that produce velocity and acceleration magnitudes greater than those realized in walking. This approach is not attractive for clinical use because it is labor intensive, e.g., each frame of data is digitized individually and requires an unacceptably long processing time.

Accelerometry. Multiaxis accelerometers can be employed to measure both linear and angular accelerations (if multiple transducers are properly configured). Velocity and position data may then be derived through numerical integration, although care must be taken with respect to the selection of initial conditions and the handling of gravitational effects.

Videocamera-Based Systems. This approach to human motion measurement involves the use of external markers that are placed on the subject's body segments and aligned with specific bony landmarks. The marker trajectories produced by the subject's ambulation through a specific measurement volume are then monitored by a system of cameras (generally from two to seven) placed around the measurement volume. In a frame-by-frame analysis, stereophotogrammetric techniques are then used to produce the instantaneous three-dimensional (3-D) coordinates of each marker (relative to a fixed laboratory coordinate system) from the set of two-dimensional camera images. The processing of the 3-D marker coordinate data is described in a later section.

The videocamera-based systems employ either passive (retroflective) or active (light-emitting diodes) markers. Passive marker camera systems use either strobe light sources (typically infrared light-emitting diodes (LEDs) configured in rings around the camera lens) or electronically shuttered cameras. The cameras then capture the light returned from the highly reflective markers (usually small spheres). Active marker camera systems record the light that is produced by small LED markers that are placed directly on the subject. Advantages and disadvantages are associated with each approach. For example, the anatomical location (or identity) of each marker used in an active marker system is immediately known because the markers are sequentially pulsed by the controlling computer. User interaction is required currently for marker identification in passive marker systems, although algorithms have been developed to expedite this process, i.e., automatic tracking. The system of cables required to power and control the LEDs of the active marker system may increase the possibility for subject distraction and gait alteration.

Ground Reaction Measurement

Force Platforms.   The three components of the ground reaction force vector, the ground reaction torque (vertical), and the point of application of the ground reaction force vector (i.e., center of pressure) are measured with force platforms embedded in the walkway. Force plates with typical measurement surface dimensions of 0.5×0.5 m are comprised of several strain gauges or piezoelectric sensor arrays rigidly mounted together.

Foot Pressure Distributions.   The dynamic distributed load that corresponds to the vertical ground reaction force can be evaluated with the use of a flat, two-dimensional array of small piezoresistive sensors. Overall resolution of the transducer is dictated by the size of the individual sensor "cell." Sensor arrays configured as shoe insole inserts and flat plates offer the clinical user two measurement alternatives. Although the currently available technology does afford the clinical practitioner better insight into the qualitative force distribution patterns across the planar surface of the foot, its quantitative capability is limited because of the challenge of calibration and signal stability (e.g., temperature-dependent sensors).

Dynamic Electromyography (EMG)

Electrodes placed on the skin's surface and fine wires inserted into muscle are used to measure the voltage potentials produced by contracting muscles. The activity of the lower limb musculature is evaluated in this way with respect to the timing and the intensity of the contraction. Data collection variables that affect the quality of the EMG signal include the placement of and distance between recording electrodes, skin surface conditions, distance between electrode and target muscle, signal amplification and filtering, and the rate of data acquisition. The phasic characteristics of the muscle activity may be estimated from the raw EMG signal. The EMG data may also be presented as a rectified and/or integrated waveform. To evaluate the intensity of the contraction, the dynamic EMG amplitudes are typically normalized by a reference value, e.g., the EMG amplitude during a maximum voluntary contraction. This latter requirement is difficult to achieve consistently for patients who have limited isolated control of individual muscles, such as children with CP.

8.2   Gait Data Reduction

The predominant approach for the collection of clinical gait data involves the placement of external markers on the surface of body segments that are aligned with particular bony landmarks. These markers are commonly attached to the subjects as either discrete units or in rigidly connected clusters (Fig. 8.1). As described briefly above, the products of the data acquisition process are arrays containing the 3-D coordinates (relative to an inertially fixed laboratory coordinate system) of the spatial trajectory of each marker over a gait cycle. If at least three markers or reference points are identified for each body segment, then the six degrees of freedom associated with the position of the segment may be determined. The following example illustrates this straightforward process.

Assume that a cluster of three markers has been attached to the thigh and shank of the test subject as shown in Fig. 8.2. A body-fixed coordinate system may be computed for each cluster. For example, for the thigh, the vector cross product of the unit vectors from markers B to A and B to C produces a vector that is perpendicular to the cluster plane. From these vectors, the unit vectors $\mathbf{T}_{TX}$ and $\mathbf{T}_{TY}$ may be determined

FIGURE 8.1 Videocamera-based motion measurement systems monitor the displacement of external markers that are placed on the subject's body segments and aligned with specific bony landmarks. These markers are commonly attached to the subject as either discrete units, e.g., the pelvis, or in rigidly connected clusters, e.g., on the thigh and shank.

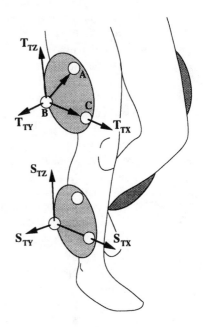

FIGURE 8.2 A body-fixed coordinate system may be computed for each cluster of three or more markers. On the thigh, for example, the vector cross product of the unit vectors from markers B to A and B to C produces a vector that is perpendicular to the cluster plane. From these vectors, the unit vectors T_{TX} and T_{TY} may be determined and used to compute the third orthogonal coordinate direction T_{TZ}.

and used to compute the third orthogonal coordinate direction T_{TZ}. In a similar manner the marker-based, or technical, coordinate system may be calculated for the shank, i.e., S_{TX}, S_{TY}, and S_{TZ}. At this point, one might use these two technical coordinate systems to provide an estimate of the absolute orientation of the thigh or shank or the relative angles between the thigh and shank. This assumes that the technical coordinate systems reasonably approximate the anatomical axes of the body segments, e.g., that T_{TZ} approximates the long axis of the thigh. A more rigorous approach incorporates the use of a subject calibration procedure to relate technical coordinate systems with pertinent anatomical directions [Cappozzo, 1984].

In a subject calibration, usually performed with the subject standing, additional data are collected by the measurement system that connects the technical coordinate systems to the underlying anatomical structure. For example, as shown in Fig. 8.3, the medial and lateral femoral condyles and the medial and lateral malleoli may be used as anatomical references with the application of additional markers. With the hip center location estimated from markers placed on the pelvis [Bell et al., 1989], and knee and ankle center locations based on the additional markers, anatomical coordinate systems may be computed, e.g., {T_A} and {S_A}. The relationship between the respective anatomical and technical coordinate system pairs as well as the location of the joint centers in terms of the appropriate technical coordinate system may be stored, to be recalled in the reduction of each frame of the walking data. In this way, the technical coordinate systems (shown in Fig. 8.3) are transformed into alignment with the anatomical coordinate systems.

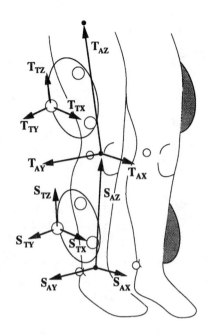

Once anatomically aligned body-fixed coordinate systems have been computed for each body segment under investigation, one may compute the angular position of the joints and segments in a number of ways. The classical approach of Euler angles is commonly used in clinical gait analysis to describe the motion of the thigh relative to the pelvis (or hip angles), the motion of the shank relative to the thigh (or knee angles), the motion of the foot relative to the shank (or ankle angles), as well as the absolute orientation of the

FIGURE 8.3 A subject calibration relates technical coordinate systems with anatomical coordinate systems, e.g., {T_T} with {T_A}, through the identification of anatomical landmarks, e.g., the medial and lateral femoral condyles and medial and lateral malleoli.

pelvis and foot in space [Grood and Suntay, 1983; Õunpuu et al., 1991]. The joint rotation sequence commonly used for the Euler angle computation is flexion–extension, adduction–abduction, and transverse plane rotation. Alternatively, joint motion has been described through the use of helical axes [Woltring et al., 1985].

The moments that soft tissue (e.g., muscle, ligaments, joint capsule) forces produce about approximate joint centers may be computed through the use of inverse dynamics, i.e., Newtonian mechanics. For example, the free-body diagram of the foot shown in Fig. 8.4 shows the various external loads to the foot as well as the reactions produced at the ankle. The mass, mass moments of inertia, and location of the center of mass may be estimated from regression-based anthropometric relationships [Dempster et al., 1959], and linear and angular velocity and acceleration may be determined by numerical differentiation. If the ground reaction loads, F_G and T, are provided by a force platform, then the unknown ankle reaction force F_A may be solved for with Newton's second law. Euler's equations of motion may then be applied to compute the net ankle reaction moment, M_A. In the application of Euler's equations of motion, care must be taken to perform the vector operations with all vectors transformed into the foot coordinate system which has been chosen to approximate the principle axes and located at the center of mass of the foot [Greenwood, 1965]. This process may then be repeated for the shank and thigh by using distal joint loads to solve for the proximal joint reactions. The mechanical power associated with a joint moment and the corresponding joint angular velocity may be computed from the vector dot product of the two vectors, e.g., ankle power is computed through $M_A \cdot \omega_A$ where ω_A is the angular velocity of the foot relative to the shank.

Although commonly referred to as *muscle moments*, these net joint reaction moments are generated by several mechanisms, e.g., ligamentous forces, passive muscle and tendon force, and active muscle contractile force, in response to external loads. Currently, clinical muscle force evaluation is not possible because of the scarcity of data related to location of the instantaneous line of muscle force as well as the

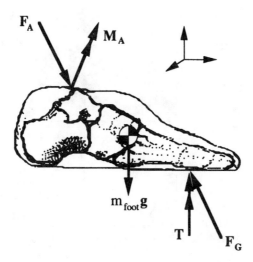

FIGURE 8.4 The moments that soft-tissue forces produce about approximate joint centers may be computed through the use of Newtonian mechanics. This free-body diagram of the foot illustrates the external loads to the foot, e.g., the ground reaction loads, F_G and T, and the weight of the foot, $m_{foot}g$, as well as the unknown reactions produced at the ankle, F_A and M_A.

overall indeterminacy of the problem. The possibility of the cocontraction of opposing muscle groups also impacts the estimation of bone-on-bone force values, i.e., the magnitude of F_A found above reflects the *minimum* estimated value of the bone-on-bone force at the ankle.

With respect to the kinematic data reduction, the body segments are assumed to be rigid, e.g., soft-tissue movement relative to underlying bony structures is small. Consequently, marker of instrumentation attachment sites must be selected carefully, e.g., over tendonous structures of the distal shank as opposed to the more proximal muscle masses of the gastrocnemius and soleus. The models described above also assume that the joint center locations remain fixed relative to the respective segmental coordinate systems, e.g., the knee center is fixed relative to the thigh coordinate system. The measurement technology and associated protocols cannot currently produce data of sufficient quality for the reliable determination of instantaneous centers of rotation.

The addition assumptions associated with the gait kinetics model described above are related to the inertial properties of the body segments. Body segment mass and mass distribution, i.e., mass moments of inertia, are generally estimated from statistical relationships derived from cadaver studies. Differences between body types are not typically incorporated into these anthropometric models. Moreover, the mass distribution changes are assumed to be negligible during motion.

8.3 Illustrative Clinical Case

As indicated above, the information available for clinical gait interpretation may include static physical examination measures, stride and temporal data, segment and joint kinematics, joint kinetics, electromyograms, and a video record. With this information the clinical team can assess the patient's gait deviations, attempt to identify the etiology of the abnormalities, and recommend treatment alternatives. In this way, clinicians are able to isolate the biomechanical insufficiency that may produce a locomotive impairment and require a compensatory response from the patient. For example, a patient may excessively elevate a hip (compensatory) in order to gain additional foot clearance in swing which is perhaps inadequate due to a weak ankle dorsiflexor (primary problem). The knee kinematics and moments and associated electromyographic data shown in Fig. 8.5 were generated by an 11-year-old male with a diagnosis of cerebral palsy, spastic diplegia. Based on the limited knee range of motion, insufficient knee flexion in the swing phase of the gait cycle, and abnormal activity of the rectus femoris in the swing

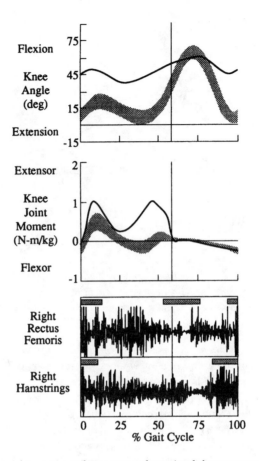

FIGURE 8.5 These are the knee kinematics and moments and associated electromyographic data for a representative subject with a diagnosis of cerebral palsy, spastic diplegia (including shaded bands/bars indicating one standard deviation about normal).

phase, a transfer of the distal end of the rectus femoris to the sartorius was recommended. Moreover, the excessive knee flexion at initial contact and throughout the stance phase due to overactivity of the hamstrings suggested intramuscular lengthenings of the medial and lateral hamstrings. Transverse plane deviations apparent from the gait analysis (not shown) led to recommendations of derotational osteotomies of both the subject's femurs and tibias.

8.4 Gait Analysis: Current Status

As indicated in the modeling discussion above, the utility of gait analysis information may be limited by sources of error such as soft tissue displacement relative to bone, estimates of joint center locations, particularly the hip, approximations of the inertial properties of the body segments, and the numerical differentiation of noisy displacement data. Other errors associated with data collection alter the results as well, for example, a marker is improperly placed or a force platform is inadvertently contacted by the swing limb. The evaluation of small subjects weakens the data because intermarker distances are reduced, thereby reducing the precision of angular computations. It is essential that the potential adverse effects of these errors on the gait information be understood and appreciated by the clinician in the interpretation process.

Controversies related to gait analysis techniques include the use of individually placed markers vs. clusters of markers, the estimation of quasi-static, body-fixed locations of joint centers (as described above) vs. the dynamic determination of the instantaneous locations, and the application of helical or screw axes vs. the use of Euler angles. Additional research and development are needed to resolve these fundamental methodological issues.

Despite these limitations, gait analysis facilitates the systematic quantitative documentation of walking patterns. With the various gait data, the clinician has the opportunity to separate the primary causes of a gait abnormality from compensatory gait mechanisms. Apparent contradictions between the different types of gait information can result in a more carefully developed understanding of the gait deviations. It provides the clinical user the capability to more precisely (than observational gait analysis alone) plan complex multilevel surgeries and evaluate the efficacy of different surgical approaches or orthotic designs. Through gait analysis, movement in planes of motion not easily observed, such as about the long axes of the lower limb segments, may be quantified. Finally, quantities that cannot be observed may be assessed, e.g., muscular activity and joint kinetics. In the future, it is anticipated that our understanding of gait will be enhanced through the application of pattern recognition strategies, coupled dynamics, and the linkage of empirical results derived through inverse dynamics with the simulations provided by forward dynamics modeling.

References

Bell AL, Pederson DR, Brand RA. 1989. Prediction of hip joint center location from external landmarks. *Human Movement Sci* 8:3.

Brand RA, Crowninshield RD. 1981. Comment on criteria for patient evaluation tools. *J Biomech* 14:655.

Cappozzo A. 1984. Gait analysis methodology. *Human Movement Sci* 3:27.

Dempster WT, Gabel WC, Felts WJL. 1959. The anthropometry of manual work space for the seated subjects. *Am J Phys Anthropometry* 17:289.

Greenwood DT. 1965. *Principles of Dynamics*, Englewood Cliffs, NJ, Prentice-Hall.

Grood ES, Suntay WJ. 1983. A joint coordinate system for the clinical description of three-dimensional motions: application to the knee. *J Biomech Eng* 105(2):136.

Õunpuu S, Gage JR, Davis RB. 1991. Three-dimensional lower extremity joint kinetics in normal pediatric gait. *J Pediatr Orthop* 11:341.

Woltring HJ, Huskies R, DeLange A. 1985. Finite centroid and helical axis estimation from noisy landmark measurement in the study of human joint kinematics. *J Biomech* 18:379.

Further Information on Gait Analysis Techniques

Allard P, Stokes IAF, Blanchi JP (Eds.) 1995. *Three-Dimensional Analysis of Human Movement*. Champaign, IL, Human Kinetics.

Berme N, Cappozzo A (Eds.) 1990. *Biomechanics of Human Movement: Applications in Rehabilitation, Sports and Ergonomics*. Worthington, OH, Bertec Corporation.

Harris GF, Smith PA (Eds.) 1996. *Human Motion Analysis*. Piscataway, NJ, IEEE Press.

Whittle M. 1991. *Gait Analysis: An Introduction*. Oxford, Butterworth-Heinemann.

Winter DA. 1990. *Biomechanics and Motor Control of Human Movement*. New York, John Wiley & Sons.

Further Information on Normal and Pathological Gait

Gage JR. 1991. *Gait Analysis in Cerebral Palsy*. London, MacKeith Press.

Perry J. 1992. *Gait Analysis: Normal and Pathological Function*. Thorofare, NJ, Slack.

Sutherland DH, Olshen RA, Biden EN et al. 1988. *The Development of Mature Walking*. London, MacKeith Press.

9

Exercise Physiology

Arthur T. Johnson
University of Maryland

Cathryn R. Dooly
University of Maryland

The study of exercise is important to medical and biological engineers. Cognizance of acute and chronic responses to exercise gives an understanding of the physiological stresses to which the body is subjected. To appreciate exercise responses requires a true systems approach to physiology, because during exercise all physiological responses become a highly integrated, total supportive mechanism for the performance of the physical stress of exercise. Unlike the study of pathology and disease, the study of exercise physiology leads to a wonderful understanding of the way the body is supposed to work while performing at its healthy best.

For exercise involving resistance, physiological and psychological adjustments begin even before the start of the exercise. The central nervous system (CNS) sizes up the task before it, assessing how much muscular force to apply and computing trial limb trajectories to accomplish the required movement. Heart rate may begin rising in anticipation of increased oxygen demands and respiration may also increase.

9.1 Muscle Energetics

Deep in muscle tissue, key components have been stored for this moment. Adenosine triphosphate (ATP), the fundamental energy source for muscle cells, is at maximal levels. Also stored are significant amounts of creatine phosphate and glycogen.

When the actinomyocin filaments of the muscles are caused to move in response to neural stimulation, ATP reserves are rapidly used, and ATP becomes adenosine diphosphate (ADP), a compound with much less energy density than ATP. Maximally contracting mammalian muscle uses approximately 1.7×10^{-5} mole of ATP per gram per second [White et al., 1959]. ATP stores in skeletal muscle tissue amount to 5×10^{-6} mole per gram of tissue, or enough to meet muscle energy demands for no more than 0.5 s.

Initial replenishment of ATP occurs through the transfer of creatine phosphate (CP) into creatine. The resting muscle contains 4 to 6 times as much CP as it does ATP, but the total supply of high-energy phosphate cannot sustain muscle activity for more than a few seconds.

Glycogen is a polysaccharide present in muscle tissues in large amounts. When required, glycogen is decomposed into glucose and pyruvic acid, which, in turn, becomes lactic acid. These reactions form ATP and proceed without oxygen. They are thus called *anaerobic*.

When sufficient oxygen is available (aerobic conditions), either in muscle tissue or elsewhere, these processes are reversed. ATP is reformed from ADP and AMP (adenosine monophosphate), CP is reformed from creatine and phosphate (P), and glycogen is reformed from glucose or lactic acid. Energy for these processes is derived from the complete oxidation of carbohydrates, fatty acids, or amino acids to form carbon dioxide and water. These reactions can be summarized by the following equations:

Anaerobic:

$$ATP \leftrightarrow ADP + P + \text{free energy} \tag{9.1}$$

$$CP + ADP \leftrightarrow \text{creatine} + ATP \tag{9.2}$$

$$\text{glycogen or glucose} + P + ADP \rightarrow \text{lactate} + ATP \tag{9.3}$$

Aerobic:

$$\text{Glycogen or fatty acids} + P + ADP + O_2 \rightarrow CO_2 + H_2O + ATP \tag{9.4}$$

All conditions:

$$2ADP \leftrightarrow ATP + AMP \tag{9.5}$$

The most intense levels of exercise occur anaerobically [Molé, 1983] and can be maintained for only a minute or two (Fig. 9.1).

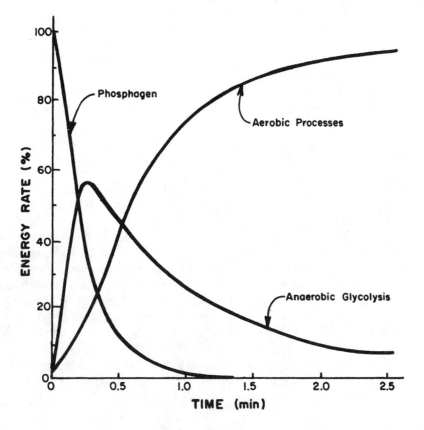

FIGURE 9.1 Muscle energy sources at the beginning of exercise. (Redrawn with permission from Molé [1983].)

9.2 Cardiovascular Adjustments

Mechanoreceptors in the muscles, tendons, and joints send information to the CNS that the muscles have begun movement, and this information is used by the CNS to increase heart rate via the sympathetic autonomic nervous system. Cardiac output, the rate of blood pumped by the heart, is the product of heart rate and stroke volume (amount of blood pumped per heart beat). Heart rate increases nearly exponentially at the beginning of exercise with a time constant of about 30 s. Stroke volume does not change immediately but lags a bit until the cardiac output completes the loop back to the heart.

During rest a large volume of blood is stored in the veins, especially in the extremities. When exercise begins, this blood is transferred from the venous side of the heart to the arterial side. Attempting to push extra blood flow through the resistance of the arteries causes a rise in both systolic (during heart ventricular contraction) and diastolic (during the pause between contractions) blood pressures. The increased blood pressures are sensed by baroreceptors located in the aortic arch and carotid sinus (Fig. 9.2).

As a consequence, small muscles encircling the entrance to the arterioles (small arteries) are caused to relax by the CNS. By Poiseuille's law:

$$R = \frac{8L\mu}{\pi r^4} \tag{9.6}$$

where R = resistance of a tube, N·s/m^5
L = length of the tube, m
μ = viscosity of the fluid, kg/(m·s) or N·s/m^2
r = radius of the tube lumen, m

increasing the arteriole radius by 19% will decrease its resistance to one-half. Thus, systolic pressure returns to its resting value, and diastolic pressure may actually fall. Increased blood pressures are called *afterload* on the heart.

To meet the oxygen demand of the muscles, blood is redistributed from tissues and organs not directly involved with exercise performance. Thus, blood flows to the gastrointestinal tract and kidneys are reduced, whereas blood flows to skeletal muscle, cardiac muscle, and skin are increased.

The heart is actually two pumping systems operated in series. The left heart pumps blood throughout the systemic blood vessels. The right heart pumps blood throughout the pulmonary system. Blood pressures in the systemic vessels are higher than blood pressures in the pulmonary system.

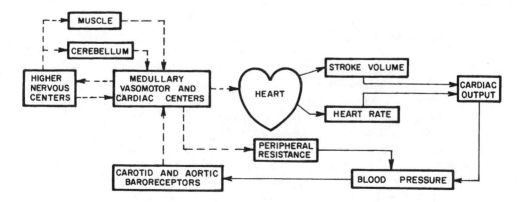

FIGURE 9.2 General scheme for blood pressure regulation. Dashed lines indicate neural communication, and solid lines indicate direct mechanical effect.

Two chambers comprise each heart. The atrium is like an assist device that produces some suction and collects blood from the veins. Its main purpose is to deliver blood to the ventricle, which is the more powerful chamber that develops blood pressure. The myocardium (heart muscle) of the left ventricle is larger and stronger than the myocardium of the right ventricle. With two hearts and four chambers in series, there could be a problem matching flow rates from each of them. If not properly matched, blood could easily accumulate downstream from the most powerful chamber and upstream from the weakest chamber.

Myocardial tissue exerts a more forceful contraction if it is stretched before contraction begins. This property (known as *Starling's law of the heart*) serves to equalize the flows between the two hearts by causing a more powerful ejection from the heart in which more blood accumulates during diastole. The amount of initial stretching of the cardiac muscle is known as *preload*.

9.3   Maximum Oxygen Uptake

The heart has been considered to be the limiting factor for delivery of oxygen to the tissues. As long as oxygen delivery is sufficient to meet demands of the working muscles, exercise is considered to be aerobic. If oxygen delivery is insufficient, anaerobic metabolism continues to supply muscular energy needs, but lactic acid accumulates in the blood. To remove lactic acid and reform glucose requires the presence of oxygen, which must usually be delayed until exercise ceases or exercise level is reduced.

Fitness of an individual is characterized by a mostly reproducible measurable quantity known as maximal oxygen uptake $(\dot{V}_{O_{2max}})$ that indicates a person's capacity for aerobic energy transfer and the ability to sustain high-intensity exercise for longer than 4 or 5 minutes (Fig. 9.3). The more fit, the higher is $\dot{V}_{O_{2max}}$. Typical values are 2.5 L/min for young male nonathletes, 5.0 L/min for well-trained male athletes; women have $\dot{V}_{O_{2max}}$ values about 70–80% as large as males. Maximal oxygen uptake declines with age steadily at 1% per year.

Exercise levels higher than those that result in $\dot{V}_{O_{2max}}$ can be sustained for various lengths of time. The accumulated difference between the oxygen equivalent of work and $\dot{V}_{O_{2max}}$ is called the *oxygen deficit* incurred by an individual (Fig. 9.4). There is a maximum oxygen deficit that cannot be exceeded by an individual. Once this maximum deficit has been reached, the person must cease exercise.

The amount of oxygen used to repay the oxygen deficit is called the *excess postexercise oxygen consumption* (EPOC). EPOC is always larger than the oxygen deficit because: (1) elevated body temperature immediately following exercise increases bodily metabolism in general, which requires more than resting levels of oxygen to service, (2) increased blood epinephrine levels increase general bodily metabolism, (3) increased respiratory and cardiac muscle activity requires oxygen, (4) refilling of body oxygen stores requires excess oxygen, and (5) there is some thermal inefficiency in replenishing muscle chemical stores. Considering only lactic acid oxygen debt, the total amount of oxygen required to return the body to its normal resting state is about twice the oxygen debt; the efficiency of anaerobic metabolism is about 50% of aerobic metabolism.

9.4   Respiratory Responses

Respiratory also increases when exercise begins, except that the time constant for respiratory response is about 45 s instead of 30 s for cardiac responses (Table 9.1). Control of respiration (Fig. 9.5) appears to begin with chemoreceptors located in the aortic arch, in the carotid bodies (in the neck), and in the ventral medula (in the brain). These receptors are sensitive to oxygen, carbon dioxide, and acidity levels but are most sensitive to carbon dioxide and acidity. Thus, the function of the respiratory system appears to be to remove excess carbon dioxide and, secondarily, to supply oxygen. Perhaps this is because excess CO_2 has narcotic effects, but insufficient oxygen does not produce severe reactions until oxygen levels in the inhaled

TABLE 9.1   Comparison of Response Time Constants for Three Major Systems of the Body

System	Dominant Time Constant, s
Heart	30
Respiratory system	45
Oxygen uptake	49
Thermal system	3600

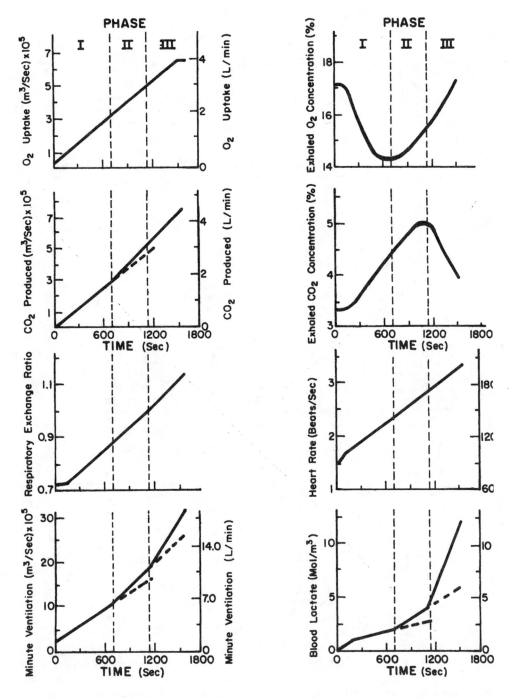

FIGURE 9.3 Concurrent typical changes in blood and respiratory parameters during exercise progressing from rest to maximum. (Adapted and redrawn from Skinner and McLellan [1980] by permission of the American Alliance for Health, Physical Education, Recreation and Dance.)

air fall to one-half of normal. There is no well-established evidence that respiration limits oxygen delivery to the tissues in normal individuals.

Oxygen is conveyed by convection in the upper airways and by diffusion in the lower airways to the alveoli (lower reaches of the lung where gas exchange with the blood occurs). Oxygen must diffuse from

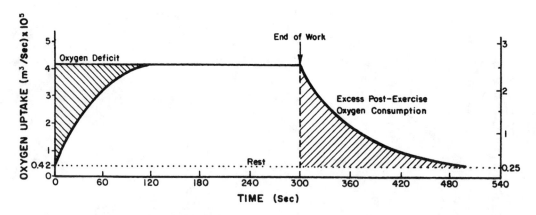

FIGURE 9.4   Oxygen uptake at the beginning of constant-load exercise increases gradually, accumulating an oxygen deficit that must be repaid at the end of exercise.

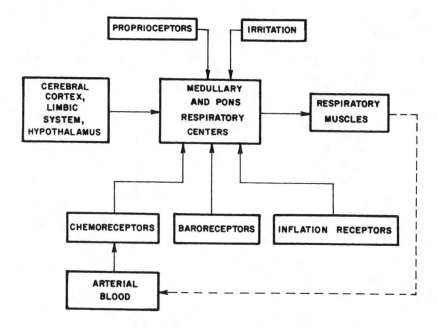

FIGURE 9.5   General scheme of respiratory control.

the alveoli, through the extremely thin alveolocapillary membrane into solution in the blood. Oxygen diffuses further into red blood cells where it is bound chemically to hemoglobin molecules. The order of each of these processes is reversed in the working muscles where the concentration gradient of oxygen is in the opposite direction. Complete equilibration of oxygen between alveolar air and pulmonary blood requires about 0.75 s. Carbon dioxide requires somewhat less, about 0.50 s. Thus, alveolar air more closely reflects levels of blood carbon dioxide than oxygen.

Both respiration rate and tidal volume (the amount of air moved per breath) increase with exercise, but above the anaerobic threshold the tidal volume no longer increases (remains at about 2–2.5 L). From that point, increases in ventilation require greater increases in respiration rate. A similar limitation occurs for stroke volume in the heart (limited to about 120 mL).

The work of respiration, representing only about 1–2% of the body's oxygen consumption at rest, increases to 8–10% or more of the body's oxygen consumption during exercise. Contributing greatly to this is the work to overcome resistance to movement of air, lung tissue, and chest wall tissue. Turbulent airflow in the upper airways (those nearest and including the mouth and nose) contributes a great deal of pressure drop. The lower airways are not as rigid as the upper airways and are influenced by the stretching and contraction of the lung surrounding them. High exhalation pressures external to the airways coupled with low static pressures inside (due to high flow rates inside) tend to close these airways somewhat and limit exhalation airflow rates. Resistance of these airways becomes very high, and the respiratory system appears like a flow source, but only during extreme exhalation.

9.5 Optimization

Energy demands during exercise are so great that optimal courses of action are followed for many physiological responses (Table 9.2). Walking occurs most naturally at a pace that represents the smallest energy expenditure; the transition from walking to running occurs when running expends less energy than walking; ejection of blood from the left ventricle appears to be optimized to minimize energy expenditure; and respiratory rate, breathing waveforms, the ratio of inhalation time to exhalation time, airways resistance, tidal volume, and other respiratory parameters all appear to be regulated to minimize energy expenditure [Johnson, 1993].

9.6 Thermal Response

When exercise extends for a long enough time, heat begins to build up in the body. In order for heat accumulation to become important, exercise must be performed at a relatively low rate. Otherwise, performance time would not be long enough for significant amounts of heat to be stored.

Muscular activities are at most 20–25% efficient, and, in general, the smaller the muscle, the less efficient it is. Heat results from the other 75–80% of the energy supplied to the muscle.

Thermal challenges are met in several ways. Blood sent to the limbs and blood returning from the limbs are normally conveyed by arteries and veins in close proximity deep inside the limb. This tends to conserve heat by countercurrent heat exchange between the arteries and veins. Thermal stress causes blood to return via surface veins rather than deep veins. Skin surface temperature increases and heat loss by convection and radiation also increases. In addition, vasodilation of cutaneous blood vessels augments surface heat loss but puts an additional burden on the heart to deliver added blood to the skin as well as the muscles. Heart rate increases as body temperature rises.

Sweating begins. Different areas of the body begin sweating earlier than others, but soon the whole body is involved. If sweat evaporation occurs on the skin surface, then the full cooling power of evaporating sweat (670 W·h/kg) is felt. If the sweat is absorbed by clothing, then the full benefit of sweat evaporation is not realized at the skin. If the sweat falls from the skin, no benefit accrues.

Sweating for a long time causes loss of plasma volume (plasma shift), resulting in some hemoconcentration (2% or more). This increased concentration increases blood viscosity, and cardiac work becomes greater.

9.7 Applications

Knowledge of exercise physiology imparts to the medical or biological engineer the ability to design devices to be used with or by humans or animals, or to borrow ideas from human physiology to apply to other situations. There is need for engineers to design the many pieces of equipment used by sports and health enthusiasts, to modify prostheses or devices for the handicapped to allow for performance of greater than light levels of work and exercise, to alleviate physiological stresses caused by personal protective equipment and other occupational ergonometric gear, to design human-powered machines

TABLE 9.2   Summary of Exercise Responses for a Normal Young Male

	Rest	Light Exercise	Moderate Exercise	Heavy Exercise	Maximal Exercise
Oxygen uptake (L/min)	0.30	0.60	2.2	3.0	3.2
Maximal oxygen uptake (%)	10	20	70	95	100
Physical work rate (watts)	0	10	140	240	430
Aerobic fraction (%)	100	100	98	85	50
Performance time (min)	∞	480	55	9.3	3.0
Carbon dioxide production (L/min)	0.18	1.5	2.3	2.8	3.7
Respiratory exchange ratio	0.72	0.84	0.94	1.0	1.1
Blood lactic acid (mMol/L)	1.0	1.8	4.0	7.2	9.6
Heart rate (beats/min)	70	130	160	175	200
Stroke volume (L)	0.075	0.100	0.105	0.110	0.110
Cardiac output (L/min)	5.2	13	17	19	22
Minute volume (L/min)	6	22	50	80	120
Tidal volume (L)	0.4	1.6	2.3	2.4	2.4
Respiration rate (breaths/min)	15	26	28	57	60
Peak flow (L/min)	216	340	450	480	480
Muscular efficiency (%)	0	5	18	20	20
Aortic hemoglobin saturation (%)	98	97	94	93	92
Inhalation time (s)	1.5	1.25	1.0	0.7	0.5
Exhalation time (s)	3.0	2.0	1.1	0.75	0.5
Respiratory work rate (watts)	0.305	0.705	5.45	12.32	20.03
Cardiac work rate (watts)	1.89	4.67	9.61	11.81	14.30
Systolic pressure (mmHg)	120	134	140	162	172
Diastolic pressure (mmHg)	80	85	90	95	100
End-inspiratory lung volume (L)	2.8	3.2	4.6	4.6	4.6
End-expiratory lung volume (L)	2.4	2.2	2.1	2.1	2.1
Gas partial pressures (mmHg)					
Arterial pCO_2	40	41	45	48	50
pO_2	100	98	94	93	92
Venous pCO_2	44	57	64	70	72
pO_2	36	23	17	10	9
Alveolar pCO_2	32	40	28	20	10
pO_2	98	94	110	115	120
Skin conductance [watts/($m^2 \cdot °C$)]	5.3	7.9	12	13	13
Sweat rate (kg/s)	0.001	0.002	0.008	0.007	0.002
Walking/running speed (m/s)	0	1.0	2.2	6.7	7.1
Ventilation/perfusion of the lung	0.52	0.50	0.54	0.82	1.1
Respiratory evaporative water loss (L/min)	1.02×10^{-5}	4.41×10^{-5}	9.01×10^{-4}	1.35×10^{-3}	2.14×10^{-3}
Total body convective heat loss (watts)	24	131	142	149	151
Mean skin temperature (°C)	34	32	30.5	29	28
Heat production (watts)	105	190	640	960	1720
Equilibrium rectal temperature (°C)	36.7	38.5	39.3	39.7	500
Final rectal temperature (°C)	37.1	38.26	39.3	37.4	37

that are compatible with the capabilities of the operators, and to invent systems to establish and maintain locally benign surroundings in otherwise harsh environments. Recipients of these efforts include athletes, the handicapped, laborers, firefighters, space explorers, military personnel, farmers, power-plant workers, and many others. The study of exercise physiology, especially in the language used by medical and biological engineers, can result in benefits to almost all of us.

Defining Terms

Anaerobic threshold: The transition between exercise levels that can be sustained through nearly complete aerobic metabolism and those that rely on at least partially anaerobic metabolism. Above the anaerobic threshold, blood lactate increases and the relationship between ventilation and oxygen uptake becomes nonlinear.

Excess postexercise oxygen consumption (EPOC): The difference between resting oxygen consumption and the accumulated rate of oxygen consumption following exercise termination.

Maximum oxygen consumption: The maximum rate of oxygen use during exercise. The amount of maximum oxygen consumption is determined by age, sex, and physical condition.

Oxygen deficit: The accumulated difference between actual oxygen consumption at the beginning of exercise and the rate of oxygen consumption that would exist if oxygen consumption rose immediately to its steady-state level corresponding to exercise level.

References

Johnson AT. 1993. How much work is expended for respiration? *Front Med Biol Eng* 5:265.

Molé PA. 1983. Exercise metabolism. In AA Bove and DT Lowenthal (Eds.), *Exercise Medicine*, pp. 43–88. New York, Academic Press.

Skinner JS, McLellan TH. 1980. The transition from aerobic to anaerobic metabolism. *Res Q Exerc Sport* 51:234.

White A, Handler P, Smith EL et al. 1959. *Principles of Biochemistry.* New York, McGraw-Hill.

Further Information

A comprehensive treatment of quantitative predictions in exercise physiology is presented in *Biomechanics and Exercise Physiology* by Arthur T. Johnson (John Wiley & Sons, 1991). There are a number of errors in the book, but an errata sheet is available from the author.

P.O. Astrand and K. Rodahl's *Textbook of Work Physiology* (McGraw-Hill, 1970) contains a great deal of exercise physiology and is probably considered to be the standard textbook on the subject.

Biological Foundations of Biomedical Engineering, edited by J. Kline (Little, Brown, Boston), is a very good textbook on physiology written for engineers.

10

Factors Affecting Mechanical Work in Humans

Arthur T. Johnson
University of Maryland

Bernard F. Hurley
University of Maryland

High technology has entered our diversions and leisure activities. Sports, exercise, and training are no longer just physical activities but include machines and techniques attuned to individual capabilities and needs. This chapter will consider several factors related to exercise and training that help in understanding human performance.

Physiological work performance is determined by energy transformations that begin with the process of photosynthesis and end with the production of biological work (Fig. 10.1). Energy in the form of nuclear transformations is converted to radiant energy, which then transforms the energy from carbon dioxide and water into oxygen and glucose through photosynthesis. In plants the glucose can also be converted to fats and proteins. Upon ingesting plants or other animals that eat plants, humans convert this energy through cellular respiration (the reverse of photosynthesis) to chemical energy in the form of adenosine triphosphate (ATP). The endergonic reactions (energy absorbed from the surroundings) that produce ATP are coupled with exergonic reactions (energy released to surroundings) that release energy from its breakdown to produce chemical and mechanical work in the human body. The steps involved in the synthesis and breakdown of carbohydrates, fats, and proteins produce chemical work and provide energy for the mechanical work produced from muscular contractions. The purpose of this chapter is to provide a brief summary of some factors that can affect mechanical work in humans.

10.1 Exercise Biomechanics

Equilibrium

Any body, including the human body, remains in stable equilibrium if the vectorial sum of all forces and torques acting on the body is zero. An unbalanced force results in linear acceleration, and an unbalanced torque results in rotational acceleration. Static equilibrium requires that:

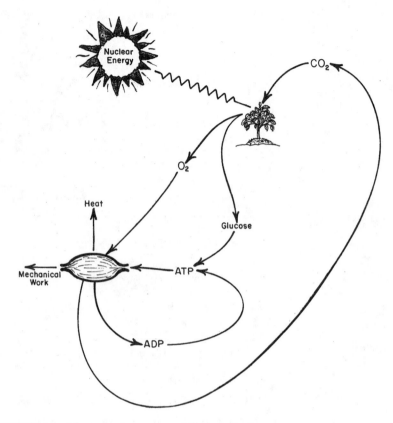

FIGURE 10.1   Schematic of energy transformations leading to muscular mechanical work.

$$\sum F = 0 \qquad\qquad (10.1)$$

$$\sum T = 0 \qquad\qquad (10.2)$$

where F = vectorial forces (N) and T = vectorial torques (N·m).

Some sport activities, such as wrestling, weightlifting, and fencing, require stability, whereas other activities, including running, jumping, and diving, cannot be performed unless there is managed instability. Shifting body position allows the proper control. The mass of the body is distributed as in Table 10.1, and the center of mass is located at approximately 56% of a person's height and midway from side-to-side and front-to-back. The center of mass can be made to shift by extending the limbs or by bending the torso.

TABLE 10.1   Fraction of Body Weights for Various Parts of the Body

Body Part	Fraction
Head and neck	0.07
Trunk	0.43
Upper Arms	0.07
Forearms and hands	0.06
Thighs	0.23
Lower legs and feet	<u>0.14</u>
	1.00

Muscular Movement

Mechanical movement results from contraction of muscles that are attached at each end to bones that can move relative to each other. The arrangement of this combination is commonly known as a class 3 lever (Fig. 10.2), where one joint acts as the fulcrum (Fig. 10.3), the other bone acts as the load, and the

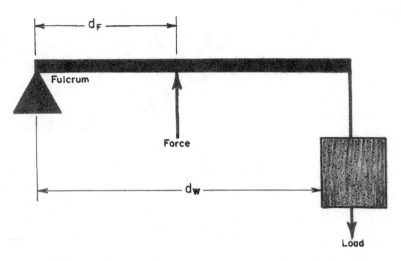

FIGURE 10.2 A class 3 lever is arranged with the applied force interposed between the fulcrum and the load. Most skeletal muscles are arranged in this fashion.

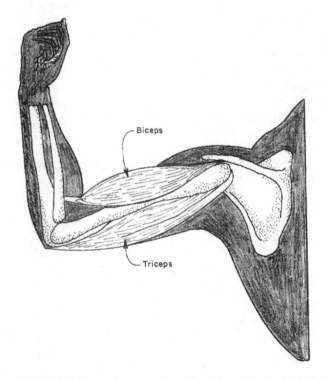

FIGURE 10.3 The biceps muscle of the arm is arranged as a class 3 lever. The load is located at the hand and the fulcrum at the elbow.

muscle provides the force interposed between fulcrum and load. This arrangement requires that the muscle force be greater than the load, sometimes by a very large amount, but the distance through which the muscle moves is made very small. These characteristics match muscle capabilities well (muscles can produce 7×10^5 N/m², but cannot move far). Since the distance is made smaller, the speed of shortening of the contracting muscle is also slower than it would be if the arrangement between the force and the load were different:

$$\frac{S_L}{S_M} = \frac{d_L}{d_M} \tag{10.3}$$

where S = speed, m/s
      d = distance from fulcrum, m
      L, M = denote load and muscle

Muscular Efficiency

Efficiency relates external, or physical, work produced to the total chemical energy consumed:

$$\eta = \frac{\text{External work produced}}{\text{Chemical energy consumed}} \tag{10.4}$$

Muscular efficiencies range from close to zero to about 20–25%. The larger numbers would be obtained for leg exercises that involve lifting body weight. In carpentry and foundry work, where both arms and legs are used, average mechanical efficiency is approximately 10% [Johnson, 1991]. For finer movements that require exquisite control, small muscles are often arranged in antagonistic fashion; that is, the final movement is produced as a result of the difference between two or more muscles working against each other. In this case, efficiencies approach zero. Isometric muscular contraction, where a force is produced but no movement results, has an efficiency of zero.

Muscles generally are able to exert the greatest force when the velocity of muscle contraction is zero. Power produced by this muscle would be zero. When the velocity of muscle contraction is about 8 m/s, the force produced by the muscle becomes zero, and the power produced by this muscle again becomes zero. Somewhere between, the power produced, and the efficiency, becomes a maximum (Fig. 10.4).

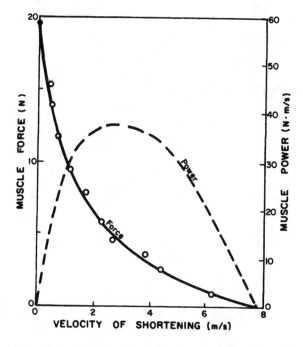

FIGURE 10.4   Force and power output of a muscle as a function of velocity. (Adapted and used with permission from Milsum [1966].)

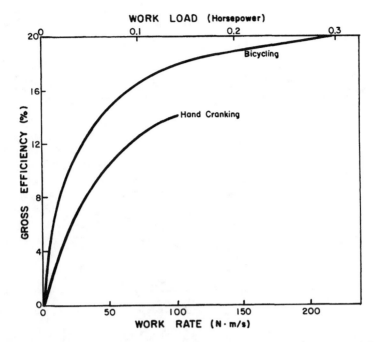

FIGURE 10.5   The gross efficiency for hand cranking or bicycling as a function of the rate of work [Goldman, 1978].

The isometric length-tension relationship of a muscle shows that the maximum force developed by a muscle is exerted at its resting length (the length of a slightly stretched muscle attached by its tendons to the skeleton) and decreases to zero at twice its resting length. Maximum force also decreases to zero at the shortest possible muscular length. Since muscular contractile force depends on the length of the muscle and length changes during contraction, muscular efficiency is always changing (Fig. 10.5).

Negative (eccentric) work is produced by a muscle when it maintains a force against an external force tending to stretch the muscle. An example of negative work is found as the action of the leg muscle during a descent of a flight of stairs. Since the body is being lowered, external work is less than zero. The muscles are using physiological energy to control the descent and prevent the body from accumulating kinetic energy as it descends.

Muscular efficiencies for walking downhill approach 120% [McMahon, 1984]. Since heat produced by the muscle is the difference between 100% and the percent efficiency, heat produced by muscles walking downhill is about 220% of their energy expenditure. Energy expenditure of muscles undergoing negative work is about one-sixth that of a muscle doing positive work [Johnson, 1991], so a leg muscle going uphill produces about twice as much heat as a leg muscle going downhill.

Locomotion

The act of locomotion involves both positive and negative work. There are four successive stages of a walking stride. In the first stage, both feet are on the ground, with one foot ahead of the other. The trailing foot pushes forward, and the front foot is pushing backward. In the second stage the trailing foot leaves the ground and the front foot applies a braking force. The center of mass of the body begins to lift over the front foot. In the third stage the trailing foot is brought forward, and the supporting foot applies a vertical force. The body center of mass is at its highest point above the supporting foot. In the last stage, the body's center of mass is lowered, and the trailing foot provides an acceleration force.

This alteration of the raising and lowering of the body center of mass, along with the pushing and braking provided by the feet, makes walking a low-efficiency maneuver. Walking has been likened to alternately applying the brakes and accelerator while driving a car. Just as the fuel efficiency of the car

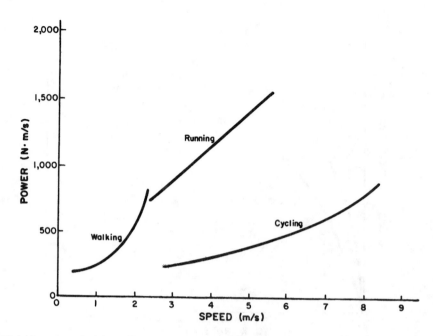

FIGURE 10.6   Power required for walking, running, and cycling by an adult male. Curves for walking and running intersect at about 2.3 m/s and show that walking is more efficient below the intersection and running is more efficient above. (Redrawn with permission from Alexander [1984].)

would suffer from this mode of propulsion, so the energy efficiency of walking suffers from the way walking is performed.

There is an optimum speed of walking. Faster than this speed, additional muscular energy is required to propel the body forward. Moving slower than the optimal speed requires additional muscular energy to retard leg movement. Thus, the optimal speed is related to the rate at which the leg can swing forward. Simple analysis of the leg as a physical pendulum shows that the optimal walking speed is related to leg length:

$$S \propto \sqrt{L} \tag{10.5}$$

Unlike walking, there is a stage of running during which both feet leave the ground. The center of mass of the body does not rise and fall as much during running as during walking, so the efficiency for running can be greater than for walking.

At a speed of about 2.5 m/s, running appears to be more energy efficient than walking, and the transition is usually made between these forms of locomotion (Fig. 10.6). Unlike walking, there does not appear to be a functional relationship between speed and leg length, so running power expenditure is linearly related to speed alone.

Why would anyone want to propel the extra weight of a bicycle in addition to body weight? On the surface it would appear that cycling would cost more energy than running or walking. However, the center of mass of the body does not move vertically as long as the cyclist sits on the seat. Without the positive and negative work associated with walking or running, cycling is a much more efficient form of locomotion than the other two (Fig. 10.6), and the cost of moving the additional weight of the bicycle can easily be supplied.

Many sports or leisure activities have a biomechanical basis. Understanding of the underlying biomechanical processes can lead to improved performance. Yet, there are limits to performance that cause frustrations for competitive athletes. Hence, additional factors are sometimes employed to expand these limits. Following is a brief discussion of some of these factors.

10.2   Exercise Training

Many compensatory reactions allow the body to adapt to minor stresses, such as mild exercise, so that homeostasis (equilibrium) can be maintained. For example, the increased energy demands of exercise stimulate an increase in heart rate, respiration, blood flow, and many other reactions that allow the body to maintain homeostasis. As the intensity of exercise increases, it becomes more difficult for compensatory mechanisms to maintain homeostasis. After exceeding about 80% of an untrained person's maximal capacity, homeostasis can no longer be maintained for more than a few minutes before exhaustion results.

Regular exercise (training) elevates the level that a single exercise session can be performed before disturbing homeostasis. It does this by elevating the maximal physiological capacity for homeostasis so that the same amount of exercise may no longer disrupt homeostasis because it is a lower percentage of maximal capacity. In addition, training produces specific adaptations during submaximal exercise that permit greater and longer amounts of work before losing homeostasis. A good example of this is when blood lactate rises with increased intensity of exercise. Prior to training, blood lactate concentration rises substantially when the intensity of exercise exceeds about 75% of maximal oxygen consumption ($\dot{V}_{O_2max}$). Following training, the same intensity of exercise results in a lower concentration of blood lactate, so exercise can now be performed at a higher fraction of $\dot{V}_{O_2max}$ before blood lactate reaches the same level as before training [Hurley et al., 1984a]. Thus, exercise training allows an individual to perform a much greater amount of work before homeostasis is disturbed to the point at which muscular exhaustion results.

10.3   Age

It is well established that the maximal capacity to exercise declines with age. However, many factors that change with age also affect an individual's maximal capacity to exercise. These include a gain in body fat, a decrease in lean body mass, onset of disease, and a decline in the level of physical activity. For this reason it is difficult to determine exactly how much of the loss in capacity to exercise with age can really be attributed solely to the effects of aging. $\dot{V}O_{2max}$ starts to decline after the age of 25 to 30 years at about the rate of 10% per decade in healthy sedentary women and men. This rate of decline is only half as great (~5%) when people maintain their physical activity levels as they age. Thus, there is no doubt that maximal work capacity declines with advancing age, but regular exercise appears to reduce this decline substantially.

10.4   Gender

Cardiovascular fitness and muscular strength are substantially higher in men compared to women. The aerobic capacity in men is about 40 to 50% higher than women. In addition, upper body strength is approximately 100% higher and lower body strength is about 50% higher in men [Lynch et al., 1999]. However, these differences are diminished substantially when body composition is taken into consideration. For example, when aerobic capacity is expressed with reference to body mass (ml O_2/kg/min) this difference decreases to about 20% and to about 10% when differences in muscle mass are taken into consideration [Lynch et al., 1999]. Similar declines in the differences in muscular strength can be demonstrated when taking lean body mass differences into consideration. There do not appear to be any differences in responsiveness to training when expressed on a relative basis (% change), but recent evidence suggests that men have greater muscle mass gains than women when expressed in absolute terms [Tracey et al., 1999].

10.5   Ergogenic Aids

Anabolic Steroids

Anabolic steroids are drugs that function similar to the male sex hormone testosterone. Upon binding with specific receptor sites, these drugs contribute greatly to male secondary sex characteristics. These drugs have been used frequently by a large number of competitive athletes, particularly those involved

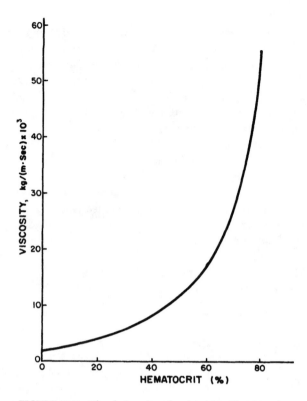

FIGURE 10.7   Blood viscosity related to blood hematocrit.

in strength and power sports. Increases in muscular strength, total body mass, and lean body mass have been reported [Bhasin et al., 1996]. It has been argued, however, that at least some of the increase in total body mass resulting from anabolic steroid use results from an increase in water retention [Casner et al., 1971]. Evidence indicates that many of these drugs act by decreasing muscle protein degradation rather than by enhancing protein synthesis [Hickson et al., 1990]. Hence, these drugs might be more appropriately named "anticatabolic" steroids. There is also some indication that these drugs work by increasing aggressive behavior and thereby promote a greater quantity and quality of training.

The factors that are responsible for transporting oxygen from the heart to skeletal muscle are enhanced following anabolic steroid administration through increases in blood volume, red blood cells (RBC), and hemoglobin. This would appear to improve cardiovascular fitness by increasing oxygen transport capacity. However, most studies show that anabolic steroids do not significantly increase cardiovascular fitness [Johnson et al., 1975]. This may be due to the increased blood viscosity resulting from the hemoconcentration effects of increasing RBCs and hemoglobin. Increases in blood viscosity may reduce blood flow to and from the heart and cause the heart to work harder in order to maintain cardiac output (Fig. 10.7).

It is clear that the use of anabolic steroids produces adverse effects on the liver and reproductive and cardiovascular systems. Effects on the liver include peliosis hepatitis (blood-filled cysts), impaired excretory function (jaundice), and liver tumors [ACSM, 1984]. Cardiovascular effects include an increase in blood pressure, abnormal alterations in cardiac tissue, and abnormal lipoprotein-lipid profiles [Hurley et al., 1984b]. Males can experience a significant reduction in sperm production, testicular size, and testosterone and gonadotrophin production, and females often experience a deepening of voice, male pattern baldness, enlargement of the clitoris, a reduction in breast size, a disruption in their menstrual cycle, and an increase in facial hair. Most of these effects are irreversible even after the drugs are discontinued. There are also many psychological effects including an increase in aggressive behavior, an increase in anger and hostility, large deviations in mood, and sudden changes in libido.

Growth Hormone

Human growth hormone, also known as somatotropic hormone, is secreted from the pituitary gland and is involved in many anabolic processes in the body including normal growth and tissue synthesis. Athletes have become more interested in the use of growth hormone (GH) in recent years because of the many reports touting its benefits, the increased awareness of the dangers of anabolic steroids, and the reduced price due to the increased availability of synthetic forms. The appeal to athletes is that GH stimulates amino acid uptake and protein synthesis in skeletal muscle and a degradation of fat in adipose tissue. It has been shown to increase fat-free mass and reduce body fat independent of diet and exercise. Nevertheless, these benefits do not come without side effects, including acromegaly (enlargement of bony structures in the extremities) and impaired glucose tolerance.

Blood Doping

This procedure is also called *red blood cell* (RBC) *reinfusion, induced erythrocythemia,* or *blood boosting.* It involves withdrawing between 1 and 4 units (1 unit = 450 ml of blood) of a person's blood and then separating the plasma (liquid portion) from the RBCs. The plasma portion is immediately reinfused, and the packed RBCs are frozen. Each unit of blood is withdrawn over a 3- to 8-week period to prevent sudden reduction in hematocrit (RBC concentration). The RBCs are then reinfused 1 to 7 days prior to an endurance event. This results in a 10 to 20% increase in RBC and hemoglobin levels as well as an increase in total blood volume. It is believed that blood doping enhances oxygen transport to the working muscles, and, since oxygen transport is considered a limiting factor for increasing $\dot{V}_{O_{2max}}$, this adaptation would appear to enhance cardiovascular endurance performance. Whether blood doping actually improves endurance performance depends on the balance between enhanced oxygen transport and increased viscosity of blood. The method of storing blood appears to be an important factor in determining the potential effects of blood doping on performance. When the proper procedures have been used it can increase $\dot{V}_{O_{2max}}$ by 5 to 13% and reduce submaximal heart rate and blood lactate during a standard exercise test [Sawka et al., 1996].

Oxygen Inhalation

Inhaling high concentrations of oxygen gas mixtures prior to athletic competition or during rest periods of sporting events are common observations. This practice is based on the notion that it will increase oxygen transport (more oxygen binding to hemoglobin), thereby increasing oxygen delivery to the working muscle and resulting in a delay in the onset of fatigue [McArdle et al., 1996]. There are at least three problems with this belief. First, hemoglobin is already almost totally saturated with oxygen during normal breathing of ambient air at rest (95 to 98%). Hence, even if this practice allowed total oxygen saturation of hemoglobin it would only add a small amount of oxygen to the arterial blood leaving the lungs (10 ml of extra oxygen for every liter of whole blood). Second, for this practice to result in a delay of fatigue one must assume that oxygen transport and delivery are limiting factors for the specific event or sport being performed. This assumption cannot be made for many of the power-type events in which oxygen inhalation is often used (e.g., football). Third, any enhancement of oxygen transport from this procedure is short-lived. Yet athletes are often observed inhaling oxygen on the sidelines for relatively long periods prior to participation. Thus, this practice does not appear to offer much more than psychological benefits when used prior to athletic competition or during prolonged rest periods.

However, breathing oxygen during prolonged steady-rate exercise results in reduced heart rates, ventilation, and blood lactate levels. Apparently even small increases in oxygen saturation of hemoglobin and some added oxygen dissolved in plasma not bound to hemoglobin results in a substantial increase in oxygen availability under conditions that approach hypoxia. It may also increase oxygen diffusion capacity across muscle capillaries by elevating the partial pressure of oxygen in the blood. Nevertheless, the timing required for oxygen inhalation to be effective makes this procedure impractical to use for most athletic events, even if it were considered ethical and legal.

Defining Terms

Anabolic steroids: Synthetic forms of sex hormones (androgens) that possess both synthesizing and masculinizing characteristics.

Biceps muscle: The major muscle group in the upper arm responsible for flexion, joined by tendons at the shoulder and elbow. Two collaborating muscles are involved in the biceps.

Endergonic: An energy-absorbing process.

Ergogenic aids: Factors that affect human performance.

Exergonic: An energy-liberating process.

Hematocrit: The concentration of red blood cells.

Homeostasis: The tendency to maintain equilibrium.

Negative work: Work where the applied force and the direction of movement are opposite.

Positive work involves an applied force in the same direction as the movement. Negative work is known as *eccentric work,* and positive work is known as *concentric* by biomechanists.

References

ASCM. 1984. The use of anabolic-androgenic steroids in sports. *Sports Med Bull* 19:13.

Alexander RM. 1984. Walking and running. *Amer Sci.* 72:348.

Bhasin S, Storer TW, Berman N, Callegari C, Clevenger B, Phillips J, Bunnell TJ, Tricker R, Shirazi A, and Casaburi R. 1996. The effects of supraphysiologic doses of testosterone on muscle size and strength in normal men. *N Eng J Med* 335:1–7.

Casner SW, Early RG, Carlson BR. 1971. Anabolic steroid effects on body composition in normal young men. *J Sports Med Phys Fitness* 11:98.

Goldman RF. 1978. Computer models in manual materials handling. In CG Drury (Ed.), *Safety in Manual Materials Handling*, pp. 110–116, Cincinnati, OH, National Institute for Occupational Safety and Health (NIOSH).

Hickson RC, Czerwinski SM, Falduto MT et al. 1990. Glucocorticoid antagonism by exercise and androgenic-anabolic steroids. *Med Sci Sports Exerc* 22:331.

Hurley BF, Hagberg JM, Allen WK et al. 1984a. Effect of training on blood lactate levels during submaximal exercise. *J Appl Physiol* 56:1260.

Hurley BF, Seals DR, Hagberg JM et al. 1984b. High-density-lipoprotein cholesterol in bodybuilders v powerlifters. *JAMA* 252:507.

Johnson AT. 1991. *Biomechanics and Exercise Physiology*, New York, John Wiley & Sons.

Johnson LC, Roundy ES, Allsen PE et al. 1975. Effect of anabolic steroid treatment on endurance. *Med Sci Sports Exerc* 7:287.

Lynch NA, Metter EJ, Lindle RS, Fozard JL, Tobin JD, Roy TA, Fleg JL, Hurley BF. 1999. Muscle quality. I. Age-associated differences between arm and leg muscle groups. *J Appl Physiol* 86:188.

McArdle WD, Katch FI, and Katch VL. 1996. Special aids to performance and conditioning. In *Exercise Physiology: Energy, Nutrition, and Human Performance*, pp. 469–470. Baltimore, MD, Williams & Wilkins.

McArdle WD, Katch FI, Katch VL. 1994. Training muscles to become stronger. In *Essentials of Exercise Physiology*, pp. 373–397, Baltimore, MD, Lea & Febiger.

McMahon TA. 1984. *Muscles, Reflexes, and Locomotion*, Princeton, NJ, Princeton University Press.

Milsum JH. 1966. *Biological Control Systems Analysis*, New York, McGraw-Hill.

Sawka MN, Joyner MJ, Miles DS, Robertson RJ, Spriets LL, Young AJ. 1996. The use of blood doping as an ergogenic aid. *Med Sci Sports Exerc* 28:i–viii.

Shapiro P, Ikedo RM, Ruebner BH, et al. 1977. Multiple hepatic tumors and peliosis hepatitis in Fanconi's anemia treated with androgens. *Am J Dis Child* 131:1104.

Tracy BL, Ivey FM, Hurlbut D, Martel GF, Lemmer JT, Siegel EL, Metter EJ, Fozard JL, Fleg JL, Hurley BF. 1999. Muscle quality. II. Effects of strength training in 65- to 75-year-old men and women. *J Appl Physiol* 86:195.

Further Information

A comprehensive treatment of quantitative predictions in exercise physiology and biomechanics is presented in *Biomechanics and Exercise Physiology* by A. T. Johnson (John Wiley & Sons, 1991). There are a number of errors in the book, but an errata sheet is available from the author.

Clear and simple explanations of the biomechanics of sports are given in *Physics in Biology and Medicine* by P. Davidovits (Prentice-Hall, 1975).

The book by Tom McMahon, *Muscles, Reflexes, and Locomotion* (Princeton University Press, 1984), is a classic not to be missed.

The article by McArdle and colleagues in *Essentials of Exercise Physiology* (Williams & Wilkens, 1996) is an excellent source for a better understanding of physiological principles of work performance.

See D. A. Winter's *Biomechanics and Motor Control of Human Movement* (John Wiley & Sons, 1990) for a good in-depth explanation of more traditional biomechanics.

11

Cardiac Biomechanics

Andrew D. McCulloch
University of California San Diego

11.1 Introduction

The primary function of the heart, to pump blood through the circulatory system, is fundamentally mechanical. In this chapter, cardiac function is discussed in the context of the mechanics of the ventricular walls from the perspective of the determinants of myocardial stresses and strains (Table 11.1). Many physiological, pathophysiological, and clinical factors are directly or indirectly affected by myocardial stress and strain (Table 11.2). Of course, the factors in Tables 11.1 and 11.2 are interrelated—most of the factors affected by myocardial stress and strain in turn affect the stress and strain in the ventricular wall. For example, changes in wall stress due to altered hemodynamic load may cause ventricular remodeling, which in turn alters geometry, structure, and material properties. This chapter is organized around the governing determinants in Table 11.1, but mention is made where appropriate of some of the factors in Table 11.2.

11.2 Cardiac Geometry and Structure

The mammalian heart consists of four pumping chambers, the left and right atria and ventricles, communicating through the atrioventricular (mitral and tricuspid) valves, which are structurally connected by chordae tendineae to papillary muscles that extend from the anterior and posterior aspects of the right and left ventricular lumens. The muscular cardiac wall is perfused via the coronary vessels that originate at the left and right coronary ostia located in the sinuses of Valsalva immediately distal to the aortic valve leaflets. Surrounding the whole heart is the collagenous parietal pericardium that fuses with the diaphragm and great vessels. These are the anatomical structures that are most commonly studied in the field of cardiac mechanics. Particular emphasis in this chapter is given to the ventricular walls, which are the most important for the pumping function of the heart. Most studies of cardiac mechanics have focused on the left ventricle, but many of the important conclusions apply equally to the right ventricle.

TABLE 11.1   Basic Determinants of Myocardial Stress and Strain

Geometry and Structure

3D shape	Wall thickness
	Curvature
	Stress-free and unloaded reference configurations
Tissue structure	Muscle fiber architecture
	Connective tissue organization
	Pericardium, epicardium, and endocardium
	Coronary vascular anatomy

Boundary/Initial Conditions

Pressure	Filling pressure (preload)
	Arterial pressure (afterload)
	Direct and indirect ventricular interactions
	Thoracic and pericardial pressure
Constraints	Effects of inspiration and expiration
	Constraints due to the pericardium and its attachments
	Valves and fibrous valve annuli, chordae tendineae
	Great vessels, lungs

Material Properties

Resting or passive	Nonlinear finite elasticity
	Quasilinear viscoelasticity
	Anisotropy
	Biphasic poroelasticity
Active dynamic	Activation sequence
	Myofiber isometric and isotonic contractile dynamics
	Sarcomere length and length history
	Cellular calcium kinetics and metabolic energy supply

TABLE 11.2   Factors Affected by Myocardial Stress and Strain

Direct Factors	Regional muscle work
	Myocardial oxygen demand and energetics
	Coronary blood flow
Electrophysiological responses	Action potential duration (QT interval)
	Repolarization (T wave morphology)
	Excitability
	Risk of arrhythmia
Development and morphogenesis	Growth rate
	Cardiac looping and septation
	Valve formation
Vulnerability to injury	Ischemia
	Arrhythmia
	Cell dropout
	Aneurysm rupture
Remodeling, repair, and adaptation	Eccentric and concentric hypertrophy
	Fibrosis
	Scar formation
Progression of disease	Transition from hypertrophy to failure
	Ventricular dilation
	Infarct expansion
	Response to reperfusion
	Aneurysm formation

TABLE 11.3 Representative Left Ventricular Minor-Axis Dimensions[a]

Species	Comments	Inner Radius (mm)	Outer Radius (mm)	Wall Thickness: Inner Radius
Dog (21 kg)	Unloaded diastole (0 mmHg)	16	26	0.62
	Normal diastole (2–12 mmHg)	19	28	0.47
	Dilated diastole (24–40 mmHg)	22	30	0.36
	Normal systole (1–9 mmHg EDP)	14	26	0.86
	Long axis, apex-equator (normal diastole)	42	47	0.12
Young rat	Unloaded diastole (0 mmHg)	1.4	3.5	1.50
Mature rat	Unloaded diastole (0 mmHg)	3.2	5.8	0.81
Human	Normal	24	32	0.34
	Compensated pressure overload	27	42	0.56
	Compensated volume overload	32	42	0.33

[a] Dog data from Ross et al. [1967] and Streeter and Hanna [1973]. Human data from Grossman et al. [1975] and Grossman [1980]. Rat data from unpublished observations in the author's laboratory.

Ventricular Geometry

From the perspective of engineering mechanics, the ventricles are three-dimensional thick-walled pressure vessels with substantial variations in wall thickness and principal curvatures both regionally and temporally through the cardiac cycle. The ventricular walls in the normal heart are thickest at the equator and base of the left ventricle and thinnest at the left ventricular apex and right ventricular free wall. There are also variations in the principal dimensions of the left ventricle with species, age, phase of the cardiac cycle, and disease (Table 11.3). But, in general, the ratio of wall thickness to radius is too high to be treated accurately by all but the most sophisticated thick-wall shell theories [Taber, 1991].

Ventricular geometry has been studied in quantitative detail in the dog heart [Streeter and Hanna, 1973; Nielsen et al., 1991]. Geometric models have been very useful in the analysis, especially the use of confocal and nonconfocal ellipses of revolution to describe the epicardial and endocardial surfaces of the left and right ventricular walls (Fig. 11.1). The canine left ventricle is reasonably modeled by a thick

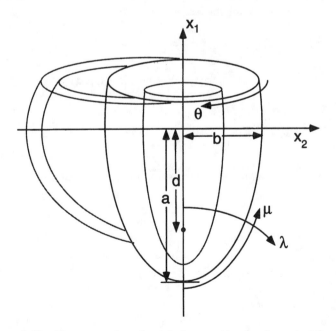

FIGURE 11.1 Truncated ellipsoid representation of ventricular geometry, showing major left ventricular radius (*a*), minor radius (*b*), focal length (*d*), and prolate spheroidal coordinates (λ, μ, θ).

ellipsoid of revolution truncated at the base. The crescentic right ventricle wraps about 180° around the heart wall circumferentially and extends longitudinally about two-thirds of the distance from the base to the apex. Using a truncated ellipsoidal model, left ventricular geometry in the dog can be defined by the major and minor radii of two surfaces, the left ventricular endocardium, and a surface defining the free wall epicardium and the septal endocardium of the right ventricle. Streeter and Hanna [1973] described the position of the basal plane using a truncation factor, f_b, defined as the ratio between the longitudinal distances from equator-to-base and equator-to-apex. Hence, the overall longitudinal distance from base to apex is $(1 + f_b)$ times the major radius of the ellipse. Since variations in f_b between diastole and systole are relatively small (0.45 to 0.51), they suggested a constant value of 0.5.

The focal length d of an ellipsoid is defined from the major and minor radii (a and b) by $d^2 = a^2 - b^2$, and varies only slightly in the dog from endocardium to epicardium between end-diastole (37.3 to 37.9 mm) and end-systole (37.7 to 37.1 mm) [Streeter and Hanna, 1973]. Hence, within the accuracy that the boundaries of the left ventricular wall can be treated as ellipsoids of revolution, the assumption that the ellipsoids are confocal appears to be a good one. This has motivated the choice of prolate spheroidal (elliptic–hyperbolic–polar) coordinates (λ, μ, θ) as a system for economically representing ventricular geometries obtained postmortem or by noninvasive tomography [Nielsen et al., 1991; Young and Axel, 1992]. The Cartesian coordinates of a point are given in terms of its prolate spheroidal coordinates by:

$$x_1 = d \cosh \lambda \cos \mu,$$
$$x_2 = d \sinh \lambda \sin \mu \cos \theta, \quad (11.1)$$
$$x_3 = d \sinh \lambda \sin \mu \sin \theta.$$

Here, the focal length d defines a family of coordinate systems that vary from spherical polar when $d = 0$ to cylindrical polar in the limit when $d \to \infty$. A surface of constant transmural coordinate λ (Fig. 11.1) is an ellipse of revolution with major radius $a = d \cosh \lambda$ and minor radius $b = d \sinh \lambda$. In an ellipsoidal model with a truncation factor of 0.5, the longitudinal coordinate μ varies from zero at the apex to 120° at the base. Integrating the Jacobian in prolate spheroidal coordinates gives the volume of the wall or cavity:

$$d^3 \int_0^{2\pi} \int_0^{\mu_2} \int_{\lambda_1}^{\lambda_2} \left[\left(\sinh^2 \lambda + \sin^2 \mu \right) \sinh \lambda \sin \mu \right] d\lambda \, d\mu \, d\theta =$$
$$\frac{2\pi d^3}{3} \left| \left(1 - \cos \mu_2 \right) \cosh^3 \lambda - \left(1 - \cos^3 \mu_2 \right) \cosh \lambda \right|_{\lambda_1}^{\lambda_2} \quad (11.2)$$

The scaling between heart mass M_H and body mass M within or between species is commonly described by the allometric formula:

$$M_H = kM^\alpha \quad (11.3)$$

Using combined measurements from a variety of mammalian species with M expressed in kilograms, the coefficient k is 5.8 g and the power α is close to unity (0.98) [Stahl, 1967]. Within individual species, the ratio of heart weight to body weight is somewhat lower in mature rabbits and rats (about 2 g/kg) than in humans (5 g/kg) and higher in horses and dogs (8 g/kg) [Rakusan, 1984]. The rate α of heart growth with body weight decreases with age in most species but not humans. At birth, left and right ventricular weights are similar, but the left ventricle is substantially more massive than the right by adulthood.

Myofiber Architecture

The cardiac ventricles have a complex three-dimensional muscle fiber architecture (for a comprehensive review see Streeter [1979]). Although the myocytes are relatively short, they are connected such that at any point in the normal heart wall there is a clear predominant fiber axis that is approximately tangent with the wall (within 3 to 5° in most regions, except near the apex and papillary muscle insertions). Each ventricular myocyte is connected via gap junctions at intercalated disks to an average of 11.3 neighbors, 5.3 on the sides, and 6.0 at the ends [Saffitz et al., 1994]. The classical anatomists dissected discrete bundles of fibrous swirls, though later investigations showed that the ventricular myocardium could be unwrapped by blunt dissection into a single continuous muscle "band" [Torrent-Guasp, 1973]. However, more modern histological techniques have shown that in the plane of the wall, the muscle fiber angle makes a smooth transmural transition from epicardium to endocardium (Fig. 11.2). Similar patterns have been described for humans, dogs, baboons, macaques, pigs, guinea pigs, and rats. In the human or dog left ventricle, the muscle fiber angle typically varies continuously from about −60° (i.e., 60° clockwise from the circumferential axis) at the epicardium to about +70° at the endocardium. The rate of change of fiber angle is usually greatest at the epicardium, so that circumferential (0°) fibers are found in the outer half of the wall, and begins to slowly approach the inner third near the trabeculata–compacta interface. There are also small increases in fiber orientation from end-diastole to systole (7 to 19°), with greatest changes at the epicardium and apex [Streeter et al., 1969].

Regional variations in ventricular myofiber orientations are generally smooth except at the junction between the right ventricular free wall and septum. A detailed study in the dog that mapped fiber angles throughout the entire right and left ventricles described the same general transmural pattern in all regions including the septum and right ventricular free wall, but with definite regional variations [Nielsen et al., 1991]. Transmural differences in fiber angle were about 120 to 140° in the left ventricular free wall, larger in the septum (160 to 180°), and smaller in the right ventricular free wall (100 to 120°). A similar study of fiber angle distributions in the rabbit left and right ventricles has recently been reported [Vetter and McCulloch, 1998]. For the most part, fiber angles in the rabbit heart were very similar to those in the dog, except on the anterior wall, where average fiber orientations were 20 to 30° counterclockwise of those in the dog.

The locus of fiber orientations at a given depth in the ventricular wall has a spiral geometry that may be modeled as a general helix by simple differential geometry. The position vector **x** of a point on a helix

Epicardium

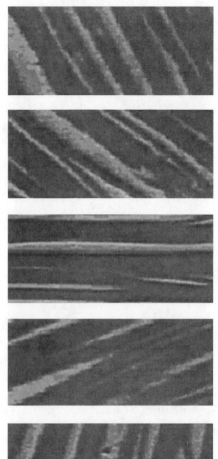

Endocardium

FIGURE 11.2 Cardiac muscle fiber orientations vary continuously through the left ventricular wall from a negative angle at the epicardium to near zero (circumferential) at the midwall and to increasing positive values toward the endocardium. (Micrographs of canine myocardium from the author's laboratory, courtesy of Dr. Deidre MacKenna).

inscribed on an ellipsoidal surface that is symmetric about the x_1 axis and has major and minor radii, a and b, is given by the parametric equation,

$$\mathbf{x} = a \sin t\ \mathbf{e}_1 + b \cos t \sin wt\ \mathbf{e}_2 + b \cos t \cos wt\ \mathbf{e}_3 \tag{11.4}$$

where the parameter is t, and the helix makes $w/4$ full turns between apex and equator. A positive w defines a left-handed helix with a positive pitch. The fiber angle or helix pitch angle η, varies along the arc length:

$$\sin \eta = \sqrt{\frac{a^2 \cos^2 t + b^2 \sin^2 t}{\left(a^2 + b^2 w^2\right)\cos^2 t + b^2 \sin^2 t}} \tag{11.5}$$

If another, deformed configuration, $\hat{\mathbf{x}}$, is defined in the same way as Eq. (11.4), the fiber segment extension ratio $\frac{d\hat{s}}{ds}$ associated with a change in the ellipsoid geometry [McCulloch et al., 1989] can be derived from:

$$\frac{d\hat{s}}{ds} = \frac{\dfrac{d\hat{s}}{dt}}{\dfrac{ds}{dt}} = \frac{\left|\dfrac{d\hat{\mathbf{x}}}{dt}\right|}{\left|\dfrac{d\mathbf{x}}{dt}\right|} \tag{11.6}$$

Although the traditional notion of discrete myofiber bundles has been revised in view of the continuous transmural variation of muscle fiber angle in the plane of the wall, there is a transverse laminar structure in the myocardium that groups fibers together in sheets an average of 4 ± 2 myocytes thick (48 ± 20 μm), separated by histologically distinct cleavage planes [Spotnitz et al., 1974; Smaill and Hunter, 1991; LeGrice et al., 1995]. LeGrice and colleagues investigated these structures in a detailed morphometric study of four dog hearts [LeGrice et al., 1995]. They describe an ordered laminar arrangement of myocytes with extensive cleavage planes running approximately radially from endocardium toward epicardium in transmural section. Like the fibers, the sheets also have a branching pattern with the number of branches varying considerably through the wall thickness.

The fibrous architecture of the myocardium has motivated models of myocardial material symmetry as transversely isotropic. The transverse laminae are the first structural evidence for material orthotropy and have motivated the development of a model describing the variation of fiber, sheet, and sheet-normal axes throughout the ventricular wall [LeGrice et al., 1997]. This has led to the idea that the laminar architecture of the ventricular myocardium affects the significant transverse shears [Waldman et al., 1985] and myofiber rearrangement [Spotnitz et al., 1974] described in the intact heart during systole. By measuring three-dimensional distributions of strain across the wall thickness using biplane radiography of radiopaque markers, LeGrice and colleagues [LeGrice et al., 1995] found that the cleavage planes coincide closely with the planes of maximum shearing during ejection, and that the consequent reorientation of the myocytes may contribute 50% or more of normal systolic wall thickening.

A detailed description of the morphogenesis of the muscle fiber system in the developing heart is not available but there is evidence of an organized myofiber pattern by day 12 in the fetal mouse heart that is similar to that seen at birth (day 20) [McLean et al., 1989]. Abnormalities of cardiac muscle fiber patterns have been described in some disease conditions. In hypertrophic cardiomyopathy, which is often familial, there is substantial myofiber disarray, typically in the interventricular septum [Maron et al., 1987; Karlon et al., 1998].

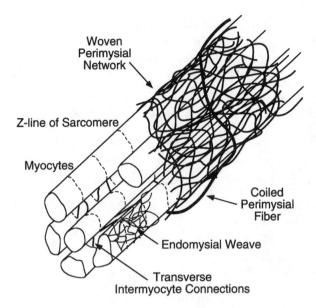

FIGURE 11.3 Schematic representation of cardiac tissue structure showing the association of endomysial and perimysial collagen fibers with cardiac myocytes. (Courtesy of Dr. Deidre MacKenna.)

Extracellular Matrix Organization

The cardiac extracellular matrix consists primarily of the fibrillar collagens, type I (85%) and III (11%), synthesized by the cardiac fibroblasts, the most abundant cell type in the heart. Collagen is the major structural protein in connective tissues but only comprises 2 to 5% of the myocardium by weight, compared with the myocytes, which make up 90% [Weber, 1989]. The collagen matrix has a hierarchical organization (Fig. 11.3), and has been classified according to conventions established for skeletal muscle into endomysium, perimysium, epimysium [Caulfield and Borg, 1979; Robinson et al., 1983]. The endomysium is associated with individual cells and includes a fine weave surrounding the cell and transverse structural connections 120 to 150 nm long connecting adjacent myocytes, with attachments localized near the *z*-line of the sarcomere. The primary purpose of the endomysium is probably to maintain registration between adjacent cells. The perimysium groups cells together and includes the collagen fibers that wrap bundles of cells into the laminar sheets described above, as well as large coiled fibers typically 1 to 3 μm in diameter which are composed of smaller collagen fibrils (40 to 50 nm) [Robinson et al., 1988]. The helix period of the coiled perimysial fibers is about 20 μm and the convolution index (ratio of fiber arclength to midline length) is approximately 1.3 in the unloaded state of the ventricle [MacKenna et al., 1996; MacKenna et al., 1997]. These perimysial fibers are most likely to be the major structural elements of the collagen extracellular matrix though they probably contribute to myocardial strain energy by uncoiling rather than stretching [MacKenna et al., 1996]. Finally, a thick epimysial collagen sheath surrounds the entire myocardium forming the protective epicardium (visceral pericardium) and endocardium.

Collagen content, organization, and ratio of types I to III change with age and in various disease conditions including myocardial ischemia and infarction, hypertension, and hypertrophy (Table 11.4). Changes in myocardial collagen content and organization coincide with alterations in diastolic myocardial stiffness [MacKenna and McCulloch, 1996]. Hence the collagen matrix plays an important role in determining the elastic material properties of the resting ventricular myocardium.

TABLE 11.4   Changes in Ventricular Collagen Structure and Mechanics with Age and Disease

Condition	Collagen Morphology	Types and Crosslinking	Passive Stiffness	Other
Pressure overload hypertrophy	Hydroxyproline: ⇑–⇑⇑⇑ [Medugorac, 1980; Weber et al., 1988] Area fraction: ⇑⇑⇑⇑ [Jalil et al., 1988; Weber et al., 1988]	Type III: ⇑ [Mukherjee and Sen, 1990] Crosslinks: no change [Harper et al., 1993]	Chamber: ⇑–⇑⇑⇑ [Jalil et al., 1988; Weber et al., 1988] Tissue: ⇑⇑⇑ [Omens et al., 1995]	Perivascular fibrosis: ⇑⇑⇑ [Weber et al., 1988] Focal scarring: [Silver et al., 1990; Contard et al., 1991]
Volume overload hypertrophy	Hydroxyproline: no change–⇓ [Michel et al., 1986; Iimoto et al., 1988] Area fraction: no change [Medugorac, 1980; Weber et al., 1990]	Crosslinks: ⇑ [Iimoto et al., 1988; Harper et al., 1993] Type III/I: ⇑ [Iimoto et al., 1988]	Chamber: ⇓ [Corin et al., 1991] Tissue: no change/⇑ [Corin et al., 1991]	Parallel changes
Acute ischemia/ stunning	Hydroxyproline: ⇓ Charney, 1992] Light microscopy: no change/⇓ [Whittaker et al., 1991] ⇓⇓ endomysial fibers [Zhao et al., 1987]		⇓ early [Forrester et al., 1972] ⇑ late [Pirzada et al., 1976]	Collagenase activity: ⇑ [Takahashi et al., 1990; Charney et al., 1992]
Chronic myocardial infarction	Hydroxyproline: ⇑⇑⇑ [Connelly et al., 1985; Jugdutt and Amy, 1986] Loss of birefringence [Whittaker et al., 1989]	Type III: ⇑ [Jensen et al., 1990]	Chamber: ⇑ early [Pfeffer et al., 1991] Chamber: ⇓ late [Pfeffer et al., 1991]	Organization: ⇑–⇑⇑⇑ [Whittaker et al., 1991; Holmes et al., 1994]
Age	Hydroxyproline: ⇑–⇑⇑⇑ [Eghbali et al., 1989; Takahashi et al., 1990] Collagen fiber diameter ⇑ [Eghbali et al., 1989]	Type III/I: ⇓ [Medugorac and Jacob, 1983] Crosslinks: ⇑ [Medugorac and Jacob, 1983]	Chamber: ⇑ [Borg et al., 1981] Papillary muscle: ⇑ [Anversa et al., 1989]	Light microscopy: fibril diameter ⇑ [Eghbali et al., 1989]

11.3   Cardiac Pump Function

Ventricular Hemodynamics

The most basic mechanical parameters of the cardiac pump are blood pressure and volume flowrate, especially in the major pumping chambers, the ventricles. From the point of view of wall mechanics, the ventricular pressure is the most important boundary condition. Schematic representations of the time-courses of pressure and volume in the left ventricle are shown in Fig. 11.4. Ventricular filling immediately following mitral valve opening (MVO) is initially rapid because the ventricle produces a diastolic suction as the relaxing myocardium recoils elastically from its compressed systolic configuration below the resting chamber volume. The later slow phase of ventricular filling (diastasis) is followed finally by atrial contraction. The deceleration of the inflowing blood reverses the pressure gradient across the valve leaflets and causes them to close (MVC). Valve closure may not, however, be completely passive, because the atrial side of the mitral valve leaflets, which unlike the pulmonic and aortic valves are cardiac in embryological origin, have muscle and nerve cells and are electrically coupled to atrial conduction [Sonnenblick et al., 1967].

Ventricular contraction is initiated by excitation, which is almost synchronous (the duration of the QRS complex of the ECG is only about 60 msec in the normal adult) and begins about 0.1 to 0.2 sec after atrial depolarization. Pressure rises rapidly during the isovolumic contraction phase (about 50 msec in adult humans), and the aortic valve opens (AVO) when the developed pressure exceeds the aortic pressure (afterload). Most of the cardiac output is ejected within the first quarter of the ejection phase before the pressure has peaked. The aortic valve closes (AVC) 20 to 30 msec after AVO when the ventricular

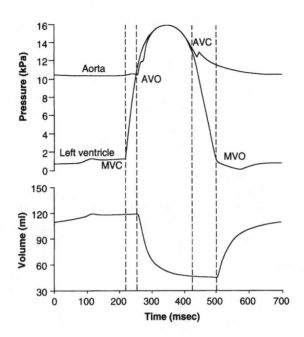

FIGURE 11.4 Left ventricular pressure, aortic pressure, and left ventricular volume during a single cardiac cycle showing the times of mitral valve closure (MVC), aortic valve opening (AVO), aortic valve closure (AVC), and mitral valve opening (MVO).

pressure falls below the aortic pressure owing to the deceleration of the ejecting blood. The dichrotic notch, a characteristic feature of the aortic pressure waveform and a useful marker of aortic valve closure, is caused by pulse wave reflections in the aorta. Since the pulmonary artery pressure against which the right ventricle pumps is much lower than the aortic pressure, the pulmonic valve opens before and closes after the aortic valve. The ventricular pressure falls during isovolumic relaxation, and the cycle continues. The rate of pressure decay from the value P_0 at the time of the peak rate of pressure falls until the mitral valve opening is commonly characterized by a single exponential time constant, i.e.,

$$P(t) = P_0 e^{-t/\tau} + P_\infty \qquad (11.7)$$

where P_∞ is the (negative) baseline pressure to which the ventricle would eventually relax if MVO were prevented [Yellin et al., 1986]. In dogs and humans, τ is normally about 40 msec, but it is increased by various factors including elevated afterload, asynchronous contraction associated with abnormal activation sequence or regional dysfunction, and slowed cytosolic calcium reuptake to the sarcoplasmic reticulum associated with cardiac hypertrophy and failure. The pressure and volume curves for the right ventricle look essentially the same, however the right ventricular and pulmonary artery pressures are only about a fifth of the corresponding pressures on the left side of the heart. The intraventricular septum separates the right and left ventricles and can transmit forces from one to the other. An increase in right ventricular volume may increase the left ventricular pressure by deformation of the septum. This direct interaction is most significant during filling [Janicki and Weber, 1980].

The phases of the cardiac cycle are customarily divided into systole and diastole. The end of diastole—the start of systole—is generally defined as the time of mitral valve closure. Mechanical end-systole is usually defined as the end of ejection, but Brutsaert and colleagues proposed extending systole until the onset of diastasis (see the review by Brutsaert and Sys [1989]) since there remains considerable myofilament interaction and active tension during relaxation. The distinction is important from the point of view of cardiac muscle mechanics: the myocardium is still active for much of diastole and may never be fully relaxed at sufficiently high heart rates (over 150 beats per minute). Here, we will retain the traditional definition of diastole, but consider the ventricular myocardium to be "passive" or "resting" only in the final slow-filling stage of diastole.

Ventricular Pressure–Volume Relations and Energetics

A useful alternative to Fig. 11.4 for displaying ventricular pressure and volume changes is the pressure-volume loop shown in Fig. 11.5a. During the last 20 years, the ventricular pressure-volume relationship has been explored extensively, particularly by Sagawa [1988], who wrote a comprehensive book on the approach. The isovolumic phases of the cardiac cycle can be recognized as the vertical segments of the loop, the lower limb represents ventricular filling, and the upper segment is the ejection phase. The difference on the horizontal axis between the vertical isovolumic segments is the stroke volume, which expressed as a fraction of the end-diastolic volume is the ejection fraction. The effects of altered loading on the ventricular pressure–volume relation have been studied in many preparations, but the best controlled experiments have used the isolated cross-circulated canine heart in which the ventricle fills and ejects against a computer-controlled volume servo-pump.

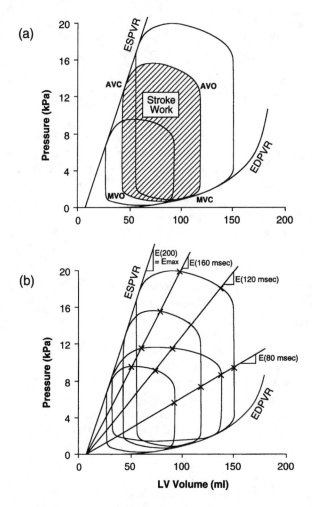

FIGURE 11.5 Schematic diagram of left ventricular pressure–volume loops: (a) End-systolic pressure–volume relation (ESPVR), end-diastolic pressure volume relation (EDPVR), and stroke work. The three P–V loops show the effects of changes in preload and afterload. (b) Time-varying elastance approximation of ventricular pump function (see text).

Changes in the filling pressure of the ventricle (preload) move the end-diastolic point along the unique end-diastolic pressure-volume relation (EDPVR), which represents the passive filling mechanics of the chamber that are determined primarily by the thick-walled geometry and nonlinear elasticity of the resting ventricular wall. Alternatively, if the afterload seen by the left ventricle is increased, stroke volume decreases in a predictable manner. The locus of end-ejection points (AVC) forms the end-systolic pressure–volume relation (ESPVR), which is approximately linear in a variety of conditions and also largely independent of the ventricular load history. Hence, the ESPVR is almost the same for isovolumic beats as for ejecting beats, although consistent effects of ejection history have been well characterized [Hunter, 1989]. Connecting pressure–volume points at corresponding times in the cardiac cycle also results in a relatively linear relationship throughout systole with the intercept on the volume axis V_0 remaining nearly constant (Fig. 11.5b). This leads to the valuable approximation that the ventricular volume $V(t)$ at any instance during systole is simply proportional to the instantaneous pressure $P(t)$ through a time-varying elastance $E(t)$:

$$P(t) = E(t)\{V(t) - V_0\} \tag{11.8}$$

The maximum elastance E_{max}, the slope of the ESPVR, has acquired considerable significance as an index of cardiac contractility that is independent of ventricular loading conditions. As the inotropic state of the myocardium increases, for example with catecholamine infusion, E_{max} increases, and with a negative inotropic effect such as a reduction in coronary artery pressure it decreases.

The area of the ventricular pressure–volume loop is the external work (EW) performed by the myocardium on the ejecting blood:

$$EW = \int_{EDV}^{ESV} P(t)\,dV \tag{11.9}$$

Plotting this stroke work against a suitable measure of preload gives a ventricular function curve, which illustrates the single most important intrinsic mechanical property of the heart pump. In 1914, Patterson and Starling performed detailed experiments on the canine heart–lung preparation, and Starling summarized their results with his famous "Law of the Heart", which states that the work output of the heart increases with ventricular filling. The so-called Frank–Starling mechanism is now well recognized to be an intrinsic mechanical property of cardiac muscle (see Section 11.4).

External stroke work is closely related to cardiac energy utilization. Since myocardial contraction is fueled by ATP, 90 to 95% of which is normally produced by oxidative phosphorylation, cardiac energy consumption is often studied in terms of myocardial oxygen consumption, VO_2 (ml $O_2 \cdot g^{-1} \cdot beat^{-1}$). Since energy is also expended during non-working contractions, Suga and colleagues [1981] defined the pressure–volume area PVA ($J \cdot g^{-1} \cdot beat^{-1}$) as the loop area (external stroke work) plus the end-systolic potential energy (internal work) which is the area under the ESPVR left of the isovolumic relaxation line (Fig. 11.5a),

$$PVA = EW + PE \tag{11.10}$$

The PVA has strong linear correlation with VO_2 independent of ejection history. Equation (11.11) has typical values for the dog heart:

$$VO_2 = 0.12(PVA) + 2.0 \times 10^{-4} \tag{11.11}$$

The intercept represents the sum of the oxygen consumption for basal metabolism and the energy associated with activation of the contractile apparatus, which is primarily used to cycle intracellular Ca^{2+} for excitation-contraction coupling [Suga et al., 1981]. The reciprocal of the slope is the contractile efficiency [Suga and Goto, 1991; Suga et al., 1993]. The VO_2–PVA relation shifts its elevation but not its slope with increments in E_{max} with most positive and negative inotropic interventions [Suga et al., 1988;

Suga, 1990; Suga and Goto, 1991; Zhao et al., 1993; Namba et al., 1994]. However, ischemic-reperfused viable but "stunned" myocardium has a smaller O_2 cost of PVA [Ohgoshi et al., 1991].

Although the PVA approach has also been useful in many settings, it is fundamentally phenomeno-logical. Because the time-varying elastance assumptions ignores the well-documented load-history dependence of cardiac muscle tension [Guccione and McCulloch, 1993; ter Keurs and de Tombe, 1993; Burkhoff et al., 1995], theoretical analyses that attempt to reconcile PVA with crossbridge mechanoen-ergetics [Taylor et al., 1993] are usually based on isometric or isotonic contractions. So that regional oxygen consumption in the intact heart can be related to myofiber biophysics, regional variations on the pressure–volume area have been proposed, such as the tension area [Goto et al., 1993], normalization of E_{max} [Sugawara et al., 1995], and the fiber stress–strain area [Delhaas et al., 1994].

In mammals, there are characteristic variations in cardiac function with heart size. In the power law relation for heart rate as a function of body mass [analogous to Eq. (11.3)], the coefficient k is 241 beats·min^{-1} and the power α is –0.25 [Stahl, 1967]. In the smallest mammals, like soricine shrews that weigh only a few grams, maximum heart rates exceeding 1000 beats·min^{-1} have been measured [Vornanen, 1992]. Ventricular cavity volume scales linearly with heart weight, and ejection fraction and blood pressure are reasonably invariant from rats to horses. Hence, stroke work also scales directly with heart size [Holt et al., 1962], and thus work rate and energy consumption would be expected to increase with decreased body size in the same manner as heart rate. However, careful studies have demonstrated only a twofold increase in myocardial heat production as body mass decreases in mammals ranging from humans to rats, despite a 4.6-fold increase in heart rate [Loiselle and Gibbs, 1979]. This suggests that cardiac energy expenditure does not scale in proportion to heart rate and that cardiac metabolism is a lower proportion of total body metabolism in the smaller species.

The primary determinants of the end-diastolic pressure–volume relation (EDPVR) are the material properties of resting myocardium, the chamber dimensions and wall thickness, and the boundary con-ditions at the epicardium, endocardium, and valve annulus [Gilbert and Glantz, 1989]. The EDPVR has been approximated by an exponential function of volume (see for example, Chapter 9 in Gaasch and LeWinter [1994]), though a cubic polynomial also works well. Therefore, the passive chamber stiffness dP/dV is approximately proportional to the filling pressure. Important influences on the EDPVR include the extent of relaxation, ventricular interaction and pericardial constraints, and coronary vascular engorgement. The material properties and boundary conditions in the septum are important since they determine how the septum deforms [Glantz et al., 1978; Glantz and Parmley, 1978]. Through septal interaction, the end-diastolic pressure–volume relationship of the left ventricle may be directly affected by changes in the hemodynamic loading conditions of the right ventricle. The ventricles also interact indirectly since the output of the right ventricle is returned as the input to the left ventricle via the pulmonary circulation. Slinker and Glantz [1986], using pulmonary artery and venae caval occlusions to produce direct (immediate) and indirect (delayed) interaction transients, concluded that the direct interaction is about half as significant as the indirect coupling. The pericardium provides a low friction mechanical enclosure for the beating heart that constrains ventricular overextension [Mirsky and Rankin, 1979]. Since the pericardium has stiffer elastic properties than the ventricles [Lee et al., 1987], it con-tributes to direct ventricular interactions. The pericardium also augments the mechanical coupling between the atria and ventricles [Maruyama et al., 1982]. Increasing coronary perfusion pressure has been seen to increase the slope of the diastolic pressure–volume relation (an "erectile" effect) [Salisbury et al., 1960; May-Newman et al., 1994].

11.4   Myocardial Material Properties

Muscle Contractile Properties

Cardiac muscle mechanics testing is far more difficult than skeletal muscle testing mainly owing to the lack of ideal test specimens like the long single fiber preparations that have been so valuable for studying the mechanisms of skeletal muscle mechanics. Moreover, under physiological conditions, cardiac muscle

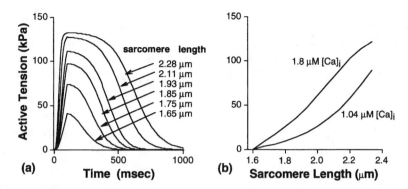

FIGURE 11.6   Cardiac muscle isometric twitch tension generated by a model of rat cardiac contraction (courtesy of Dr. Julius Guccione): (a) Developed twitch tension as a function of time and sarcomere length; (b) peak isometric twitch tension vs. sarcomere length for low and high calcium concentration.

cannot be stimulated to produce sustained tetanic contractions due to the absolute refractory period of the myocyte cell membrane. Cardiac muscle also exhibits a mechanical property analogous to the relative refractory period of excitation. After a single isometric contraction, some recovery time is required before another contraction of equal amplitude can be activated. The time constant for this mechanical restitution property of cardiac muscle is about 1 sec [Bers, 1991].

Unlike skeletal muscle, in which maximal active force generation occurs at a sarcomere length that optimizes myofilament overlap (~2.1 μm), the isometric twitch tension developed by isolated cardiac muscle continues to rise with increased sarcomere length in the physiological range (1.6 to 2.4 μm) (Fig. 11.6a). Early evidence for a descending limb of the cardiac muscle isometric length–tension curve was found to be caused by shortening in the central region of the isolated muscle at the expense of stretching at the damaged ends where the specimen was tethered to the test apparatus. If muscle length is controlled so that sarcomere length in the undamaged part of the muscle is indeed constant, or if the developed tension is plotted against the instantaneous sarcomere length rather than the muscle length, the descending limb is eliminated [ter Keurs et al., 1980]. Thus, the increase with chamber volume of end-systolic pressure and stroke work is reflected in isolated muscles as a monotonic increase in peak isometric tension with sarcomere length (Fig. 11.6b). Note that the active tension shown in Fig. 11.6 is the total tension minus the resting tension which, unlike in skeletal muscle, becomes very significant at sarcomere lengths over 2.3 μm. The increase in slope of the ESPVR associated with increased contractility is mirrored by the effects of increased calcium concentration in the length–tension relation. The duration, as well as the tension developed in the active cardiac twitch, also increases substantially with sarcomere length (Fig. 11.6a).

The relationship between cytosolic calcium concentration and isometric muscle tension has mostly been investigated in muscle preparations in which the sarcolemma has been chemically permeabilized. Because there is evidence that this chemical "skinning" alters the calcium sensitivity of myofilament interaction, recent studies have also investigated myofilament calcium sensitivity in intact muscles tetanized by high frequency stimulation in the presence of a compound such as ryanodine that opens calcium release sites in the sarcoplasmic reticulum. Intracellular calcium concentration was estimated using calcium-sensitive optical indicators such as Fura. The myofilaments are activated in a graded manner by micromolar concentrations of calcium, which binds to troponin-C according to a sigmoidal relation [Rüegg, 1988]. Half-maximal tension in cardiac muscle is developed at intracellular calcium concentrations of 10^{-6} to 10^{-5} M (the C_{50}) depending on factors such as species and temperature [Bers, 1991]. Hence, relative isometric tension T_0 / T_{max} may be modeled using [Tözeren, 1985; Hunter et al., 1998]:

$$\frac{T_0}{T_{max}} = \frac{\left[Ca\right]^n}{\left[Ca\right]^n + C_{50}^n} .$$

(11.12)

The Hill coefficient, n, governs the steepness of the sigmoidal curve. A wide variety of values have been reported but most have been in the range 3 to 6 [Kentish et al., 1986; Yue et al., 1986; Gao et al., 1994; Backx et al., 1995]. The steepness of the isometric length-tension relation (Fig. 11.6b), compared with that of skeletal muscle is due to length-dependent calcium sensitivity. That is, the C_{50} (M) and n both change with sarcomere length, L (µm). Hunter et al. [1998] used the following approximations to fit the data of Kentish et al. [1986] from rat right ventricular trabeculae:

$$n = 4.25\left\{1 + 1.95\left(L/L_{\text{ref}} - 1\right)\right\}, \qquad \text{pC}_{50} = -\log_{10} C_{50} = 5.33\left\{1 + 0.31\left(L/L_{\text{ref}} - 1\right)\right\} \quad (11.13)$$

where the reference sarcomere length L_{ref} was taken to be 2.0 µm.

The isotonic force–velocity relation of cardiac muscle is similar to that of skeletal muscle, and A.V. Hill's well-known hyperbolic relation is a good approximation except at larger forces greater than about 85% of the isometric value. The maximal (unloaded) velocity of shortening is essentially independent of preload, but does change with time during the cardiac twitch and is affected by factors that affect contractile ATPase activity and hence crossbridge cycling rates. De Tombe and colleagues [de Tombe and ter Keurs, 1992] using sarcomere length-controlled isovelocity release experiments found that viscous forces impose a significant internal load opposing sarcomere shortening. If the isotonic shortening response is adjusted for the confounding effects of passive viscoelasticity, the underlying crossbridge force-velocity relation is found to be linear.

Cardiac muscle contraction also exhibits other significant length-history-dependent properties. An important example is "deactivation" associated with length transients. The isometric twitch tension redevelops following a brief length transient that dissociates crossbridges that reach the original isometric value when the transient is imposed early in the twitch before the peak tension is reached. But following transients applied at times after the peak twitch tension has occurred, the fraction of tension redeveloped declines progressively since the activator calcium has fallen to levels below that necessary for all cross-bridges to reattach [ter Keurs et al., 1980].

There have been many model formulations of cardiac muscle contractile mechanics, too numerous to summarize here. In essence they may be grouped into three categories. *Time-varying elastance models* include the essential dependence of cardiac active force development on muscle length and time. These models would seem to be well suited to the continuum analysis of whole heart mechanics [Arts et al., 1979; Chadwick, 1982; Taber, 1991] by virtue of the success of the time-varying elastance concept of ventricular function (see Section 11.3 above). In *Hill models*, the active fiber stress development is modified by shortening or lengthening according to the force–velocity relation, so that fiber tension is reduced by increased shortening velocity [Arts et al., 1982; Nevo and Lanir, 1989]. *Fully history-dependent models* are more complex and are generally based on A.F. Huxley's crossbridge theory [Panerai, 1980; Guccione et al., 1993; Landesberg and Sideman, 1994; Landesberg et al., 1996]. A statistical approach known as the *distribution moment model* has also been shown to provide an excellent approximation to the crossbridge theory [Ma and Zahalak, 1991]. An alternative, more phenomenological approach, is *Hunter's fading memory theory*, which captures the complete length-history dependence of cardiac muscle contraction without requiring all the biophysical complexity of crossbridge models [Hunter et al., 1998]. The appropriate choice of model will depend on the purpose of the analysis. For many models of global ventricular function, a time-varying elastance model will suffice, but for an analysis of sarcomere dynamics in isolated muscle or the ejecting heart, a history-dependent analysis is more appropriate.

Although Hill's basic assumption that resting and active muscle fiber tension are additive is axiomatic in one-dimensional tests of isolated cardiac mechanics, there remains little experimental information on how the passive and active material properties of myocardium superpose in two dimensions or three. The simplest and most common assumption is that active stress is strictly one dimensional and adds to the fiber component of the three-dimensional passive stress. However, even this addition will indirectly affect all the other components of the stress response, since myocardial elastic deformations are finite, nonlinear, and approximately isochoric (volume conserving). In an interesting and important new development, biaxial testing of tetanized and barium-contracted ventricular myocardium has shown that developed systolic stress also has a large component in directions transverse to the mean myofiber axis that can exceed 50% of the axial fiber component [Lin and Yin, 1998]. The magnitude of this transverse active stress depends significantly on the biaxial loading conditions. Moreover, evidence from osmotic swelling and other studies suggests that transverse strain can affect contractile tension development along the fiber axis by altering myofibril lattice spacing [Schoenberg, 1980; Zahalak, 1996]. The mechanisms of transverse active stress development remain unclear but two possible contributors are the geometry of the crossbridge head itself which is oriented oblique to the myofilament axis [Schoenberg, 1980], and the dispersion of myofiber orientation, which tends to be distributed in intact myocardium with a standard deviation of about 12° [Karlon et al., 1998].

Resting Myocardial Properties

Since, by the Frank–Starling mechanism, end-diastolic volume directly affects systolic ventricular work, the mechanics of resting myocardium also have fundamental physiological significance. Most biomechanical studies of passive myocardial properties have been conducted in isolated, arrested whole heart or tissue preparations. Passive cardiac muscle exhibits most of the mechanical properties characteristic of soft tissues in general [Fung, 1993]. In cyclic uniaxial loading and unloading, the stress–strain relationship is nonlinear with small but significant hysteresis. Depending on the preparation used, resting cardiac muscle typically requires from 2 to 10 repeated loading cycles to achieve a reproducible (preconditioned) response. Intact cardiac muscle experiences finite deformations during the normal cardiac cycle, with maximum Lagrangian strains (which are generally radial and endocardial) that may easily exceed 0.5 in magnitude. Hence, the classical linear theory of elasticity is quite inappropriate for resting myocardial mechanics. The hysteresis of the tissue is consistent with a viscoelastic response, which is undoubtedly related to the substantial water content of the myocardium (about 80% by mass). Changes in water content, such as edema, can cause substantial alterations in the passive stiffness and viscoelastic properties of myocardium. The viscoelasticity of passive cardiac muscle has been characterized in creep and relaxation studies of papillary muscle from cat and rabbit. In both species, the tensile stress in response to a step in strain relaxes 30 to 40% in the first 10 sec [Pinto and Patitucci, 1977; Pinto and Patitucci, 1980]. The relaxation curves exhibit a short exponential time constant (<0.02 sec) and a long one (about 1000 sec), and are largely independent of the strain magnitude, which supports the approximation that myocardial viscoelasticity is quasilinear. Myocardial creep under isotonic loading is 2 to 3% of the original length after 100 sec of isotonic loading and is also quasilinear with an exponential timecourse. There is also evidence that passive ventricular muscle exhibits other anelastic properties such as maximum strain-dependent "strain softening" [Emery et al., 1997a,b], a well-known property in elastomers first described by Mullins [1947].

Since the hysteresis of passive cardiac muscle is small and only weakly affected by changes in strain rate, the assumption of pseudoelasticity [Fung, 1993] is often appropriate. That is, the resting myocardium is considered to be a finite elastic material with different elastic properties in loading vs. unloading. Although various preparations have been used to study resting myocardial elasticity, the most detailed and complete information has come from biaxial and multiaxial tests of isolated sheets of cardiac tissue, mainly from the dog [Demer and Yin, 1983; Halperin et al., 1987; Humphrey et al., 1990]. These experiments have shown that the arrested myocardium exhibits significant anisotropy with substantially greater stiffness in the muscle fiber direction than transversely. In equibiaxial tests of muscle sheets cut from planes parallel to the ventricular wall, fiber stress was greater than the transverse stress (Fig. 11.7) by an average factor of close to 2.0 [Yin et al., 1987]. Moreover, as suggested by the structural organization of the myocardium described in Section 11.2, there may be also be significant anisotropy in the plane of the tissue transverse to the fiber axis.

The biaxial stress-strain properties of passive myocardium display some heterogeneity. Novak et al. [1994] measured regional variations of biaxial mechanics in the canine left ventricle. Specimens from the inner and outer thirds of the left ventricular free wall were stiffer than those from the midwall and

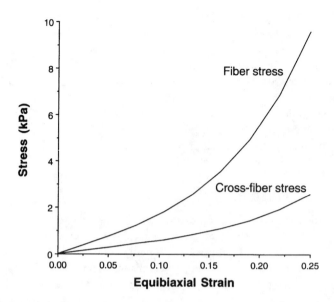

FIGURE 11.7   Representative stress–strain curves for passive rat myocardium computed using Eqs. (11.17) and (11.19). Fiber and crossfiber stress are shown for equibiaxial strain. (Courtesy of Dr. Jeffrey Omens.)

interventricular septum, but the degree of anisotropy was similar in each region. Significant species variations in myocardial stiffness have also been described. Using measurements of two-dimensional regional strain during left ventricular inflation in the isolated whole heart, a parameter optimization approach showed that canine cardiac tissue was several times stiffer than that of the rat, though the nonlinearity and anisotropy were similar [Omens et al., 1993]. Biaxial testing of the collagenous parietal pericardium and epicardium have shown that these tissues have distinctly different properties than the myocardium, being very compliant and isotropic at low biaxial strains (<0.1 to 0.15), but rapidly becoming very stiff and anisotropic as the strain is increased [Lee et al., 1987; Humphrey et al., 1990].

Various constitutive models have been proposed for the elasticity of passive cardiac tissues. Because of the large deformations and nonlinearity of these materials, the most useful framework has been provided by the pseudostrain-energy formulation for hyperelasticity. For a detailed review of the material properties of passive myocardium and approaches to constitutive modeling, see Chapters 1–6 of Glass et al. [1991]. In hyperelasticity, the components of the stress[1] are obtained from the strain energy W as a function of the Lagrangian (Green's) strain E_{RS}.

The myocardium is generally assumed to be an incompressible material, which is a good approximation in the isolated tissue, although in the intact heart significant redistribution of tissue volume is sometimes associated with phasic changes in regional coronary blood volume. Incompressibility is included as a kinematic constraint in the finite elasticity analysis, which introduces a new pressure variable that is added as a Lagrange multiplier in the strain energy. The examples that follow are various strain–energy functions, with representative parameter values (for W in kPa, i.e., mJ·ml^{-1}), that have been suggested for cardiac tissues. For the two-dimensional properties of canine myocardium, Yin and colleagues [1987] obtained reasonable fits to experimental data with an exponential function,

$$W = 0.47\, e^{\left(35 E_{11}^{1.2} + 20 E_{22}^{1.2} \right)} \tag{11.14}$$

---

[1]In a hyperelastic material, the *second Piola Kirchhoff* stress tensor is given by $P_{RS} = \dfrac{1}{2}\left(\dfrac{\partial W}{\partial E_{RS}} + \dfrac{\partial W}{\partial E_{SR}} \right)$.

where E_{11} is the fiber strain and E_{22} is the crossfiber in-plane strain. Humphrey and Yin [1987] proposed a three-dimensional form for W as the sum of an isotropic exponential function of the first principal invariant I_1 of the right Cauchy–Green deformation tensor and another exponential function of the fiber stretch ratio λ_F:

$$W = 0.21\left(e^{9.4\left(I_1 - 3\right)} - 1\right) + 0.35\left(e^{66\left(\lambda_F - 1\right)^2} - 1\right) \tag{11.15}$$

The isotropic part of this expression has also been used to model the myocardium of the embryonic chick heart during the ventricular looping stages, with coefficients of 0.02 kPa during diastole and 0.78 kPa at end-systole, and exponent parameters of 1.1 and 0.85, respectively [Lin and Taber, 1994]. Another, related transversely isotropic strain–energy function was used by Guccione et al. [1991] and Omens et al. [1991] to model material properties in the isolated mature rat and dog hearts:

$$W = 0.6\left(e^Q - 1\right) \tag{11.16}$$

where, in the dog,

$$Q = 26.7E_{11}^2 + 2.0\left(E_{22}^2 + E_{33}^2 + E_{23}^2 + E_{32}^2\right) + 14.7\left(E_{12}^2 + E_{21}^2 + E_{13}^2 + E_{31}^2\right) \tag{11.17}$$

and, in the rat,

$$Q = 9.2E_{11}^2 + 2.0\left(E_{22}^2 + E_{33}^2 + E_{23}^2 + E_{32}^2\right) + 3.7\left(E_{12}^2 + E_{21}^2 + E_{13}^2 + E_{31}^2\right) \tag{11.18}$$

In Eqs. (11.17) and (11.18), normal and shear strain components involving the radial (x_3) axis are included. Humphrey and colleagues [1990] determined a new polynomial form directly from biaxial tests. Novak et al. [1994] gave representative coefficients for canine myocardium from three layers of the left ventricular free wall. The outer third follows as:

$$W = 4.8\left(\lambda_F - 1\right)^2 + 3.4\left(\lambda_F - 1\right)^3 + 0.77\left(I_1 - 3\right) - 6.1\left(I_1 - 3\right)\left(\lambda_F - 1\right) + 6.2\left(I_1 - 3\right)^2 \tag{11.19}$$

The midwall region follows as:

$$W = 5.3\left(\lambda_F - 1\right)^2 + 7.5\left(\lambda_F - 1\right)^3 + 0.43\left(I_1 - 3\right) - 7.7\left(I_1 - 3\right)\left(\lambda_F - 1\right) + 5.6\left(I_1 - 3\right)^2 \tag{11.20}$$

The inner layer of the wall follows as

$$W = 0.51\left(\lambda_F - 1\right)^2 + 27.6\left(\lambda_F - 1\right)^3 + 0.74\left(I_1 - 3\right) - 7.3\left(I_1 - 3\right)\left(\lambda_F - 1\right) + 7.0\left(I_1 - 3\right)^2 \tag{11.21}$$

A power law strain–energy function expressed in terms of circumferential, longitudinal, and transmural extension ratios (λ_1, λ_2, and λ_3) was used [Gupta et al., 1994] to describe the biaxial properties of sheep myocardium 2 weeks after experimental myocardial infarction, in the scarred infarct region:

$$W = 0.36\left(\frac{\lambda_1^{32}}{32} + \frac{\lambda_2^{30}}{30} + \frac{\lambda_3^{31}}{31} - 3\right) \tag{11.22}$$

and in the remote, non-infarcted tissue:

$$W = 0.11\left(\frac{\lambda_1^{22}}{22} + \frac{\lambda_2^{26}}{26} + \frac{\lambda_3^{24}}{24} - 3\right) \tag{11.23}$$

Finally, based on the observation that resting stiffness rises steeply at strains that extend coiled collagen fibers to the limit of uncoiling, Hunter and colleagues have proposed a pole-zero constitutive relation in which the stresses rise asymptotically as the strain approaches a limiting elastic strain [Hunter et al., 1998].

The strain in the constitutive equation, must generally be referred to the stress-free state of the tissue. However, the unloaded state of the passive left ventricle is not stress free; residual stress exists in the intact, unloaded myocardium, as shown by Omens and Fung [1990]. Cross-sectional equatorial rings from potassium-arrested rat hearts spring open elastically when the left ventricular wall is resected radially. The average opening angle of the resulting curved arc is 45 ± 10° in the rat. Subsequent radial cuts produce no further change. Hence, a slice with one radial cut is considered to be stress free, and there is a nonuniform distribution of residual strain across the intact wall, being compressive at the endocardium and tensile at the epicardium, with some regional differences. Stress analyses of the diastolic left ventricle show that residual stress acts to minimize the endocardial stress concentrations that would otherwise be associated with diastolic loading [Guccione et al., 1991]. An important physiological consequence of residual stress is that sarcomere length is nonuniform in the unloaded resting heart. Rodriguez et al. [1993] showed that sarcomere length is about 0.13 μm greater at epicardium than endocardium in the unloaded rat heart, and this gradient vanishes when residual stress is relieved. Three-dimensional studies have also revealed the presence of substantial transverse residual shear strains [Costa et al., 1997]. Residual stress and strain may have an important relationship to cardiac growth and remodeling. Theoretical studies have shown that residual stress in tissues can arise from growth fields that are kinematically incompatible [Skalak et al., 1982; Rodriguez et al., 1994].

11.5 Regional Ventricular Mechanics: Stress and Strain

Although ventricular pressures and volumes are valuable for assessing the global pumping performance of the heart, myocardial stress and strain distributions are need to characterize regional ventricular function, especially in pathological conditions, such as myocardial ischemia and infarction, where profound localized changes may occur. The measurement of stress in the intact myocardium involves resolving the local forces acting on defined planes in the heart wall. Attempts to measure local forces [Feigl et al., 1967; Huisman et al., 1980] have had limited success because of the large deformations of the myocardium and the uncertain nature of the mechanical coupling between the transducer elements and the tissue. Efforts to measure intramyocardial pressures using miniature implanted transducers have been more successful but have also raised controversy over the extent to which they accurately represent changes in interstitial fluid pressure. In all cases, these methods provide an incomplete description of three-dimensional wall stress distributions. Therefore, the most common approach for estimating myocardial stress distributions is the use of mathematical models based on the laws of continuum mechanics [Hunter and Smaill, 1989]. Although there is not room to review these analyses here, the important elements of such models are the geometry and structure, boundary conditions and material properties, described in the foregoing sections. An excellent review of ventricular wall stress analysis is given by Yin [1981]. The most versatile and powerful method for ventricular stress analysis is the finite element method, which has been used in cardiac mechanics for over 20 years [Yin, 1985]. However, models must also be validated with experimental measurements. Since the measurement of myocardial stresses is not yet reliable, the best experimental data for model validation are measurements of strains in the ventricular wall.

The earliest myocardial strain gauges were mercury-in-rubber transducers sutured to the epicardium. Today, local segment length changes are routinely measured with various forms of the piezoelectric crystal sonomicrometer. However, since the ventricular myocardium is a three-dimensional continuum, the local strain is only fully defined by all the normal and shear components of the myocardial strain tensor. Villarreal et al. [1988] measured two-dimensional midwall strain components by arranging three piezo-electric crystals in a small triangle so that three segment lengths could be measured simultaneously. They showed that the principal axis of greatest shortening is not aligned with circumferential midwall fibers, and that this axis changes with altered ventricular loading and contractility. Therefore, uniaxial segment measurements do not reveal the full extent of alterations in regional function caused by an experimental intervention. Another approach to measuring regional myocardial strains is the use of clinical imaging techniques, such as contrast ventriculography, high-speed x-ray tomography, magnetic resonance imaging (MRI), or two-dimensional echocardiography. But the conventional application of these techniques is not suitable for measuring regional strains because they can not be used to identify the motion of distinct myocardial points. They only produce a profile or silhouette of a surface, except in the unusual circumstance when radiopaque markers are implanted in the myocardium during cardiac surgery or transplantation [Ingels et al., 1975]. Hunter and Zerhouni [1989] describe the prospects for non-invasive imaging of discrete points in the ventricular wall. The most promising method is the use of MRI tagging methods, which are now being used to map three-dimensional ventricular strain fields in conscious subjects [Young and Axel, 1992].

In experimental research, implantable radiopaque markers are used for tracking myocardial motions with high spatial and temporal resolution. Meier et al. [1980a,b] placed triplets of metal markers 10–15 mm apart near the epicardium of the canine right ventricle and reconstructed their positions from biplane cinéradiographic recordings. By polar decomposition, they obtained the two principal epicardial strains, the principal angle, and the local rotation in the region. The use of radiopaque markers was extended to three dimensions by Waldman and colleagues [1985], who implanted three closely separated columns of 5 to 6 metal beads in the ventricular wall. With this technique, it is possible to find all six components of strain and all three rigid-body rotation angles at sites through the wall. For details of this method, see the review by Waldman in Chapter 7 of Glass et al. [1991]. An enhancement to this method uses high-order finite element interpolation of the marker positions to compute continuous transmural distributions of myocardial deformation [McCulloch and Omens, 1991].

Studies and models like these are producing an increasingly detailed picture of regional myocardial stress and strain distributions. Of the many interesting observations, there are some useful generalizations, particularly regarding the strain. Myocardial deformations are large and three dimensional, hence the nonlinear finite strain tensors are more appropriate measures than the linear infinitesimal Cauchy strain. During filling in the normal heart, the wall stretches biaxially but nonuniformly in the plane of the wall, and thins in the transmural direction. During systole, shortening is also two-dimensional and the wall thickens. There are substantial regional differences in the time-course, magnitude, and pattern of myocardial deformations. In humans and dogs, in-plane systolic myocardial shortening and diastolic lengthening vary with longitudinal position on the left and right ventricular free walls generally increasing in magnitude from base to apex.

Both during systole and diastole, there are significant shear strains in the wall. In-plane (torsional) shears are negative during diastole, consistent with a small left-handed torsion of the left ventricle during filling, and positive as the ventricular twist reverses during ejection. Consequently, the principal axes of greatest diastolic segment lengthening and systolic shortening are not circumferential or longitudinal but at oblique axes, that are typically rotated 10 to 60° clockwise from circumferential. There are circumferential variations in the regional left ventricular strain. The principal axes of greatest diastolic lengthening and systolic shortening tend to be more longitudinal on the posterior wall and more circumferentially oriented on the anterior wall. Perhaps the most significant regional variations are transmural. In-plane and transmural, normal or principal strains, are usually significantly greater in magnitude at the endocardium than the epicardium both in filling and ejection. However, when the strain is resolved in the local muscle fiber direction, the transmural variation of fiber strain becomes insignificant. The combination of torsional deformation and the transmural variation in fiber direction means that systolic shortening and diastolic lengthening tend to be maximized in the fiber direction at the epicardium and minimized at the endocardium. Hence, whereas maximum shortening and lengthening are closely aligned with muscle fibers at the subepicardium, they are almost perpendicular to the fibers at the subendocardium. In the left ventricular wall there are also substantial transverse shear strains (i.e., in the circumferential–radial and longitudinal–radial planes) during systole, though during filling they are smaller. Their functional significance remains unclear, though they change substantially during acute myocardial ischemia or ventricular pacing and are apparently associated with the transverse laminae described earlier [LeGrice et al., 1995].

Sophisticated continuum mechanics models are needed to determine the stress distributions associated with these complex myocardial deformations. With modern finite element methods it is now possible to include in the analysis the three-dimensional geometry and fiber architecture, finite deformations, nonlinear material properties, and muscle contraction of the ventricular myocardium. Some models have included other factors such as viscoelasticity, poroelasticity, coronary perfusion, growth and remodeling, regional ischemia, and residual stress. To date, continuum models have provided some valuable insight into regional cardiac mechanics. These include the importance of muscle fiber orientation, torsional deformations and residual stress, and the substantial inhomogeneities associated with regional variations in geometry and fiber angle or myocardial ischemia and infarction.

Acknowledgments

I am indebted to many colleagues and students, past and present, for their input and perspective on cardiac biomechanics. Owing to space constraints, I have relied on much of their work without adequate citation, especially in the final section. Special thanks to Drs. Peter Hunter, Bruce Smaill, Lewis Waldman, Y.-C. Fung, James Covell, Jeffrey Omens, Wilbur Lew, and Francisco Villarreal.

References

Anversa P, Puntillo E, Nikitin P et al. 1989. Effects of age on mechanical and structural properties of myocardium of Fischer 344 rats. *Am J Physiol* 256:H1440.

Arts T, Reneman RS, Veenstra PC. 1979. A model of the mechanics of the left ventricle. *Ann Biomed Eng* 7:299.

Arts T, Veenstra PC, Reneman RS. 1982. Epicardial deformation and left ventricular wall mechanics during ejection in the dog. *Am J Physiol* 243:H379.

Backx PH, Gao WD, Azan-Backx MD et al. 1995. The relationship between contractile force and intracellular [Ca^{2+}] in intact rat cardiac trabeculae. *J Gen Physiol* 105:1.

Bers DM. 1991. *Excitation-Contraction Coupling and Cardiac Contractile Force*. Dordrecht, Kluwer.

Borg TK, Ranson WF, Moslehy FA et al. 1981. Structural basis of ventricular stiffness. *Lab Invest* 44:49.

Brutsaert DL, Sys SU. 1989. Relaxation and diastole of the heart. *Physiol Rev* 69:1228.

Burkhoff D, Schnellbacher M, Stennett RA et al. 1995. Explaining load-dependent ventricular performance and energetics based on a model of E-C coupling. In: *Cardiac Energetics: From E_{max} to Pressure–Volume Area*, LeWinter MM, Suga H, Watkins MW (Eds.). Boston, Kluwer Academic.

Caulfield JB, Borg TK. 1979. The collagen network of the heart. *Lab Invest* 40:364.

Chadwick RS. 1982. Mechanics of the left ventricle. *Biophys J* 39:279.

Charney RH, Takahashi S, Zhao M et al. 1992. Collagen loss in the stunned myocardium. *Circulation* 85:1483.

Connelly CM, Vogel WM, Wiegner AW et al. 1985. Effects of reperfusion after coronary artery occlusion on post-infarction scar tissue. *Circ Res* 57:562.

Contard F, Koteliansky V, Marotte F et al. 1991. Specific alterations in the distribution of extracellular matrix components within rat myocardium during the development of pressure overload. *Lab Invest* 64:65.

Corin WJ, Murakami T, Monrad ES et al. 1991. Left ventricular passive diastolic properties in chronic mitral regurgitation. *Circulation* 83:797.

Costa K, May-Newman K, Farr D et al. 1997. Three-dimensional residual strain in canine mid-anterior left ventricle. *Am J Physiol* 273:H1968.

de Tombe PP, ter Keurs HE. 1992. An internal viscous element limits unloaded velocity of sarcomere shortening in rat myocardium. *J Physiol (Lond)* 454:619.

Delhaas T, Arts T, Prinzen FW et al. 1994. Regional fibre stress-fibre strain area as an estimate of regional blood flow and oxygen demand in the canine heart. *J Physiol (Lond)* 477:481.

Demer LL, Yin FCP. 1983. Passive biaxial mechanical properties of isolated canine myocardium. *J Physiol* 339:615.

Eghbali M, Robinson TF, Seifter S et al. 1989. Collagen accumulation in heart ventricles as a function of growth and aging. *Cardiovascular Res* 23:723.

Emery JL, Omens JH, McCulloch AD. 1997a. Biaxial mechanics of the passively overstretched left ventricle. *Am J Physiol* 272:H2299.

Emery JL, Omens JH, McCulloch AD. 1997b. Strain softening in rat left ventricular myocardium. *J Biomech Eng* 119:6.

Feigl EO, Simon GA, Fry DL. 1967. Auxotonic and isometric cardiac force transducers. *J Appl Physiol* 23:597.

Forrester JS, Diamond G, Parmley WW et al. 1972. Early increase in left ventricular compliance after myocardial infarction. *J Clin Invest* 51:598.

Fung YC. 1993. *Biomechanics: Mechanical Properties of Living Tissues.* New York, Springer-Verlag Inc.

Gaasch WH, LeWinter MM. 1994. *Left Ventricular Diastolic Dysfunction and Heart Failure.* Philadelphia, Lea & Febiger.

Gao WD, Backx PH, Azan-Backx M et al. 1994. Myofilament Ca^{2+} sensitivity in intact versus skinned rat ventricular muscle. *Circ Res* 74:408.

Gilbert JC, Glantz SA. 1989. Determinants of left ventricular filling and of the diastolic pressure–volume relation. *Circ Res* 64:827.

Glantz SA, Parmley WW. 1978. Factors which affect the diastolic pressure–volume curve. *Circ Res* 42:171.

Glantz SA, Misbach GA, Moores WY et al. 1978. The pericardium substantially affects the left ventricular diastolic pressure-volume relationship in the dog. *Circ Res* 42:433.

Glass L, Hunter P, McCulloch AD, Eds. 1991. *Theory of Heart: Biomechanics, Biophysics and Nonlinear Dynamics of Cardiac Function.* New York, Springer-Verlag.

Goto Y, Futaki S, Kawaguchi O et al. 1993. Coupling between regional myocardial oxygen consumption and contraction under altered preload and afterload. *J Am Coll Cardiol* 21:1522.

Grossman W. 1980. Cardiac hypertrophy: useful adaptation or pathologic process? *Am J Med* 69:576.

Grossman W, Jones D, McLaurin LP. 1975. Wall stress and patterns of hypertrophy in the human left ventricle. *J Clin Invest* 56:56.

Guccione JM, McCulloch AD. 1993. Mechanics of active contraction in cardiac muscle: Part I—Constitutive relations for fiber stress that describe deactivation. *J Biomech Eng* 115:72.

Guccione JM, McCulloch AD, Waldman LK. 1991. Passive material properties of intact ventricular myocardium determined from a cylindrical model. *J Biomech Eng* 113:42.

Guccione JM, Waldman LK, McCulloch AD. 1993. Mechanics of active contraction in cardiac muscle: Part II—Cylindrical models of the systolic left ventricle. *J Biomech Eng* 115:82.

Gupta KB, Ratcliff MB, Fallert MA et al. 1994. Changes in passive mechanical stiffness of myocardial tissue with aneurysm formation. *Circulation* 89:2315.

Halperin HR, Chew PH, Weisfeldt ML et al. 1987. Transverse stiffness: a method for estimation of myocardial wall stress. *Circ Res* 61:695.

Harper J, Harper E, Covell JW. 1993. Collagen characterization in volume-overload- and pressure-overload-induced cardiac hypertrophy in minipigs. *Am J Physiol* 265:H434.

Holmes JW, Yamashita H, Waldman LK et al. 1994. Scar remodeling and transmural deformation after infarction in the pig. *Circulation* 90:411.

Holt JP, Rhode EA, Peoples SA et al. 1962. Left ventricular function in mammals of greatly different size. *Circ Res* 10:798.

Huisman RM, Elzinga G, Westerhof N et al. 1980. Measurement of left ventricular wall stress. *Cardiovasc Res* 14:142.

Humphrey JD, Yin FCP. 1987. A new constitutive formulation for characterizing the mechanical behavior of soft tissues. *Biophys J* 52:563.

Humphrey JD, Strumpf RK, Yin FCP. 1990a. Biaxial mechanical behavior of excised ventricular epicardium. Am J Physiol 259:H101.

Humphrey JD, Strumpf RK, Yin FCP. 1990b. Determination of a constitutive relation for passive myocardium: I. A new functional form. *J Biomech Eng* 112:333.

Hunter WC. 1989. End-systolic pressure as a balance between opposing effects of ejection. *Circ Res* 64:265.

Hunter PJ, Smaill BH. 1989. The analysis of cardiac function: a continuum approach. *Prog Biophys Mol Biol* 52:101.

Hunter WC, Zerhouni EA. 1989. Imaging distinct points in left ventricular myocardium to study regional wall deformation. In: *Innovations in Diagnostic Radiology.* Anderson JH (Ed.). New York, Springer-Verlag, p. 169.

Hunter PJ, McCulloch AD, ter Keurs HEDJ. 1998. Modeling the mechanical properties of cardiac muscle. *Prog Biophys Mol Biol* 69:289.

Iimoto DS, Covell JW, Harper E. 1988. Increase in crosslinking of type I and type III collagens associated with volume overload hypertropy. *Circ Res* 63:399.

Ingels NB, Jr, Daughters GT, II, Stinson EB et al. 1975. Measurement of midwall myocardial dynamics in intact man by radiography of surgically implanted markers. *Circulation* 52:859.

Jalil JE, Doering CW, Janicki JS et al. 1988. Structural vs. contractile protein remodeling and myocardial stiffness in hypertrophied rat left ventricle. *J Mol Cell Cardiol* 20:1179.

Janicki JS, Weber KT. 1980. The pericardium and ventricular interaction, distensibility and function. *Am J Physiol* 238:H494.

Jensen LT, Hørslev-Petersen K, Toft P et al. 1990. Serum aminoterminal type III procollagen peptide reflects repair after acute myocardial infarction. *Circulation* 81:52.

Jugdutt BI, Amy RW. 1986. Healing after myocardial infarction in the dog: changes in infarct hydroxyproline and topography. *J Am Coll Cardiol* 7:91.

Karlon WJ, Covell JW, McCulloch AD et al. 1998. Automated measurement of myofiber disarray in transgenic mice with ventricular expression of ras. *Anatom Rec* 252:612.

Kentish JC, Ter Keurs HEDJ, Ricciari L et al. 1986. Comparisons between the sarcomere length-force relations of intact and skinned trabeculae from rat right ventricle. *Circ Res* 58:755.

Landesberg A, Sideman S. 1994. Coupling calcium binding to troponin C and cross-bridge cycling in skinned cardiac cells. *Am J Physiol* 266:H1260.

Landesberg A, Markhasin VS, Beyar R et al. 1996. Effect of cellular inhomogeneity on cardiac tissue mechanics based on intracellular control mechanisms. *Am J Physiol* 270:H1101.

Lee MC, Fung YC, Shabetai R et al. 1987. Biaxial mechanical properties of human pericardium and canine comparisons. *Am J Physiol* 253:H75.

LeGrice IJ, Smaill BH, Chai LZ et al. 1995a. Laminar structure of the heart: ventricular myocyte arrangement and connective tissue architecture in the dog. *Am J Physiol* 269:H571.

LeGrice IJ, Takayama Y, Covell JW. 1995b. Transverse shear along myocardial cleavage planes provides a mechanism for normal systolic wall thickening. *Circ Res* 77:182.

LeGrice IJ, Hunter PJ, Smaill BH. 1997. Laminar structure of the heart: a mathematical model. *Am J Physiol* 272:H2466.

Lin DHS, Yin FCP. 1998. A multiaxial constitutive law for mammalian left ventricular myocardium in steady-state barium contracture or tetanus. *J Biomech Eng* 120:504.

Lin I-E, Taber LA. 1994. Mechanical effects of looping in the embryonic chick heart. *J Biomech* 27:311.

Loiselle DS, Gibbs CL. 1979. Species differences in cardiac energetics. *Am J Physiol* 237.

Ma SP, Zahalak GI. 1991. A distribution-moment model of energetics in skeletal muscle [see comments]. *J Biomech* 24:21.

MacKenna DA, McCulloch AD. 1996. Contribution of the collagen extracellular matrix to ventricular mechanics. In: *Systolic and Diastolic Function of the Heart*, Ingels NB, Daughters GT, Baan J et al. (Eds.). Amsterdam, IOS Press, p. 35.

MacKenna DA, Omens JH, Covell JW. 1996. Left ventricular perimysial collagen fibers uncoil rather than stretch during diastolic filling. *Basic Res Cardiol* 91:111.

MacKenna DA, Vaplon SM, McCulloch AD. 1997. Microstructural model of perimysial collagen fibers for resting myocardial mechanics during ventricular filling. *Am J Physiol* 273:H1576.

Maron BJ, Bonow RO, Cannon RO et al. 1987. Hypertrophic cardiomyopathy. Interrelations of clinical manifestations, pathophysiology, and therapy (1). *N Engl J Med* 316:780.

Maruyama Y, Ashikawa K, Isoyama S et al. 1982. Mechanical interactions between the four heart chambers with and without the pericardium in canine hearts. *Circ Res* 50:86.

May-Newman KD, Omens JH, Pavelec RS et al. 1994. Three-dimensional transmural mechanical interaction between the coronary vasculature and passive myocardium in the dog. *Circ Res* 74:1166.

McCulloch AD, Omens JH. 1991. Non-homogeneous analysis of three-dimensional transmural finite deformations in canine ventricular myocardium. *J Biomech* 24: 539.

McCulloch AD, Smaill BH, Hunter PJ. 1989. Regional left ventricular epicardial deformation in the passive dog heart. *Circ Res* 64:721.

McLean M, Ross MA, Prothero J. 1989. Three-dimensional reconstruction of the myofiber pattern in the fetal and neonatal mouse heart. *Anat Rec* 224:392.

Medugorac I. 1980. Myocardial collagen in different forms of hypertrophy in the rat. *Res Exp Med (Berl)* 177:201.

Medugorac I, Jacob R. 1983. Characterisation of left ventricular collagen in the rat. *Cardiovascular Res* 17:15.

Meier GD, Bove AA, Santamore WP et al. 1980a. Contractile function in canine right ventricle. *Am J Physiol* 239:H794.

Meier GD, Ziskin MC, Santamore WP et al. 1980b. Kinematics of the beating heart. *IEEE Trans Biomed Eng* 27:319.

Michel JB, Salzmann JL, Ossondo Nlom M et al. 1986. Morphometric analysis of collagen network and plasma perfused capillary bed in the myocardium of rats during evolution of cardiac hypertrophy. *Basic Res Cardiol* 81:142.

Mirsky I, Rankin JS. 1979. The effects of geometry, elasticity, and external pressures on the diastolic pressure-volume and stiffness-stress relations: how important is the pericardium? *Circ Res* 44:601.

Mukherjee D, Sen S. 1990. Collagen phenotypes during development and regression of myocardial hypertrophy in spontaneously hypertensive rats. *Circ Res* 67:1474.

Mullins L. 1947. Effect of stretching on the properties of rubber. *J Rubber Res* 16:275.

Namba T, Takaki M, Araki J et al. 1994. Energetics of the negative and positive inotropism of pentobarbitone sodium in the canine left ventricle. *Cardiovascular Res* 28:557.

Nevo E, Lanir Y. 1989. Structural finite deformation model of the left ventricle during diastole and systole. *J Biomech Eng* 111:342.

Nielsen PMF, Le Grice IJ, Smaill BH et al. 1991. Mathematical model of geometry and fibrous structure of the heart. *Am J Physiol* 260:H1365.

Novak VP, Yin FCP, Humphrey JD. 1994. Regional mechanical properties of passive myocardium. *J Biomech* 27:403.

Ohgoshi Y, Goto Y, Futaki S et al. 1991. Increased oxygen cost of contractility in stunned myocardium of dog. *Circ Res* 69:975.

Omens JH, Fung YC. 1990. Residual strain in rat left ventricle. *Circ Res* 66:37.

Omens JH, MacKenna DA, McCulloch AD. 1991. Measurement of two-dimensional strain and analysis of stress in the arrested rat left ventricle. *Adv Bioeng* BED-20:635.

Omens JH, MacKenna DA, McCulloch AD. 1993. Measurement of strain and analysis of stress in resting rat left ventricular myocardium. *J Biomech* 26:665.

Omens JH, Milkes DE, Covell JW. 1995. Effects of pressure overload on the passive mechanics of the rat left ventricle. *Ann Biomed Eng* 23:152.

Panerai RB. 1980. A model of cardiac muscle mechanics and energetics. *J Biomech* 13:929.

Patterson SW, Starling EH. 1914. On the mechanical factors which determine the output of the ventricles. *J Physiol* 48:357.

Pfeffer JM, Pfeffer MA, Fletcher PJ et al. 1991. Progressive ventricular remodeling in rat with myocardial infarction. *Am J Physiol* 260:H1406.

Pinto JG, Patitucci PJ. 1977. Creep in cardiac muscle. *Am J Physiol* 232:H553.

Pinto JG, Patitucci PJ. 1980. Visco-elasticity of passive cardiac muscle. *J Biomech Eng* 102:57.

Pirzada FA, Ekong EA, Vokonas PS et al. 1976. Experimental myocardial infarction XIII. Sequential changes in left ventricular pressure-length relationships in the acute phase. *Circulation* 53:970.

Rakusan K. 1984. Cardiac growth, maturation and aging. In: *Growth of the Heart in Health and Disease*, Zak R (Ed.). New York, Raven Press, p. 131.

Robinson TF, Cohen-Gould L, Factor SM. 1983. Skeletal framework of mammalian heart muscle: arrangement of inter- and pericellular connective tissue structures. *Lab Invest* 49:482.

Robinson TF, Geraci MA, Sonnenblick EH et al. 1988. Coiled perimysial fibers of papillary muscle in rat heart: morphology, distribution, and changes in configuration. *Circ Res* 63:577.

Rodriguez EK, Omens JH, Waldman LK et al. 1993. Effect of residual stress on transmural sarcomere length distribution in rat left ventricle. *Am J Physiol* 264:H1048.

Rodriguez E, Hoger A, McCulloch A. 1994. Stress-dependent finite growth in soft elastic tissues. *J Biomech* 27:455.

Ross J, Jr, Sonnenblick EH, Covell JW et al. 1967. The architecture of the heart in systole and diastole: technique of rapid fixation and analysis of left ventricular geometry. *Circ Res* 21:409.

Rüegg JC. 1988. *Calcium in Muscle Activation: A Comparative Approach*. Berlin, Springer-Verlag.

Saffitz JE, Kanter HL, Green KG et al. 1994. Tissue-specific determinants of anisotropic conduction velocity in canine atrial and ventricular myocardium. *Circ Res* 74:1065.

Sagawa K. 1988. *Cardiac Contraction and the Pressure–Volume Relationship*. New York, Oxford University Press.

Salisbury PF, Cross CE, Rieben PA. 1960. Influence of coronary artery pressure upon myocardial elasticity. *Circ Res* 8:794.

Schoenberg M. 1980a. Geometrical factors influencing muscle force development. I. The effect of filament spacing upon axial forces. *Biophys J* 30:51.

Schoenberg M. 1980b. Geometrical factors influencing muscle force development. II. Radial forces. *Biophys J* 30:69.

Silver MA, Pick R, Brilla CG et al. 1990. Reactive and reparative fibrillar collagen remodeling in the hypertrophied rat left ventricle: two experimental models of myocardial fibrosis. *Cardiovascular Res* 24:741.

Skalak R, Dasgupta G, Moss M et al. 1982. Analytical description of growth. *J Theor Biol* 94:555.

Slinker BK, Glantz SA. 1986. End-systolic and end-diastolic ventricular interaction. *Am J Physiol* 251:H1062.

Smaill BH, Hunter PJ. 1991. Structure and function of the diastolic heart. In: *Theory of Heart*, Glass L, Hunter PJ, McCulloch AD (Eds.). New York, Springer-Verlag, p. 1.

Sonnenblick EH, Napolitano LM, Daggett WM et al. 1967. An intrinsic neuromuscular basis for mitral valve motion in the dog. *Circ Res* 21:9.

Spotnitz HM, Spotnitz WD, Cottrell TS et al. 1974. Cellular basis for volume related wall thickness changes in the rat left ventricle. *J Mol Cell Cardiol* 6:317.

Stahl WR. 1967. Scaling of respiratory variable in mammals. *J Appl Physiol* 22:453.

Streeter DD, Jr. 1979. Gross morphology and fiber geometry of the heart. In: *Handbook of Physiology*, Section 2: *The Cardiovascular System*, Chapter 4. Bethesda, MD, American Physiological Society, p. 61.

Streeter DD, Jr, Hanna WT. 1973. Engineering mechanics for successive states in canine left ventricular myocardium: I. Cavity and wall geometry. *Circ Res* 33:639.

Streeter DD, Jr, Spotnitz HM, Patel DP et al. 1969. Fiber orientation in the canine left ventricle during diastole and systole. *Circ Res* 24:339.

Suga H. 1990. Ventricular energetics. *Physiol Rev* 70:247.

Suga H, Goto Y. 1991. Cardiac oxygen costs of contractility (E_{max}) and mechanical energy (PVA): new key concepts in cardiac energetics. In: *Recent Progress in Failing Heart Syndrome*, Sasayama S, Suga H (Eds.). Tokyo, Springer-Verlag, p. 61.

Suga H, Hayashi T, Shirahata M. 1981. Ventricular systolic pressure-volume area as predictor of cardiac oxygen consumption. *Am J Physiol* 240:H39.

Suga H, Goto Y, Yasumura Y et al. 1988. O_2 consumption of dog heart under decreased coronary perfusion and propranolol. *Am J Physiol* 254:H292.

Suga H, Goto Y, Kawaguchi O et al. 1993. Ventricular perspective on efficiency. *Basic Res Cardiol* 88(Suppl 2):43.

Sugawara M, Kondoh Y, Nakano K. 1995. Normalization of E_{max} and PVA. In: *Cardiac Energetics: From E_{max} to Pressure–Volume Area.* LeWinter MM, Suga H, Watkins MW (Eds.). Boston, Kluwer Academic, p. 65.

Taber LA. 1991. On a nonlinear theory for muscle shells. Part II. Application to the beating left ventricle. *ASME J Biomech Eng* 113:63.

Takahashi S, Barry AC, Factor SM. 1990. Collagen degradation in ischaemic rat hearts. *Biochem J* 265:233.

Taylor TW, Goto Y, Suga H. 1993. Variable cross-bridge cycling-ATP coupling accounts for cardiac mechanoenergetics. *Am J Physiol* 264:H994.

ter Keurs HE, de Tombe PP. 1993. Determinants of velocity of sarcomere shortening in mammalian myocardium. *Adv Exp Med Biol* 332:649.

ter Keurs HEDJ, Rijnsburger WH, van Heuningen R. 1980a. Restoring forces and relaxation of rat cardiac muscle. *Eur Heart J* 1:67.

ter Keurs HEDJ, Rijnsburger WH, van Heuningen R et al. 1980b. Tension development and sarcomere length in rat cardiac trabeculae: evidence of length-dependent activation. *Circ Res* 46:703.

Torrent-Guasp F. 1973. *The Cardiac Muscle.* Madrid, Juan March Foundation.

Tözeren A. 1985. Continuum rheology of muscle contraction and its application to cardiac contractility. *Biophys J* 47:303.

Vetter F, McCulloch A. 1998. Three-dimensional analysis of regional cardiac function: a model of the rabbit ventricular anatomy. *Prog Biophys Mol Biol* 69:157.

Villarreal FJ, Waldman LK, Lew WYW. 1988. Technique for measuring regional two-dimensional finite strains in canine left ventricle. *Circ Res* 62:711.

Vornanen M. 1992. Maximum heart rate of sorcine shrews: correlation with contractile properties and myosin composition. *Am J Physiol* 31:R842.

Waldman LK, Fung YC, Covell JW. 1985. Transmural myocardial deformation in the canine left ventricle: normal *in vivo* three-dimensional finite strains. *Circ Res* 57:152.

Weber KT. 1989. Cardiac interstituim in health and disease: the fibrillar collagen network. *J Am Coll Cardiol* 13:1637.

Weber KT, Janicki JS, Shroff SG et al. 1988. Collagen remodeling of the pressure-overloaded, hypertrophied nonhuman primate myocardium. *Circ Res* 62:757.

Weber KT, Pick R, Silver MA et al. 1990. Fibrillar collagen and remodeling of dilated canine left ventricle. *Circulation* 82:1387.

Whittaker P, Boughner DR, Kloner RA. 1989. Analysis of healing after myocardial infarction using polarized light microscopy. *Am J Pathol* 134:879.

Whittaker P, Boughner DR, Kloner RA. 1991a. Role of collagen in acute myocardial infarct expansion. *Circulation* 84:2123.

Whittaker P, Boughner DR, Kloner RA et al. 1991b. Stunned myocardium and myocardial collagen damage: differential effects of single and repeated occlusions. *Am Heart J* 121:434.

Yellin EL, Hori M, Yoran C et al. 1986. Left ventricular relaxation in the filling and nonfilling intact canine heart. *Am J Physiol* 250:H620.

Yin FCP. 1981. Ventricular wall stress. *Circ Res* 49:829.

Yin FCP. 1985. Applications of the finite-element method to ventricular mechanics. *CRC Crit Rev Biomed Eng* 12:311.

Yin FCP, Strumpf RK, Chew PH et al. 1987. Quantification of the mechanical properties of noncontracting canine myocardium under simultaneous biaxial loading. *J Biomech* 20:577.

Young AA, Axel L. 1992. Three-dimensional motion and deformation in the heart wall: estimation from spatial modulation of magnetization—a model-based approach. *Radiology* 185:241.

Yue DT, Marban E, Wier WG. 1986. Relationship between force and intracellular [Ca^{2+}] in tetanized mammalian heart muscle. *J Gen Physiol* 87:223.

Zahalak GI. 1996. Non-axial muscle stress and stiffness. *J Theor Biol* 182:59.

Zhao M, Zhang H, Robinson TF et al. 1987. Profound structural alterations of the extracellular collagen matrix in postischemic dysfunctional ("stunned") but viable myocardium. *J Am Coll Cardiol* 10:1322.

Zhao DD, Namba T, Araki J et al. 1993. Nipradilol depresses cardiac contractility and O_2 consumption without decreasing coronary resistance in dogs. *Acta Medica Okayama* 47:29.

12

Heart Valve Dynamics

Ajit P. Yoganathan
Georgia Institute of Technology

Jack D. Lemmon
Georgia Institute of Technology

Jeffrey T. Ellis
Georgia Institute of Technology

The heart has four valves that control the direction of blood flow through the heart, permitting forward flow and preventing back flow. On the right side of the heart, the tricuspid and pulmonic valves regulate the flow of blood that is returned from the body to the lungs for oxygenation. The mitral and aortic valves control the flow of oxygenated blood from the left side of the heart to the body. The aortic and pulmonic valves allow blood to be pumped from the ventricles into arteries on the left and right side of the heart, respectively. Similarly, the mitral and tricuspid valves lie between the atria and ventricles of the left and right sides of the heart, respectively. The aortic and pulmonic valves open during systole when the ventricles are contracting, and close during diastole when the ventricles are filling through the open mitral and tricuspid valves. During isovolumic contraction and relaxation, all four valves are closed (Fig. 12.1).

When closed, the pulmonic and tricuspid valves must withstand a pressure of approximately 30 mmHg. However, the closing pressures on the left side of the heart are much higher. The aortic valve withstands pressures of approximately 100 mmHg, while the mitral valve closes against pressures up to 150 mmHg. Since diseases of the valves on the left side of the heart are more prevalent, most of this chapter will focus on the aortic and mitral valves. Where pertinent, reference will be made to the pulmonic and tricuspid valves.

12.1 Aortic and Pulmonic Valves

The aortic valve is composed of three semilunar cusps, or leaflets, contained within a connective tissue sleeve. The valve cusps are attached to a fibrous ring embedded in the fibers of the ventricular septum and the anterior leaflet of the mitral valve. Each of the leaflets is lined with endothelial cells and has a dense collagenous core adjacent to the high pressure aortic side. The side adjacent to the aorta is termed the fibrosa and is the major fibrous layer within the belly of the leaflet. The layer covering the ventricular side of the valve is called the ventricularis and is composed of both collagen and elastin. The ventricularis is thinner than the fibrosa and presents a very smooth surface to the flow of blood [Christie, 1990]. The central portion of the valve, called the spongiosa, contains variable loose connective tissue and proteins and is normally not vascularized. The collagen fibers within the fibrosa and ventricularis are unorganized in the unstressed state. When a stress is applied, they become oriented primarily in the circumferential direction with a lower concentration of elastin and collagen in the radial direction [Christie, 1990; Thubrikar, 1990].

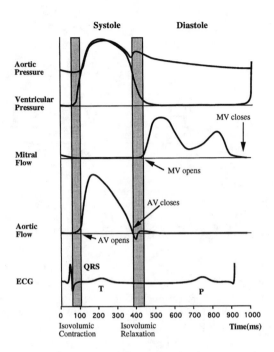

FIGURE 12.1   Typical pressure and flow curves for the aortic and mitral valves.

The fibrous annular ring of the aortic valve separates the aorta from the left ventricle and superior to this ring is a structure called the sinus of Valsalva, or aortic sinus. The sinus is comprised of three bulges at the root of the aorta, with each bulge aligned with the belly or central part of the specific valve leaflet. Each valve cusp and corresponding sinus is named according to its anatomical location within the aorta. Two of these sinuses give rise to coronary arteries that branch off the aorta, providing blood flow to the heart itself. The right coronary artery is based at the right or right anterior sinus, the left coronary artery exits the left or left posterior sinus, and the third sinus is called the non-coronary or right posterior sinus. Figure 12.2 shows the configuration of the normal aortic sinuses and valve in the closed position. Because the length of the aortic valve cusps is greater than the annular radius, a small overlap of tissue from each leaflet protrudes and forms a coaptation surface within the aorta when the valve is closed [Emery and Arom, 1991]. This overlapped tissue, called the lunula, may help to ensure that the valve is sealed. When the valve is open, the leaflets extend to the upper edge of the sinuses of Valsalva. The anatomy of the pulmonic valve is similar to that of the aortic valve, but the surrounding structure is slightly different. The main differences are that the sinuses are smaller in the pulmonary artery and the pulmonic valve annulus is slightly larger than that of the aortic valve.

The dimensions of the aortic and pulmonic valves and their leaflets have been measured in a number of ways. Before noninvasive measurement techniques such as echocardiography became available, valve measurements were recorded from autopsy specimens. An examination of 160 pathologic specimens revealed the aortic valve diameter to be 23.2 ± 3.3 mm, whereas the diameter of the pulmonic valve was measured at 24.3 ± 3.0 mm [Westaby et al., 1984]. However, according to M-mode echocardiographic measurements, the aortic root diameter at the end of systole was 35 ± 4.2 mm and 33.7 ± 4.4 mm at the end of diastole [Gramiak and Shah, 1970]. The differences in these measurements reflect the fact that the autopsy measurements were not performed under physiologic pressure conditions and that intrinsic differences in the measurement techniques exist. On average, pulmonic leaflets are thinnner than aortic leaflets: 0.49 mm vs. 0.67 mm [David et al., 1994], although the leaflets of the aortic valve show variable dimensions depending on the respective leaflet. For example, the posterior leaflet tends to be thicker, have a larger surface area, and weigh more than the right or left leaflet [Silver and Roberts, 1985; Sahasakul

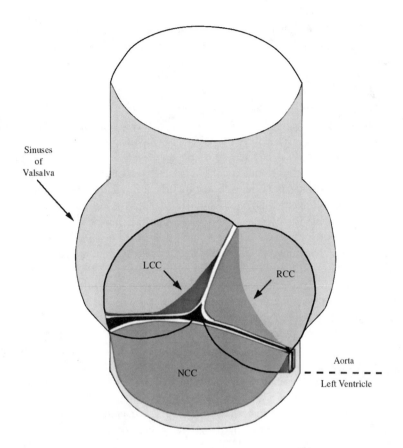

FIGURE 12.2 The aortic sinuses and valve in the closed position. The noncoronary cusp (NCC) is in front. The left and right coronary cusps (LCC and RCC) are positioned as marked. The aorta is above the closed valve in this orientation and the left ventricle is below the dashed line.

et al., 1988], and the average width of the right aortic leaflet is greater than that of the other two [Vollebergh and Becker, 1977].

Mechanical Properties

Due to the location and critical function of the aortic valve, it is difficult to obtain measurements of its mechanical properties *in vivo*; however, reports are available from a small number of animal studies. This section will reference the *in vivo* data whenever possible and defer to the *in vitro* data when necessary. Since little mathematical modeling of the aortic valve's material properties has been reported, it will be sufficient to describe the known mechanical properties of the valve. Like most biological tissues, the aortic valve is anisotropic, inhomogeneous, and viscoelastic. The collagen fibers within each valve cusp are aligned along the circumferential direction. Vesely and Noseworthy [1992] found that both the ventricularis and fibrosa were stiffer in the circumferential direction than in the radial direction. However, the ventricularis was more extensible radially than circumferentially, while the fibrosa had uniform extensibility in both directions.

There are also elastin fibers, at a lesser concentration, that are oriented orthogonal to the collagen. It is this fiber structure that accounts for the anisotropic properties of the valve. The variation in thickness and composition across the leaflets is responsible for their inhomogeneous material properties. Although the aortic valve leaflet as a whole is asymmetric in its distensibility, the basal region tends to be relatively isotropic while the central region shows the greatest degree of anisotropy [Lo and Vesely, 1995]. The role and morphology of elastin and how elastin is coupled to collagen remain points of investigation. Scott and Vesely [1996] have shown that the elastin in the ventricularis consists of continuous amorphous sheets or compact meshes, while elastin in the fibrosa consists of complex arrays of large tubes that extend circumferentially across the leaflet. These tubes may surround the large circumferential collagen bundles in the fibrosa. Mechanical testing of elastin structures from the fibrosa and ventricularis separately have shown that the purpose of elastin in the aortic valve leaflet is to maintain a specific collagen fiber configuration and return the fibers to that state during cyclic loading [Vesely, 1998]. The valve's viscoelastic properties are actually dominated by the elastic component (over the range of *in vitro* testing) so that the viscous effects, which are largely responsible for energy losses, are small [Thubrikar, 1990]. In addition to the collagen and elastin, clusters of lipids have been observed in the central spongiosa of porcine aortic valves. Vesely et al. [1994] have shown that the lipids tend to be concentrated at the base of the valve leaflets, while the coaptation regions and free edges of the leaflets tend to be devoid of these lipids. In addition, the spatial distribution of the lipids within the spongiosal layer of the aortic leaflets corresponded to areas in which calcification is commonly observed on bioprosthetic valves suggesting that these lipid clusters may be potential nucleation sites for calcification. In contrast, pulmonic leaflets showed a substantially lower incidence of lipids [Dunmore-Buyze et al., 1995]. The aortic valve leaflets have also been shown to be slightly stiffer than pulmonary valve leaflets, although the extensibilities and relaxation rates of the two tissues are similar [Leeson-Dietrich et al., 1995].

Using marker fluoroscopy, in which the aortic valve leaflets were surgically tagged with radio-opaque markers and imaged with high-speed x-rays, the leaflets have been shown to change length during the cardiac cycle [Thubrikar, 1990]. The cusps are longer during diastole than systole in both the radial and circumferential direction. The variation in length is greatest in the radial direction, approximately 20%, while the strain in the circumferential direction is about 10% of the normal systolic length [Lo and Vesely, 1995]. The difference in strain is due to the presence of the compliant elastin fibers aligned in this radial direction. The length change in both directions results in an increased valve surface area during diastole. During systole, the shortening of the valve leaflets helps to reduce obstruction of the aorta during the systolic ejection of blood. It should be noted that this change in area is by no means an active mechanism within the aortic valve; the valve simply reacts to the stresses it encounters in a passive manner.

In addition to this change in surface area, the aortic valve leaflets also undergo bending in the circumferential direction during the cardiac cycle. In diastole when the valve is closed, each leaflet is convex toward the ventricular side. During systole when the valve is open, the curvature changes and each leaflet is concave toward the ventricle. This bending is localized on the valve cusp near the wall of the aorta. This location is often thicker than the rest of the leaflet. The total diastolic stress in a valve leaflet has been estimated at 2.5×10^6 dynes/cm^2 for a strain of 15% [Thubrikar, 1990] The stress in the circumferential direction was found to be the primary load bearing element in the aortic valve. Due to the collagen fibers oriented circumferentially, the valve is relatively stiff in this direction. The strain that does occur circumferentially is primarily due to scissoring of the fibrous matrix and straightening of collagen fibers that are kinked or crimped in the presence of no external stress. However, in the radial direction, because elastin is the primary element, the valve can undergo a great deal of strain, ranging from 20 to 60% in tissue specimens [Christie, 1990; Lo and Vesely, 1995]. In the closed position, the radial stress levels are actually small compared to those in the circumferential direction. This demonstrates the importance of the lunula in ensuring that the valve seals tightly to prevent leakage. Because of their anatomical location, the lunula cause these high circumferential stress levels by enabling the aortic pressure to pull each leaflet in the circumferential direction towards the other leaflets.

The composition, properties, and dimensions of the aortic valve change with age and in the presence of certain diseases. The valve leaflets become thicker, the lunula become fenestrated, or mesh-like, and in later stages of disease the central portion of the valve may become calcified [Davies, 1980]. This thickening of the valve typically occurs on the ventricular side of the valve, in the region where the tips of the leaflets come together. Another site of calcification and fibrosis is the point of maximum cusp flexion and is thought to be a response to fatigue in the normal valve tissue.

Valve Dynamics

The aortic valve opens during systole when the ventricle is contracting and then closes during diastole as the ventricle relaxes and fills from the atrium. Systole lasts about one third of the cardiac cycle and begins when the aortic valve opens, which typically takes only 20 to 30 msec [Bellhouse, 1969]. Blood rapidly accelerates through the valve and reaches a peak velocity after the leaflets have opened to their full extent and start to close again. Peak velocity is reached during the first third of systole and the flow begins to decelerate rapidly after the peak is reached, albeit not as fast as its initial acceleration. The adverse pressure gradient that is developed affects the low momentum fluid near the wall of the aorta more than that at the center; this causes reverse flow in the sinus region [Reul and Talukdar, 1979]. Figure 12.2 illustrates the pressure and flow relations across the aortic valve during the cardiac cycle. During systole, the pressure difference required to drive the blood through the aortic valve is on the order of a few millimeters of mercury; however, the diastolic pressure difference reaches 80 mmHg in normal individuals. The valve closes near the end of the deceleration phase of systole with very little reverse flow through the valve.

During the cardiac cycle the heart undergoes translation and rotation due to its own contraction pattern. As a result, the base of the aortic valve varies in size and also translates, mainly along the axis of the aorta. Using marker fluoroscopy to study the base of the aortic valve in dogs, Thubrikar et al. [1993] found that the base perimeter is at its largest at end diastole and decreases in size during systole; it then reaches a minimum at the end of systole and increases again during diastole. The range of this perimeter variation during the cardiac cycle was 22% for an aortic pressure variation of 120/80 mmHg. The valve annulus also undergoes translation, primarily parallel to the aortic axis. The aortic annulus moves downward toward the ventricle during systole and then recoils back toward the aorta as the ventricle fills during diastole. The annulus also experiences a slight side-to-side translation with its magnitude approximately one half the displacement along the aortic axis.

During systole, vortices develop in all three sinuses behind the leaflets of the aortic valve. The function of these vortices was first described by Leonardo da Vinci in 1513, and they have been researched extensively in this century primarily through the use of *in vitro* models [Bellhouse, 1969; Reul and Talukdar, 1979]. It has been hypothesized that the vortices help to close the aortic valve so that blood is prevented from returning to the ventricle during the closing process. These vortices create a transverse pressure difference that pushes the leaflets toward the center of the aorta and each other at the end of systole, thus minimizing any possible closing volume. However, as shown *in vitro* by Reul and Talukdar [1979], the axial pressure difference alone is enough to close the valve. Without the vortices in the sinuses, the valve still closes but its closure is not as quick as when the vortices are present. The adverse axial pressure difference within the aorta causes the low inertia flow within the developing boundary layer along the aortic wall to decelerate first and to reverse direction. This action forces the belly of the leaflets away from the aortic wall and toward the closed position. When this force is coupled with the vortices that push the leaflet tips toward the closed position, a very efficient and fast closure is obtained. Closing volumes have been estimated to be less than 5% of the forward flow [Bellhouse and Bellhouse, 1969].

The parameters that describe the normal blood flow through the aortic valve are the velocity profile, time course of the blood velocity or flow, and magnitude of the peak velocity. These are determined in part by the pressure difference between the ventricle and aorta and by the geometry of the aortic valve complex. As seen in Fig. 12.3, the velocity profile at the level of the aortic valve annulus is relatively flat. However there is usually a slight skew toward the septal wall (less than 10% of the center-line velocity) which is caused by the orientation of the aortic valve relative to the long axis of the left ventricle. This skew in the velocity profile has been shown by many experimental techniques, including hot film anemometry, Doppler ultrasound, and MRI [Paulsen and Hasenkam, 1983; Rossvol et al., 1991; Kilner et al., 1993]. In healthy individuals, blood flows through the aortic valve at the beginning of systole and then rapidly accelerates to its peak value of 1.35 ± 0.35 m/s; for children this value is slightly higher at 1.5 ± 0.3 m/s [Hatle and Angelson, 1985]. At the end of systole there is a very short period of reverse flow that can be measured with Doppler ultrasound. This reverse flow is probably either a small closing

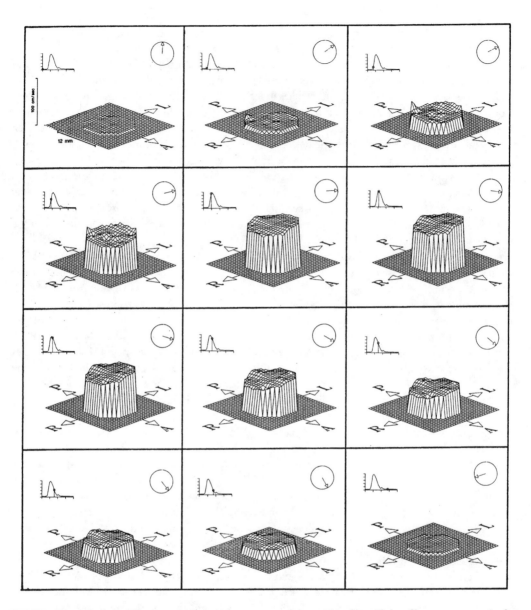

FIGURE 12.3   Velocity profiles measured 2 cm downstream of the aortic valve with hot film anemometry in dogs [Paulsen and Hasenkam, 1983]. The timing of the measurements during the cardiac cycle is shown by the marker on the aortic flow curve.

volume or the velocity of the valve leaflets as they move toward their closed position. The flow patterns just downstream of the aortic valve are of particular interest because of their complexity and relation to arterial disease. Highly skewed velocity profiles and corresponding helical flow patterns have been observed in the human aortic arch using magnetic resonance phase velocity mapping [Kilner et al., 1993].

The pulmonic valve flow behaves similarly to that of the aortic valve, but the magnitude of the velocity is not as great. Typical peak velocities for healthy adults are 0.75 ± 0.15 m/s; again, these values are slightly higher for children at 0.9 ± 0.2 m/s [Weyman, 1994]. As seen in Fig. 12.4, a rotation of the peak velocity can be observed in the pulmonary artery velocity profile. During acceleration, the peak velocity is observed inferiorly with the peak rotating counterclockwise throughout the remainder of the ejection phase [Sloth et al., 1994]. The mean spatial profile is relatively flat, however, although there is a region

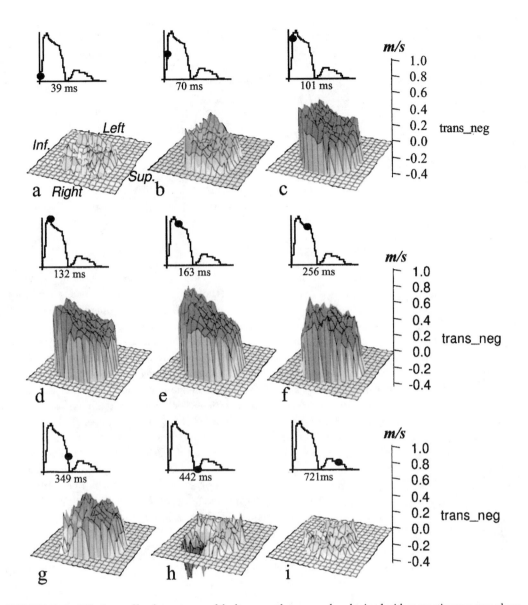

FIGURE 12.4 Velocity profiles downstream of the human pulmonary valve obtained with magnetic resonance phase velocity mapping [Sloth, 1994]. Again, the timing of the measurements is shown by the marker on the flow curve.

of reverse flow that occurs in late systole which may be representative of flow separation. Typically, there is only a slight skew to the profile. The peak velocity is generally within 20% of the spatial mean throughout the cardiac cycle. Secondary flow patterns can also be observed in the pulmonary artery and its bifurcation. *In vitro* laser Doppler anemometry experiments have shown that these flow patterns are dependent on the valve geometry and thus can be used to evaluate function and fitness of the heart valve [Sung and Yoganathan, 1990].

12.2   Mitral and Tricuspid Valves

The mitral (Fig. 12.5) and tricuspid valves are similar in structure with both valves composed of four primary elements: (1) the valve annulus, (2) the valve leaflets, (3) the papillary muscles, and (4) the chordae tendineae. The base of the mitral leaflets form the mitral annulus, which attaches to the atrial

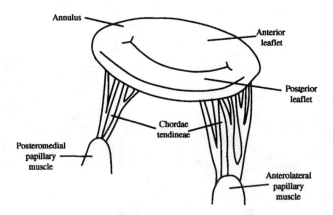

FIGURE 12.5   Schematic of the mitral valve showing the valve leaflets, papillary muscles, and chordae tendineae.

and ventricular walls, and aortic root. At the free edge of the leaflets, the chordae tendinae insert at multiple locations and extend to the tips of the papillary muscles. This arrangement provides continuity between the valve and ventricular wall to enhance valvular function. The valvular apparatus, or complex, requires an intricate interplay between all components throughout the cardiac cycle.

The mitral annulus is an elliptical ring composed of dense collagenous tissue surrounded by muscle. It goes through dynamic changes during the cardiac cycle by not only changing in size, but also by moving three-dimensionally. The circumference of the mitral annulus ranges from 8 to 12 cm during diastole. Recent studies involving the measurement of annular shape have also shown that the mitral annulus is not planar, but instead has a three-dimensional form. The annulus actually forms a saddle, or ski-slope shape [Glasson et al., 1996; Pai et al., 1995; Komoda et al., 1994; Levine et al., 1987]. This three-dimensional shape must be taken into account when non-invasively evaluating valvular function.

The mitral valve is a bileaflet valve comprised of an anterior and posterior leaflet. The leaflet tissue is primarily collagen-reinforced endothelium, but also contains striated muscle cells, non-myelinated nerve fibers, and blood vessels. The anterior and posterior leaflets of the valve are actually one continuous piece of tissue, as shown in Fig. 12.6. The free edge of this tissue shows several indentations of which two are regularly placed, called the commisures. The commisures separate the tissue into the anterior and posterior leaflets. The location of the commisures can be identified by the fan-like distribution of chordae tendinae and the relative positioning of the papillary muscles. The combined surface area of both leaflets is approximately twice the area of the mitral orifice; this extra surface area permits a large

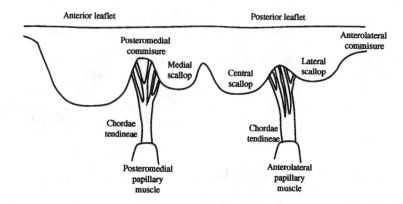

FIGURE 12.6   Diagram of the mitral valve as a continuous piece of tissue. The posterior and anterior leaflets are indicated, as are the scallops, chordae tendineae, and papillary muscles.

line of coaptation and ample coverage of the mitral orifice during normal function and provides compensation in cases of disease [He, 1997, 1999]. The posterior leaflet encircles roughly two-thirds of the mitral annulus and is essentially an extension of the mural endocardium from the free walls of the left atrium. The anterior leaflet portion of the annulus is a line of connection for the leaflet, the wall of the ascending aorta, the aortic valve, and the atrial septum. The anterior leaflet is slightly larger than the posterior leaflet, and is roughly semi-lunar in shape as opposed to the quadrangular-shaped posterior leaflet. The normal width and height of the anterior leaflet are approximately 3.3 cm and 2.3 cm, respectively. The height of the posterior leaflet is 1.3 cm, while the commisure height is less than 1.0 cm. The posterior leaflet typically has indentations, called scallops, that divide the leaflet into three regions: (1) the medial, (2) central, and (3) lateral scallop [Kunzelman et al., 1994; Barlow, 1987; Roberts, 1983; Raganathan et al., 1970; Silverman and Hurst, 1968].

The mitral leaflet tissue can be divided into both a rough and clear zone. The rough zone is the thicker part of the leaflet and is defined from the free edge of the valve to the valve's line of closure. The term "rough" is used to denote the texture of the leaflet due to the insertion of the chordae tendineae in this area. The clear zone is thinner and translucent and extends from the line of closure to the annulus in the anterior leaflet and to the basal zone in the posterior leaflet. Unlike the mitral valve, the tricuspid valve has three leaflets: (1) an anterior leaflet, (2) a posterior leaflet with a variable number of scallops, and (3) a septal leaflet. The tricuspid valve is larger and structurally more complicated than the mitral valve and the separation of the valve tissue into distinct leaflets is less pronounced than with the mitral valve. The surface of the leaflets is similar to that of the mitral valve; however, the basal zone is present in all of the leaflets [Silver et al., 1971].

Chordae tendineae from both leaflets attach to each of the papillary muscles. The chordae tendineae consist of an inner core of collagen surrounded by loosely meshed elastin and collagen fibers with an outer layer of endothelial cells. In the mitral complex structure, there are marginal and basal chordae that insert into the mitral leaflets. From each papillary muscle, several chordae originate and branch into the marginal and basal chordae. The thinner marginal chordae insert into the leaflet free edge at multiple insertion points, while the thicker basal chordae insert into the leaflets at a higher level towards the annulus. The marginal chordae function to keep the leaflets stationary while the basal chordae seem to act more as supports [Kunzelman, 1994].

The left side of the heart has two papillary muscles, called anterolateral and posteromedial, that attach to the ventricular free wall and tether the mitral valve in place via the chordae tendinae. This tethering prevents the mitral valve from prolapsing into the atrium during ventricular ejection. On the right side of the heart, the tricuspid valve has three papillary muscles. The largest one, the anterior papillary muscle, attaches to the valve at the commissure between the anterior and posterior leaflets. The posterior papillary muscle is located between the posterior and septal leaflets. The smallest papillary muscle, called the septal muscle, is sometimes not even present. Improper tethering of the leaflets will result in valve prolapse during ventricular contraction, permitting the valve leaflets to extend into the atrium. This incomplete apposition of the valve leaflets can cause regurgitation, which is leaking of the blood being ejected back into the atrium.

Mechanical Properties

Studies on the mechanical behavior of the mitral leaflet tissue have been conducted to determine the key connective tissue components which influence the valve function. Histological studies have shown that the tissue is composed of three layers which can be identified by differences in cellularity and collagen density. Analysis of the leaflets under tension indicated that the anterior leaflet would be more capable of supporting larger tensile loads than the posterior leaflet. The differences between the mechanical properties between the two leaflets may require different material selection for repair or replacement of the individual leaflets [Kunzelman et al., 1993a, b].

Studies have also been done on the strength of the chordae tendinae. The tension of chordae tendineae in dogs was monitored throughout the cardiac cycle by Salisbury and co-workers [1963]. They found that the tension only paralleled the left ventricular pressure tracings during isovolumic contraction, indicating slackness at other times in the cycle. Investigation of the tensile properties of the chordae tendineae at different strain rates by Lim and Bouchner [1975] found that the chordae had a non-linear stress–strain relationship. They found that the size of the chordae had a more significant effect on the development of the tension than did the strain rate. The smaller chordae with a cross-sectional area of 0.001 to 0.006 cm^2 had a modulus of 2×10^9 dynes/cm^2, while larger chordae with a cross-sectional area of 0.006 to 0.03 cm^2 had a modulus of 1×10^9 dynes/cm^2.

A theoretical study of the stresses sustained by the mitral valve was performed by Ghista and Rao [1972]. This study determined that the stress level can reach as high as 2.2×10^6 dynes/cm^2 just prior to the opening of the aortic valve, with the left ventricular pressure rising to 150 mmHg. A mathematical model has also been created for the mechanics of the mitral valve. It incorporates the relationship between chordae tendineae tension, left ventricular pressure, and mitral valve geometry [Arts et al., 1983]. This study examined the force balance on a closed valve, and determined that the chordae tendinae force was always more than half the force exerted on the mitral valve orifice by the transmitral pressure gradient. During the past 10 years, computational models of mitral valve mechanics have been developed, with the most advanced modeling being three-dimensional finite element models (FEMs) of the complete mitral apparatus. Kunzelman and co-workers [1993, 1998] have developed a model of the mitral complex that includes the mitral leaflets, chordae tendinae, contracting annulus, and contracting papillary muscles. From these studies, the maximum principal stresses found at peak loading (120 mmHg) were 5.7 $\times 10^6$ dynes/cm^2 in the annular region, while the stresses in the anterior leaflet ranged from 2×10^6 to 4×10^6 dynes/cm^2. This model has also been used to evaluate mitral valve disease, repair in chordal rupture, and valvular annuloplasty.

Valve Dynamics

The valve leaflets, chordae tendineae, and papillary muscles all participate to ensure normal functioning of the mitral valve. During isovolumic relaxation, the pressure in the left atrium exceeds that of the left ventricle, and the mitral valve cusps open. Blood flows through the open valve from the left atrium to the left ventricle during diastole. The velocity profiles at both the annulus and the mitral valve tips have been shown to be skewed [Kim et al., 1994] and therefore are not flat as is commonly assumed. This skewing of the inflow profile is shown in Fig. 12.7. The initial filling is enhanced by the active relaxation of the ventricle, maintaining a positive transmitral pressure. The mitral velocity flow curve shows a peak in the flow curve, called the E-wave, which occurs during the early filling phase. Normal peak E-wave velocities in healthy individuals range from 50 to 80 cm/s [Samstad et al., 1989; Oh et al., 1997]. Following active ventricular relaxation, the fluid begins to decelerate and the mitral valve undergoes partial closure. Then the atrium contracts and the blood accelerates through the valve again to a secondary peak, termed the A-wave. The atrium contraction plays an important role in additional filling of the ventricle during late diastole. In healthy individuals, velocities during the A-wave are typically lower than those of the E-wave, with a normal E/A velocity ratio ranging from 1.5 to 1.7 [Oh et al., 1997]. Thus, normal diastolic filling of the left ventricle shows two distinct peaks in the flow curve with no flow leaking back through the valve during systole.

The tricuspid flow profile is similar to that of the mitral valve, although the velocities in the tricuspid valve are lower because it has a larger valve orifice. In addition, the timing of the valve opening is slightly different. Since the peak pressure in the right ventricle is less than that of the left ventricle, the time for right ventricular pressure to fall below the right atrial pressure is less than the corresponding time period for the left side of the heart. This leads to a shorter right ventricular isovolumic relaxation and thus an earlier tricuspid opening. Tricuspid closure occurs after the mitral valve closes since the activation of the left ventricle precedes that of the right ventricle [Weyman, 1994].

A primary focus in explaining the fluid mechanics of mitral valve function has been understanding the closing mechanism of the valve. Bellhouse [1972] first suggested that the vortices generated by ventricular filling were important for the partial closure of the mitral valve following early diastole. Their

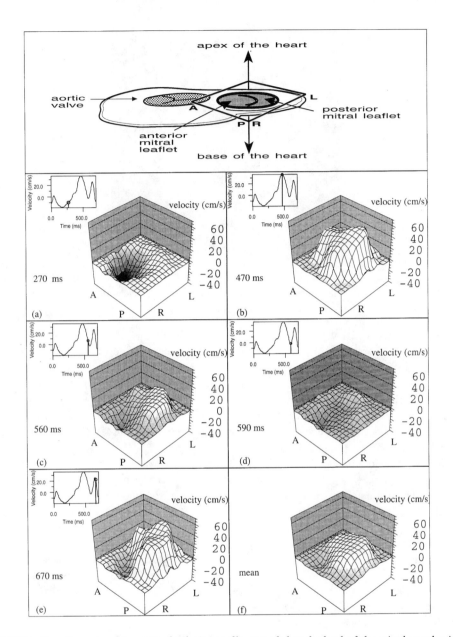

FIGURE 12.7 Two-dimensional transmitral velocity profiles recorded at the level of the mitral annulus in a pig [Kim et al., 1994]: (a) systole; (b) peak E-wave; (c) deceleration phase of early diastole; (d) mid-diastolic period (diastasis); (e) peak A-wave; (f) time averaged diastolic cross-sectional mitral velocity profile. (Reprinted with permission from the American College of Cardiology, *J. Am. Coll. Cardiol.* 24:532–545.)

in vitro experiments suggested that without the strong outflow tract vortices, the valve would remain open at the onset of ventricular contraction, thus resulting in a significant amount of mitral regurgitation before complete closure. Later *in vitro* experiments by Reul and Talukdar [1981] in a left ventricle model made from silicone suggested that an adverse pressure differential in mid-diastole could explain both the flow deceleration and the partial valve closure, even in the absence of a ventricular vortex. Thus, the studies by Reul and Talukdar suggest that the vortices may provide additional closing effects at the initial stage; however, the pressure forces are the dominant effect in valve closure. A more unified theory of valve closure put forth by Yellin and co-workers [1981] includes the importance of chordal tension, flow

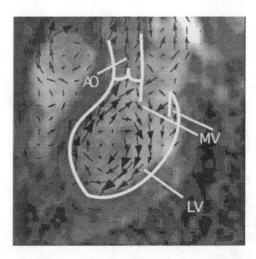

FIGURE 12.8 Magnetic resonance image of a healthy individual during diastole. An outline of the interior left ventricle (LV) is indicated in white as are the mitral valve leaflets (MV) and the aorta (AO). Velocity vectors were obtained from MRI phase velocity mapping and superimposed on the anatomical image.

deceleration, and ventricular vortices, with chordal tension being a necessary condition for the other two. Their animal studies indicated that competent valve closure can occur even in the absence of vortices and flow deceleration. Recent studies using magnetic resonance imaging to visualize the three-dimensional flow field in the left ventricle showed that in normal individuals a large anterior vortex is present at initial partial closure of the valve, as well as following atrial contraction [Kim et al., 1995]. Studies conducted in our laboratory using magnetic resonance imaging of healthy individuals clearly show the vortices in the left ventricle [Walker et al., 1996], which may be an indication of normal diastolic function. An example of these vortices is presented in Fig. 12.8.

The rarea of interest has been the motion of the mitral valve complex. The heart moves throughout the cardiac cycle; similarly, the mitral apparatus moves and changes shape. Recent studies have been conducted which examined the three-dimensional dynamics of the mitral annulus during the cardiac cycle. [Pai et al., 1995; Glasson et al., 1996; Komoda et al., 1994; Ormiston et al., 1981]. These studies have shown that during systole the annular circumference decreases from the diastolic value due to the contraction of the ventricle, and this reduction in area ranges from 10 to 25%. This result agrees with an animal study of Tsakiris and co-workers [1971] that looked at the difference in the size, shape, and position of the mitral annulus at different stages in the cardiac cycle. They noted an eccentric narrowing of the annulus during both atrial and ventricular contractions that reduced the mitral valve area by 10 to 36% from its peak diastolic area. This reduction in the annular area during systole is significant, resulting in a smaller orifice area for the larger leaflet area to cover. Not only does the annulus change size, but it also translates during the cardiac cycle. The movement of the annulus towards the atrium has been suggested to play a role in ventricular filling, possibly increasing the efficiency of blood flow into the ventricle. During ventricular contraction, there is a shortening of the left ventricular chamber along its longitudinal axis, and the mitral and tricuspid annuli move toward the apex [Pai et al., 1995; Alam and Rosenhamer, 1992; Hammarström et al., 1991; Simonson and Schiller, 1989].

The movement of the papillary muscles is also important in maintaining proper mitral valve function. The papillary muscles play an important role in keeping the mitral valve in position during ventricular contraction. Abnormal strain on the papillary muscles could cause the chordae to rupture, resulting in mitral regurgitation. It is necessary for the papillary muscles to contract and shorten during systole to prevent mitral prolapse; therefore, the distance between the apex of the heart to the mitral apparatus is important. The distance from the papillary muscle tips to the annulus was measured in normal individuals during systole and was found to remain constant [Sanfilippo et al., 1992]. In patients with mitral valve prolapse, however, this distance decreased, corresponding to a superior displacement of the papillary muscle towards the annulus.

The normal function of the mitral valve requires a balanced interplay between all of the components of the mitral apparatus, as well as the interaction of the atrium and ventricle. Engineering studies into mitral valve function have provided some insight into its mechanical properties and function. Further fundamental and detailed studies are needed to aid surgeons in repairing the diseased mitral valve and in understanding the changes in function due to mitral valve pathologies. In addition, these studies are crucial for improving the design of prosthetic valves that more closely replicate native valve function.

References

Alam, M. and Rosenhamer, G. 1992. Atrioventricular plane displacement and left ventricular function. *J. Am. Soc. Echocardiogr.* 5:427–433.

Arts, T., Meerbaum, S., Reneman, R., and Corday, E. 1983. Stresses in the closed mitral valve: a model study. *J. Biomechanics* 16:539–547.

Barlow, J.B. 1987. *Perspectives on the Mitral Valve.* F.A. Davis Company, Philadelphia, PA.

Bellhouse, B.J. 1969. Velocity and pressure distributions in the aortic valve. *J. Fluid. Mech.* 37:587–600.

Bellhouse, B.J. 1972. The fluid mechanics of a model mitral valve and left ventricle. *Cardiovasc. Res.* 6:199–210.

Bellhouse, B.J. and Bellhouse, F. 1969. Fluid mechanics of model normal and stenosed aortic valves. *Circ. Res.* 25:693–704.

Christie, G.W. 1990. Anatomy of aortic heart valve leaflets: the influence of glutaraldehyde fixation on function. *Eur. J. Cardio-Thorac. Surg.* 6(suppl. 1):S25–S33.

David, H., Boughner, D.R., Vesely, I., and Gerosa, G. 1994. The pulmonary valve: is it mechanically suitable for use as an aortic valve replacement? *ASAIO J.* 40:206–212.

Davies, M.J. 1980. *Pathology of Cardiac Valves.* Butterworths, London.

Dunmore-Buyze, J., Boughner, D.R., Macris, N., and Vesely, I. 1995. A comparison of macroscopic lipid content within porcine pulmonary and aortic valves: implications for bioprosthetic valves. *J. Thorac. Cardiovasc. Surg.* 110:1756–1761.

Emery, R.W. and Arom, K.V. 1991. *The Aortic Valve.* Hanley & Belfus, Philadelphia, PA.

Ghista, D.N. and Rao, A.P. 1972. Structural mechanics of the mitral valve: stresses sustained by the valve; non-traumatic determination of the stiffness of the *in vivo* valve. *J. Biomechanics* 5:295–307.

Glasson, J.R., Komeda, M., Daughters, G.T., Niczyporuk, M.A., Bolger, A.F., Ingels, N.B., and Miller, D.C. 1996. Three-dimensional regional dynamics of the normal mitral annulus during left ventricular ejection. *J. Thorac. Cardiovasc. Surg.* 111:574–585.

Gramiak, R. and Shah, P.M. 1970. Echocardiography of the normal and diseased aortic valve. *Radiology* 96:1.

Hammarström, E., Wranne, B., Pinto, F.J., Puryear, J., and Popp, R.L. 1991. Tricuspid annular motion. *J. Am. Soc. Echocardiogr.* 4:131–139.

Hatle, L. and Angelsen, B. 1985. *Doppler Ultrasound in Cardiology Physical Principals and Clinical Applications.* Lea & Febiger, Philadelphia, PA.

He, S., Fontaine, A.A., Schwammenthal, E., Yoganathan, A.P., and Levine, R.A. 1997. Integrated mechanism for functional mitral regurgitation: leaflet elongation versus coapting force: *in vitro* study. *Circulation* 96(6):1826–1834.

He, S., Lemmon, J.D., Weston, M.W., Jensen, M.O., Levine, R.A., and Yoganathan, A.P. 1999. Mitral valve compensation for annular dilatation: *in vitro* study into the mechanisms of functional mitral regurgitation with an adjustable annulus model. *J. Heart Valve Dis.* 8:294–302.

Kilner, P.J., Yang, G.Z., Mohiaddin, R.H., Firmin, D.N., and Longmore, D.B. 1993. Helical and retrograde secondary flow patterns in the aortic arch studied by three-directional magnetic resonance velocity mapping. *Circulation* 88(part I):2235–2247.

Kim, W.Y., Bisgaard, T., Nielsen, S.L., Poulsen, J.K., Pederson, E.M., Hasenkam, J.M., and Yoganathan, A.P. 1994. Two-dimensional mitral flow velocity profiles in pig models using epicardial Doppler echocardiography. *J. Am. Coll. Cardiol.* 24(2):532–545.

Kim, W.Y., Walker, P.G., Pederson, E.M., Poulsen, J.K., Oyre, S. Houlind, K.C., and Yoganathan, A.P. 1995. Left ventricular blood flow patterns in normal subjects: a quantitative analysis by three-dimensional magnetic resonance velocity mapping. *J. Am. Coll. Cardiol.* 26(1):224–230.

Komoda, T., Hetzer, R., Uyama, C., Siniawski, H., Maeta, H., Rosendahl, P., and Ozaki, K. 1994. Mitral ammular function assessed by 3D imaging for mitral valve surgery. *J. Heart Valve Dis.* 3:483–490.

Kunzelman, K.S., Cochran, R.P., Murphee, S.S., Ring, W.S., Verrier, E.D., and Eberhart, R.C. 1993a. Differential collagen distribution in the mitral valve and its influence on biomechanical behaviour. *J. Heart Valve Dis.* 2:236–244.

Kunzelman, K.S., Cochran, R.P., Chuong, C., Ring, W.S., Verner, E.D., and Eberhart, R.D. 1993b. Finite element analysis of the mitral valve. *J. Heart Valve Dis.* 2:326–340.

Kunzelman, K.S., Cochran, R.P., Verner, E.D., and Eberhart, R.D. 1994. Anatomic basis for mitral valve modeling. *J. Heart Valve Dis.* 3:491–496.

Kunzelman, K.S., Reimink, M.S., and Cochran, R.P. 1998. Flexible versus rigid ring annuloplasty for mitral valve annular dilatation: a finite element model. *J. Heart Valve Dis.* 7:108–116.

Leeson-Dietrich, J., Boughner, D., and Vesely, I. 1995. Porcine pulmonary and aortic valves: a comparison of their tensile viscoelastic properties at physiological strain rates. *J. Heart Valve Dis.* 4:88–94.

Levine, R.A., Trivizi, M.O., Harrigan, P., and Weyman, A.E. 1987. The relationship of mitral annular shape to the diagnosis of mitral valve prolapse. *Circulation* 75:756–767.

Lim, K.O. and Bouchner, D.P. 1975. Mechanical properties of human mitral valve chordae tendineae: variation with size and strain rate. *Can. J. Physiol. Pharmacol.* 53:330–339.

Lo, D. and Vesely, I. 1995. Biaxial strain analysis of the porcine aortic valve. *Ann. Thorac. Surg.* 60(suppl. II):S374–378.

Oh, J.K., Appleton, C.P., Hatle, L.K., Nishimura, R.A., Seward, J.B., and Tajik, A.J. 1997. The noninvasive assessment of left ventricular diastolic function with two-dimensional and Doppler echocardiography. *J. Am. Soc. Echocardiol.* 10:246–270.

Ormiston, J.A., Shah, P.M., Tei, C., and Wong, M. 1981. Size and motion of the mitral valve annulus in man: a two-dimensional echocardiographic method and findings in normal subjects. *Circulation* 64:113–120.

Pai, R.G., Tanimoto, M., Jintapakorn, W., Azevedo, J., Pandian, N.G., and Shah, P.M. 1995. Volume-rendered three-dimensional dynamic anatomy of the mitral annulus using a transesophageal echocardiographic technique. *J. Heart Valve Dis.* 4:623–627.

Paulsen, P.K. and Hasenkam, J.M. 1983. Three-dimensional visualization of velocity profiles in the ascending aorta in dogs, measured with a hot film anemometer. *J. Biomechanics* 16:201–210.

Raganathan, N., Lam, J.H.C., Wigle, E.D., and Silver, M.D. 1970. Morphology of the human mitral valve: the valve leaflets. *Circulation* 41:459–467.

Reul, H. and Talukdar, N. 1979. Heart valve mechanics. In *Quantitative Cardiovascular Studies: Clinical and Research Applications of Engineering Principles.* Hwang, N.H.C., Gross, D.R. and Patel, D.J., Eds. University Park Press, Baltimore, MD, 527–564.

Roberts, W.C. 1983. Morphologic features of the normal and abnormal mitral valve. *Am. J. Cardiol.* 51:1005–1028.

Rossvoll, O., Samstad, S., Torp, H.G., Linker, D.T., Skjærpe, T., Angelsen, B.A.J., and Hatle, L. 1991. The velocity distribution in the aortic annulus in normal subjects: a quantitative analysis of two-dimensional Doppler flow maps. *J. Am. Soc. Echocardiogr.* 4:367–378.

Sahasakul, Y., Edwards, W.D., Naessens, J.M., and Tajik, A.J. 1988. Age-related changes in aortic and mitral valve thickness: implications for two-dimensional echocardiography based on an autopsy study of 200 normal human hearts. *Am. J. Cardiol.* 62:424–430.

Salisbury, P.F., Cross, C.E., and Rieben, P.A. 1963. Chordae tendinea tension. *Am. J. Physiol.* 25:385–392.

Samstad, O., Gorp, H.G., Linker, D.T., Rossvoll, O., Skjaerpe, T., Johansen, E., Kristoffersen, K., Angelson, B.A.J., and Hatle, L. 1989. Cross-sectional early mitral flow velocity profiles from colour Doppler. *Br. Heart J.* 62:177–184.

Sanfilippo, A.J., Harrigan, P., Popovic, A.D., Weyman, A.E., and Levine, R.A. 1992. Papillary muscle traction in mitral valve prolapse: quantitation by two-dimensional echocardiography. *J. Am. Coll. Cardiol.* 19:564–571.

Scott, M.J. and Vesely, I. 1996. Morphology of porcine aortic valve cusp elastin. *J. Heart Valve Dis.* 5:464–471.

Silver, M.A. and Roberts, W.C. 1985. Detailed anatomy of the normally functioning aortic valve in hearts of normal and increased weight. *Am. J. Cardiol.* 55:454–461.

Silver, M.D., Lam, J.H.C., Raganathan, N., and Wigle, E.D. 1971. Morphology of the human tricuspid valve. *Circulation* 43:333–348.

Silverman, M.E. and Hurst, J.W. 1968. The mitral complex: Interaction of the anatomy, physiology, and pathology of the mitral annulus, mitral valve leaflets, chordae tendineae and papillary muscles. *Am. Heart J.* 76:399–418.

Simonson, J.S. and Schiller, N.B. 1989. Descent of the base of the left ventricle: an echocardiographic index of left ventricular function. *J. Am. Soc. Echocardiogr.* 2:25–35.

Sloth, E., Houlind, K.C., Oyre, S., Kim, Y.K., Pedersen, E.M., Jørgensen, H.S., and Hasenkam, J.M. 1994. Three-dimensional visualiztion of velocity profiles in the human main pulmonary artery using magnetic resonance phase velocity mapping. *Am. Heart J.* 128:1130–1138.

Sung, H.W. and Yoganathan, A.P. 1990. Axial flow velocity patterns in a normal human pulmonary artery model: pulsatile *in vitro* studies. *J. Biomechanics* 23(3):210–214.

Thubrikar, M. 1990. *The Aortic Valve.* CRC Press, Boca Raton, FL.

Thubrikar, M., Heckman, J.L., and Nolan, S.P. 1993. High speed cine-radiographic study of aortic valve leaflet motion. *J. Heart Valve Dis.* 2:653–661.

Tsakiris, A.G., von Bernuth, G., Rastelli, G.C., Bourgeois, M.J., Titus, J.L., and Wood, E.H. 1971. Size and motion of the mitral valve annulus in anesthetized intact dogs. *J. Appl. Physiol.* 30:611–618.

Vesely, I. 1998. The role of elastin in aortic valve mechanics. *J. Biomech.* 31:115–123.

Vesely, I. and Noseworthy, R. 1992. Micromechanics of the fibrosa and the ventricularis in aortic valve leaflets. *J. Biomech.* 25:101–113.

Vesely, I., Macris, N., Dunmore, P.J., and Boughner, D. 1994. The distribution and morphology of aortic valve cusp lipids. *J. Heart Valve Dis.* 3:451–456.

Vollebergh, F.E.M.G. and Becker, A.E. 1977. Minor congenital variations of cusp size in tricuspid aortic valves: possible link with isolated aortic stenosis. *Br. Heart J.* 39:106–111.

Walker, P.G., Cranney, G.B., Grimes, R.Y., Delatore, J., Rectenwald, J., Pohost, G.M., and Yoganathan, A.P. 1996. Three-dimensional reconstruction of the flow in a human left heart by magnetic resonance phase velocity encoding. *Ann. Biomed. Eng.* 24:139–147.

Westaby, S., Karp, R.B., Blackstone, E.H., and Bishop, S.P. 1984. Adult human valve dimensions and their surgical significance. *Am. J. Cardiol.* 53:552–556.

Weyman, A.E. 1994. *Principles and Practices of Echocardiography.* Lea & Febiger, Philadelphia, PA.

Yellin, E.L., Peskin, C., Yoran, C., Koenigsberg, M., Matsumoto, M., Laniado, S., McQueen, D., Shore, D., and Frater, R.W.M. 1981. Mechanisms of mitral valve motion during diastole. *Am. J. Physiol.* 241:H389–H400.

13

Arterial Macrocirculatory Hemodynamics

Baruch B. Lieber
State University of New York at Buffalo

The arterial circulation is a multiply branched network of compliant tubes. The geometry of the network is complex, and the vessels exhibit nonlinear *viscoelastic* behavior. Flow is pulsatile, and the blood flowing through the network is a suspension of red blood cells and other particles in plasma which exhibits complex *non-Newtonian* properties. Whereas the development of an exact biomechanical description of arterial hemodynamics is a formidable task, surprisingly useful results can be obtained with greatly simplified models.

The geometrical parameters of the canine *systemic* and *pulmonary* circulations are summarized in Table 13.1. Vessel diameters vary from a maximum of 19 mm in the proximal aorta to 0.008 mm (8 microns) in the capillaries. Because of the multiple branching, the total cross-sectional area increases from 2.8 cm² in the proximal aorta to 1357 cm² in the capillaries. Of the total blood volume, approximately 83% is in the systemic circulation, 12% is in the pulmonary circulation, and the remaining 5% is in the heart. Most of the systemic blood is in the venous circulation, where changes in compliance are used to control mean circulatory blood pressure. This chapter will be concerned with flow in the larger arteries, classes 1–5 in the systemic circulation and 1–3 in the pulmonary circulation in Table 13.1. Flow in the microcirculation is discussed in Chapter 14, and venous hemodynamics is covered in Chapter 16.

13.1 Blood Vessel Walls

The detailed properties of blood vessels were described earlier in this section, but a few general observations are made here to facilitate the following discussion. Blood vessels are composed of three layers, the intima, media, and adventitia. The inner layer, or intima, is composed primarily of *endothelial* cells, which line the vessel and are involved in control of vessel diameter. The media, composed of *elastin*, *collagen*, and smooth muscle, largely determines the elastic properties of the vessel. The outer layer, or adventitia, is composed mainly of connective tissue. Unlike in structures composed of passive elastic materials, vessel diameter and elastic modulus vary with smooth-muscle tone. Dilation in response to increases in flow and *myogenic* constriction in response to increases in pressure have been observed in

TABLE 13.1   Model of Vascular Dimensions in 20-kg Dog

Class	Vessels	Mean Diam. (mm)	Number of Vessels	Mean Length (mm)	Total Cross-Section (cm^2)	Total Blood Volume (ml)	Percentage of Total Volume (%)
			Systemic				
1	Aorta	(19–4.5)	1		(2.8–0.2)	60	
2	Arteries	4.000	40	150.0	5.0	75	
3	Arteries	1.300	500	45.0	6.6	30	
4	Arteries	0.450	6000	13.5	9.5	13	11
5	Arteries	0.150	110,000	4.0	19.4	8	
6	Arterioles	0.050	2.8×10^6	1.2	55.0	7	
7	Capillaries	0.008	2.7×10^9	0.65	1357.0	88	5
8	Venules	0.100	1.0×10^7	1.6	785.4	126	
9	Veins	0.280	660,000	4.8	406.4	196	
10	Veins	0.700	40,000	13.5	154.0	208	
11	Veins	1.800	2,100	45.0	53.4	240	
12	Veins	4.500	110	150.0	17.5	263	67
13	Venae cavae	(5–14)	2		(0.2–1.5)	92	
Total						1406	
			Pulmonary				
1	Main artery	1.600	1	28.0	2.0	6	
2	Arteries	4.000	20	10.0	2.5	25	3
3	Arteries	1.000	1550	14.0	12.2	17	
4	Arterioles	0.100	1.5×10^6	0.7	120.0	8	
5	Capillaries	0.008	2.7×10^9	0.5	1357.0	68	4
6	Venules	0.110	2.0×10^6	0.7	190.0	13	
7	Veins	1.100	1650	14.0	15.7	22	
8	Veins	4.200	25	100.0		35	5
9	Main veins	8.000	4	30.0		6	
Total						200	
			Heart				
	Atria		2			30	
	Ventricles		2			54	5
Total						84	
Total circulation						1690	100

Source: Milnor, W.R. 1989. *Hemodynamics*, 2nd ed., p. 45. Baltimore, Williams & Wilkins. With permission.

some arteries. Smooth-muscle tone is also affected by circulating *vasoconstrictors* such as norepinephrine and *vasodilators* such as nitroprusside. Blood vessels, like other soft biological tissues, generally do not obey Hooke's law, becoming stiffer as pressure is increased. They also exhibit viscoelastic characteristics such as hysteresis and creep. Fortunately, for many purposes a linear elastic model of blood vessel behavior provides adequate results.

13.2   Flow Characteristics

Blood is a complex substance containing water, inorganic ions, proteins, and cells. Approximately 50% is plasma, a nearly Newtonian fluid consisting of water, ions, and proteins. The balance contains erythrocytes (red blood cells), leukocytes (white blood cells), and platelets. Whereas the behavior of blood in vessels smaller than approximately 100 μm exhibits significant non-Newtonian effects, flow in larger vessels can be described reasonably accurately using the Newtonian assumption. There is some evidence

TABLE 13.2 Normal Average Hemodynamics Values in Human and Dog

	Dog (20 kg)			Human (70 kg, 1.8 m²)		
	N_w	Velocity (cm/s)	N_R	N_w	Velocity (cm/s)	N_R
Systemic vessels						
Ascending aorta	16	15.8 (89/0)[a]	870 (4900)[b]	21	18 (112/0)[a]	1500 (9400)[a]
Abdominal aorta	9	12 (60.0)	370 (1870)	12	14 (75/0)	640 (3600)
Renal artery	3	41 (74/26)	440 (800)	4	40 (73/26)	700 (1300)
Femoral artery	4	10 (42/1)	130 (580)	4	12 (52/2)	200 (860)
Femoral vein	5	5	92	7	4	104
Superior vena cava	10	8 (20/0)	320 (790)	15	9 (23/0)	550 (1400)
Inferior vena cava	11	19 (40/0)	800 (1800)	17	21 (46/0)	1400 (3000)
Pulmonary vessels						
Main artery	14	18 (72/0)	900 (3700)	20	19 (96/0)	1600 (7800)
Main vein[c]	7	18 (30/9)	270 (800)	10	19 (38/10)	800 (2200)

[a] Mean (systolic/diastolic)
[b] Mean (peak)
[c] One of the usually four terminal pulmonary veins
Source: Milnor, W.R. 1989. *Hemodynamics*, 2nd ed., p. 148. Baltimore, Williams & Wilkins. With permission.

suggesting that in blood analog fluids wall shear stress distributions may differ somewhat from Newtonian values [Liepsch et al., 1991].

Flow in the arterial circulation is predominantly laminar with the possible exception of the proximal aorta and main pulmonary artery. In steady flow, transition to turbulence occurs at Reynolds numbers (N_R) above approximately 2300:

$$N_R = \frac{2rV}{v} \tag{13.1}$$

where r = vessel radius, V = velocity, v = kinematic viscosity = viscosity/density.

Flow in the major systemic and pulmonary arteries is highly pulsatile. Peak-to-mean flow amplitudes as high as 6 to 1 have been reported in both human and dog [Milnor, 1989, p. 149]. Womersley's analysis of incompressible flow in rigid and elastic tubes [1957] showed that the importance of pulsatility in the velocity distributions depended on the parameter:

$$N_w = r\sqrt{\frac{\omega}{v}} \tag{13.2}$$

where ω = frequency.

This is usually referred to as the *Womersley number* (N_w) or α-*parameter*. Womersley's original report is not readily available; however, Milnor provides a reasonably complete account [Milnor, 1989, pp. 106–121].

Mean and peak Reynolds numbers in human and dog are given in Table 13.2, which also includes mean, peak, and minimum velocities as well as the Womersley number. Mean Reynolds numbers in the entire systemic and pulmonary circulations are below 2300. Peak systolic Reynolds numbers exceed 2300 in the aorta and pulmonary artery, and some evidence of transition to turbulence has been reported. In dogs, distributed flow occurs at Reynolds numbers as low as 1000, with higher Womersley numbers increasing the transition Reynolds number [Nerem and Seed, 1972]. The values in Table 13.2 are typical for individuals at rest. During exercise, cardiac output and hence Reynolds numbers can increase severalfold. The Womersley number also affects the shape of the instantaneous velocity profiles as discussed below.

TABLE 13.3   Pressure Wave Velocities in Arteries[a,b]

Artery	Species	Wave Velocity (cm/s)
Ascending Aorta	Man	440–520
	Dog	350–472
Thoracic aorta	Man	400–650
	Dog	400–700
Abdominal aorta	Man	500–620
	Dog	550–960
Iliac	Man	700–880
	Dog	700–800
Femoral	Man	800–1800
	Dog	800–1300
Popliteal	Dog	1220–1310
Tibial	Dog	1040–1430
Carotid	Man	680–830
	Dog	610–1240
Pulmonary	Man	168–182
	Dog	255–275
	Rabbit	100
	Pig	190

[a] All data are apparent pressure wave velocities (although the average of higher frequency harmonics approximates the true velocity in many cases), from relatively young subjects with normal cardiovascular systems, at approximately normal distending pressures.
[b] Ranges for each vessel and species taken from Table 9.1 of source.
Source: Milnor, W.R. 1989. *Hemodynamics*, 2nd ed., p. 235. Baltimore, Williams & Wilkins. With permission.

13.3   Wave Propagation

The viscoelasticity of blood vessels affects the hemodynamics of arterial flow. The primary function of arterial elasticity is to store blood during systole so that forward flow continues when the aortic valve is closed. Elasticity also causes a finite wave propagation velocity, which is given approximately by the Moens–Korteweg relationship:

$$c = \sqrt{\frac{Eh}{2\rho r}} \tag{13.3}$$

where E = wall elastic modulus, h = wall thickness, ρ = blood density, r = vessel radius.

Although Moens [1878] and Korteweg [1878] are credited with this formulation, Fung [1984, p. 107] has pointed out that the formula was first derived much earlier [Young, 1808]. Wave speeds in arterial blood vessels from several species are given in Table 13.3. In general, wave speeds increase toward the periphery as vessel radius decreases and are considerably lower in the main pulmonary artery than in the aorta owing primarily to the lower pressure and consequently lower elastic modulus.

Wave reflections occur at branches where there is not perfect *impedance* matching of parent and daughter vessels. The input impedance of a network of vessels is the ratio of pressure to flow. For rigid vessels with laminar flow and negligible inertial effects, the input impedance is simply the resistance and is independent of pressure and flow rate. For elastic vessels, the impedance is dependent on the frequency of the fluctuations in pressure and flow. The impedance can be described by a complex function expressing the amplitude ratio of pressure to flow oscillations and the phase difference between the peaks.

$$\bar{Z}_i(\omega) = \frac{\bar{P}(\omega)}{\bar{Q}(\omega)}$$

$$\left|\bar{Z}_i(\omega)\right| = \left|\frac{\bar{P}(\omega)}{\bar{Q}(\omega)}\right| \tag{13.4}$$

$$\theta_i(\omega) = \theta\left[\left(\bar{P}(\omega)\right)\right] - \theta\left[\left(\bar{Q}(\omega)\right)\right]$$

where $\bar{Z}_i$ is the complex impedance, $|\bar{Z}_i|$ is the amplitude, and θ_i is the phase.

For an infinitely long straight tube with constant properties, input impedance will be independent of position in the tube and dependent only on vessel and fluid properties. The corresponding value of input impedance is called the *characteristic impedance Z_o* given by:

$$Z_0 = \frac{\rho c}{A} \tag{13.5}$$

where A = vessel cross-sectional area.

In general, the input impedance will vary from point to point in the network because of variations in vessel sizes and properties. If the network has the same impedance at each point (perfect impedance matching), there will be no wave reflections. Such a network will transmit energy most efficiently. The reflection coefficient R, defined as the ratio of reflected to incident wave amplitude, is related to the relative characteristic impedance of the vessels at a junction. For a parent tube with characteristic impedance Z_0 branching into two daughter tubes with characteristic impedances Z_1 and Z_2, the reflection coefficient is given by:

$$R = \frac{Z_0^{-1} - \left(Z_1^{-1} + Z_2^{-1}\right)}{Z_0^{-1} + \left(Z_1^{-1} + Z_2^{-1}\right)} \tag{13.6}$$

and perfect impedance matching requires:

$$\frac{1}{Z_0} = \frac{1}{Z_1} + \frac{1}{Z_2} \tag{13.7}$$

The arterial circulation exhibits partial impedance matching; however, wave reflections do occur. At each branch point, local reflection coefficients typically are less than 0.2. Nonetheless, global reflection coefficients, which account for all reflections distal to a given site, can be considerably higher [Milnor, 1989, p. 217].

In the absence of wave reflections, the input impedance is equal to the characteristic impedance. Womersley's analysis predicts that impedance modulus will decrease monotonically with increasing frequency, whereas the phase angle is negative at low frequency and becomes progressively more positive with increasing frequency. Typical values calculated from Womersley's analysis are shown in Fig. 13.1. In the actual circulation, wave reflections cause oscillations in the modulus and phase. Figure 13.2 shows input impedance measured in the ascending aorta of a human. Measurements of input resistance, characteristic impedance, and the frequency of the first minimum in the input impedance are summarized in Table 13.4.

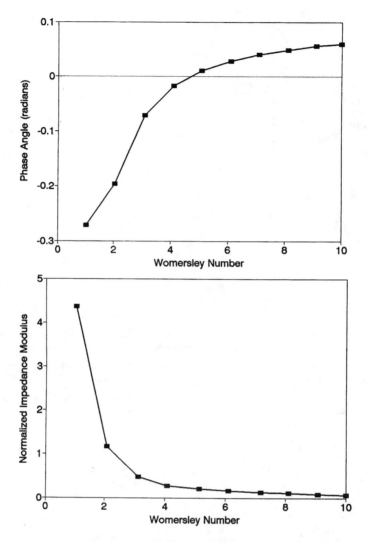

FIGURE 13.1 Characteristic impedance calculated from Womersley's analysis. The top panel contains the phase of the impedance and the bottom panel the modulus, both plotted as a function of the Womersley number N_w, which is proportional to frequency. The curves shown are for an unconstrained tube and include the effects of wall viscosity. The original figure has an error in the scale of the ordinate which has been corrected. *Source:* Milnor, W.R. 1989. *Hemodynamics*, 2nd ed., p. 172. Baltimore, Williams & Wilkins. With permission.

13.4   Velocity Profiles

Typical pressure and velocity fluctuations throughout the cardiac cycle in man are shown in Fig. 13.3. Although mean pressure decreases slightly toward the periphery due to viscous effects, peak pressure shows small increases in the distal aorta due to wave reflection and vessel taper. A rough estimate of mean pressure can be obtained as 1/3 of the sum of systolic pressure and twice the diastolic pressure. Velocity peaks during systole, with some backflow observed in the aorta early in diastole. Flow in the aorta is nearly zero through most of the diastole; however, more peripheral arteries such as the iliac and renal show forward flow throughout the cardiac cycle. This is a result of capacitive discharge of the central arteries as arterial pressure decreases.

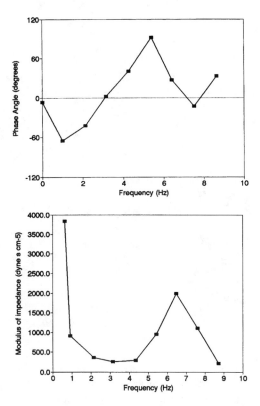

FIGURE 13.2 Input impedance derived from the pressure and velocity data in the ascending aorta of Fig. 13.3. The top panel contains the modulus and the bottom panel the phase, both plotted as functions of frequency. The peripheral resistance (DC impedance) for this plot was 16470 dyne s/cm^5. *Source:* Mills, C.J, Gabe, I.T, Gault, J.N. et al. 1970. Pressure-flow relationships and vascular impedance in man. *Cardiovasc. Res.* 4:405. With permission.

TABLE 13.4 Characteristic Arterial Impedances in Some Mammals: Average ($\pm$SE)[a,b]

Species	Artery	R_{in}	Z_o	f_{min}
Dog	Aorta	2809–6830	125–288	6–8
Dog	Pulmonary	536–807	132–295	2–3.5
Dog	Femoral	110–162[c]	4.5–15.8[c]	8–13
Dog	Carotid	69[c]	7.0–9.4[c]	8–11
Rabbit	Aorta	20–50[c]	1.8–2.1[c]	4.5–9.8
Rabbit	Pulmonary		1.1[c]	3.0
Rat	Aorta	153[c]	11.2[c]	12

Abbreviations: R_{in}, input resistance (mean arterial pressure/flow) in dyn s/cm^5; Z_o, characteristic impedance, in dyn s/cm^5, estimated by averaging high-frequency input impedances in aorta and pulmonary artery; value at 5Hz for other arteries, f_{min}, frequency of first minimum of Z_i.

[a] Values estimated from published figures if averages were not reported.
[b] Ranges for each species and vessel taken from values in Table 7.2 of source.
[c] 10^3 dyn s/cm^5.

Source: Milnor, W.R. 1989. *Hemodynamics*, 2nd ed., p. 183. Baltimore, Williams & Wilkins. With permission.

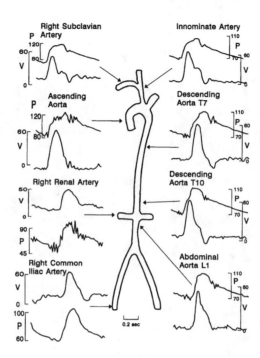

FIGURE 13.3 Simultaneous pressure and blood velocity patterns recorded at points in the systemic circulation of a human. Velocities were recorded with a catheter-tip electromagnetic flowmeter probe. The catheter included a lumen for simultaneous pressure measurement. V = velocity (cm/s), P = pressure (mmHg). *Source:* Mills, C.J, Gabe, I.T, Gault, J.N. et al. 1970. Pressure-flow relationships and vascular impedance in man. *Cardiovasc. Res.* 40:405. With permission.

Velocity varies across the vessel due to viscous and inertial effects as mentioned earlier. The velocities in Fig. 13.3 were measured at one point in the artery. Velocity profiles are complex because the flow is pulsatile and vessels are elastic, curved, and tapered. Profiles measured in the thoracic aorta of a dog at normal arterial pressure and cardiac output are shown in Fig. 13.4. Backflow occurs during diastole, and profiles are flattened even during peak systolic flow. The shape of the profiles varies considerably with mean aortic pressure and cardiac output [Ling et al., 1973].

In more peripheral arteries the profiles are resembling parabolic ones as in fully developed laminar flow. The general features of these fully developed flow profiles can be modeled using Womersley's approach, although nonlinear effects may be important in some cases. The qualitative features of the profile depend on the Womersley number N_w. Unsteady effects become more important as N_w increases. Below a value of about 2 the instantaneous profiles are close to the steady parabolic shape. Profiles in the aortic arch are skewed due to curvature of the arch.

13.5   Pathology

Atherosclerosis is a disease of the arterial wall which appears to be strongly influenced by hemodynamics. The disease begins with a thickening of the intimal layer in locations which correlate with the shear stress distribution on the endothelial surface [Friedman et al., 1993]. Over time the lesion continues to grow until a significant portion of the vessel lumen is occluded. The peripheral circulation will dilate to compensate for the increase in resistance of the large vessels, compromising the ability of the system to respond to increases in demand during exercise. Eventually the circulation is completely dilated, and resting flow begins to decrease. A blood clot may form at the site or lodge in a narrowed segment, causing

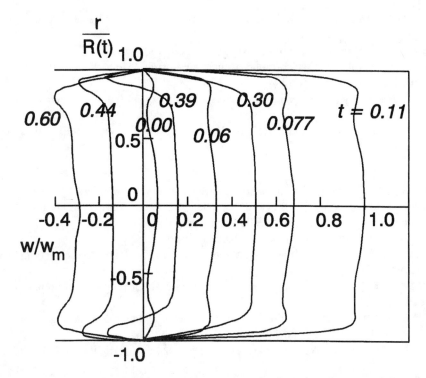

FIGURE 13.4 Velocity profiles obtained with a hot-film anemometer probe in the descending thoracic aorta of a dog at normal arterial pressure and cardiac output. The velocity at t = time/(cardiac period) is plotted as a function of radial position. Velocity w is normalized by the maximum velocity w_m and radial position at each time by the instantaneous vessel radius $R(t)$. The aortic valve opens at $t = 0$. Peak velocity occurs 11% of the cardiac period after aortic valve opening. *Source:* Ling, S.C., Atabek, W.G., Letzing, W.G. et al. 1973. Nonlinear analysis of aortic flow in living dogs. *Circ. Res.* 33:198. With permission.

an acute loss of blood flow. The disease is particularly dangerous in the coronary and carotid arteries due to the critical oxygen requirements of the heart and brain.

In addition to intimal thickening, the arterial wall properties also change with age. Most measurements suggest that arterial elastic modulus increases with age (hardening of the arteries); however, in some cases arteries do become more compliant (inverse of elasticity) [Learoyd and Taylor, 1966]. Local weakening of the wall may also occur, particularly in the descending aorta, giving rise to an aneurysm, which, if it ruptures, can cause sudden death.

Defining Terms

Aneurysm: A ballooning of a blood vessel wall caused by weakening of the elastic material in the wall.
Atherosclerosis: A disease of the blood vessels characterized by thickening of the vessel wall and eventual occlusion of the vessel.
Collagen: A protein found in blood vessels which is much stiffer than elastin.
Elastin: A very elastic protein found in blood vessels.
Endothelial: The inner lining of blood vessels.
Impedance: A (generally) complex number expressing the ratio of pressure to flow.
Myogenic: A change in smooth-muscle tone due to stretch or relaxation, causing a blood vessel to resist changes in diameter.
Newtonian: A fluid whose stress–rate-of-strain relationship is linear, following Newton's law. The fluid will have a viscosity whose value is independent of rate of strain.

Pulmonary: The circulation which delivers blood to the lungs for reoxygenation and carbon dioxide removal.

Systemic: The circulation which supplies oxygenated blood to the tissues of the body.

Vasoconstrictor: A substance which causes an increase in smooth-muscle tone, thereby constricting blood vessels.

Vasodilator: A substance which causes a decrease in smooth-muscle tone, thereby dilating blood vessels.

Viscoelastic: A substance which exhibits both elastic (solid) and viscous (liquid) characteristics.

References

Chandran KB. 1992. *Cardiovascular Biomechanics.* New York, New York University Press.

Friedman MH, Brinkman AM, Qin JJ et al. 1993. Relation between coronary artery geometry and the distribution of early sudanophilic lesions. *Atherosclerosis* 98:193.

Fung YC. 1984. *Biodynamics: Circulation.* New York, Springer-Verlag.

Korteweg DJ. 1878. Über die Fortpflanzungsgeschwindigkeit des Schalles in elastischen Rohren. *Ann Phys Chem (NS)* 5:525.

Learoyd BM, Taylor MG. 1966. Alterations with age in the viscoelastic properties of human arterial walls. *Circ Res* 18:278.

Liepsch D, Thurston G, Lee M. 1991. Studies of fluids simulating blood-like rheological properties and applications in models of arterial branches. *Biorheology* 28:39.

Ling SC, Atabek WG, Letzing WG et al. 1973. Nonlinear analysis of aortic flow in living dogs. *Circ Res* 33:198.

Mills CJ, Gabe IT, Gault JN et al. 1970. Pressure-flow relationships and vascular impedance in man. *Cardiovasc Res* 4:405.

Milnor WR. 1989. *Hemodynamics,* 2nd ed. Baltimore, Williams & Wilkins.

Moens AI. 1878. *Die Pulskurve,* Leiden.

Nerem RM, Seed WA. 1972. An *in-vivo* study of aortic flow disturbances. *Cardiovasc Res* 6:1.

Womersley JR. 1957. The mathematical analysis of the arterial circulation in a state of oscillatory motion. Wright Air Development Center Technical Report WADC-TR-56-614.

Young T. 1808. Hydraulic investigations, subservient to an intended Croonian lecture on the motion of the blood. *Phil Trans Roy Soc London* 98:164.

Further Information

A good introduction to cardiovascular biomechanics, including arterial hemodynamics, is provided by K. B. Chandran in *Cardiovascular Biomechanics.* Y. C. Fung's *Biodynamics: Circulation* is also an excellent starting point, somewhat more mathematical than Chandran. Perhaps the most complete treatment of the subject is in *Hemodynamics* by W. R. Milnor, from which much of this chapter was taken. Milnor's book is quite mathematical and may be difficult for a novice to follow.

Current work in arterial hemodynamics is reported in a number of engineering and physiological journals, including the *Annals of Biomedical Engineering, Journal of Biomechanical Engineering, Circulation Research,* and *The American Journal of Physiology, Heart and Circulatory Physiology.* Symposia sponsored by the American Society of Mechanical Engineers, Biomedical Engineering Society, American Heart Association, and the American Physiological Society contain reports of current research.

14

Mechanics and Transport in the Microcirculation

Aleksander S. Popel
The Johns Hopkins University

Roland N. Pittman
Virginia Commonwealth University

14.1   Introduction

Microcirculation comprises blood vessels—arterioles, capillaries, and venules—with diameters of less than approximately 150 µm. The importance of the microcirculation is underscored by the fact that most of the hydrodynamic resistance of the circulatory system lies in the microvessels, especially in arterioles, and most of the exchange of nutrients and waste products occurs at the level of the smallest microvessels. The subjects of microcirculatory research are: (1) blood flow and molecular transport in microvessels, (2) mechanical interactions and molecular exchange between these vessels and the surrounding tissue, and (3) regulation of blood flow and molecular transport. Quantitative knowledge of microcirculatory mechanics and mass transport has accumulated primarily in the past 30 years owing to significant innovations in methods and techniques to measure microcirculatory parameters and methods to analyze microcirculatory data. The development of these methods has required joint efforts of physiologists and biomedical engineers. Key innovations include: (1) significant improvements in intravital microscopy, (2) the dual-slit method (Wayland–Johnson) for measuring velocity in microvessels, (3) the servo-null method (Wiederhielm–Intaglietta) for measuring pressure in microvessels, (4) the recessed oxygen microelectrode method (Whalen) for polarographic measurements of partial pressure of oxygen, and (5) the microspectrophotometric method (Pittman–Duling) for measuring oxyhemoglobin saturation in microvessels. The single-capillary cannulation method (Landis–Michel) has provided a powerful tool for studies of the transport of water and solutes through the capillary endothelium. In the last decade, new experimental techniques have appeared, many adapted from cell biology and modified for *in vivo* studies, that are having a tremendous impact on the field. Examples include (1) confocal microscopy for better three-dimensional resolution of microvascular structures, (2) methods of optical imaging using fluorescent labels (e.g., labeling blood cells for velocity measurements), and fluorescent dyes (e.g., calcium ion sensitive dyes for measuring the dynamics of calcium ion concentration in arteriolar smooth muscle and endothelial cells *in vivo*), (3) development of sensors (glass filaments and optical and magnetic tweezers) for measuring forces in the nanonewton range that are characteristic of cell–cell interactions, (4) phosphorescence decay measurements as an indicator of oxygen tension, and (5) methods of manipulating receptors on the surfaces of blood cells and endothelial cells. In addition to the dramatic developments in experimental techniques, quantitative knowledge and understanding of the microcirculation have been significantly enhanced by theoretical studies, perhaps having a larger impact than in other areas of physiology. Extensive theoretical work has been conducted on (1) the mechanics of the red blood cell (RBC) and leukocyte, (2) mechanics of blood flow in single microvessels and microvascular networks, (3) oxygen (O_2) and carbon dioxide (CO_2) exchange between microvessels and surrounding tissue, and (4) water and solute transport through capillary endothelium and the surrounding tissue. These theoretical studies not only aid in the interpretation of experimental data, but in many cases also serve as a framework for quantitative testing of working hypotheses and as a guide in designing and conducting further experiments. The accumulated knowledge has led to significant progress in our understanding of mechanisms of regulation of blood flow and molecular exchange in the microcirculation in many organs and tissues under a variety of physiological and pathological conditions (e.g., hypoxia, hypertension, sickle-cell anemia, diabetes, inflammation, sepsis, and cancer). Discussions are under way to start organizing the enormous amount of information on the microcirculation in the form of a database or a network of databases encompassing anatomical and functional data, and conceptual (pathway) and computational models [Popel et al., 1998]. This effort is called the Microcirculation Physiome project.

The goal of this chapter is to give an overview of the current status of research on the systemic microcirculation. Issues of pulmonary microcirculation are not discussed. Because of space limitations, it is not possible to recognize numerous important contributions to the field of microcirculatory mechanics and mass transport. In most cases reference is made to recent reviews, when available, and journal articles where earlier references can be found. A discussion of experimental and theoretical findings is presented along with the gaps in the understanding of microcirculatory flow phenomena.

14.2   Mechanics of Microvascular Blood Flow

Vessel dimensions in the microcirculation are small enough so that the effects of the particulate nature of blood are significant [Lipowsky, 1987]. Blood is a suspension of formed elements (red blood cells, white blood cells [leukocytes] and platelets) in plasma. Plasma is an aqueous solution of mostly proteins (albumins, globulins, and fibrinogen) and electrolytes. Under static conditions, human RBCs are biconcave discs with diameter ~7 to 9 μm. The chief function of the RBC is delivery of O_2 to tissue. Most of the O_2 carried by the blood is chemically bound to hemoglobin inside the RBCs. The mammalian RBC comprises a viscoelastic membrane filled with a viscous fluid, concentrated hemoglobin solution. The membrane consists of the plasma membrane and underlying cytoskeleton. The membrane can undergo deformations without changing its surface area, which is nearly conserved locally. RBCs are so easily deformable that they can flow through small pores with diameter <3 μm. Leukocytes (grouped into three categories: granulocytes, monocytes, and lymphocytes) are spherical cells with diameter ~10 to 20 μm. They are stiffer than RBCs. The main function of these cells is immunologic, i.e., protection of the body against microorganisms causing disease. In contrast to mammalian RBCs, leukocytes are nucleated and are endowed with an internal structural cytoskeleton. Leukocytes are capable of active ameboid motion, the property which allows their migration from the blood stream into the tissue. Platelets are disc-shaped blood elements with diameter ~2 to 3 μm. Platelets play a key role in thrombogenic processes and blood coagulation. The normal volume fraction of RBCs (hematocrit) in humans is 40 to 45%. The total volume of RBCs in blood is approximately 500 times greater than the volume of leukocytes and platelets. Rheological properties of blood in arterioles and venules and larger vessels are determined primarily by RBCs; however, leukocytes play an important mechanical role in capillaries and small venules.

Blood plasma is a Newtonian fluid with viscosity of approximately 1.2 cP. The viscosity of whole blood in a rotational viscometer or a large-bore capillary viscometer exhibits shear-thinning behavior, i.e., decreases when shear rate increases. At shear rates >100 s^{-1} and a hematocrit of 40% typical viscosity values are 3 to 4 cP. The dominant mechanism of the non-Newtonian behavior is RBC aggregation and the secondary mechanism is RBC deformation under shear forces. The cross-sectional distribution of RBCs in vessels is nonuniform, with a core of concentrated RBC suspension and a cell-free or cell-depleted marginal layer, typically 2 to 5 μm thick, adjacent to the vessel wall. The nonuniform RBC distribution results in the *Fahraeus effect* (the microvessel hematocrit is smaller than the feed or discharge hematocrit) due to the fact that, on the average, RBCs move with a higher velocity than blood plasma, and the concomitant *Fahraeus–Lindqvist effect* (the apparent viscosity of blood is lower than the bulk viscosity measured with a rotational viscometer or a large-bore capillary viscometer at high shear rate). The *apparent viscosity* of a fluid flowing in a cylindrical vessel of radius R and length L under the influence of a pressure difference ΔP is defined as:

$$\eta_a = \frac{\pi \Delta P R^4}{8QL}$$

where Q is the volumetric flow rate. For a Newtonian fluid the apparent viscosity becomes the dynamic viscosity of the fluid and the above equation represents *Poiseuille's law*. The apparent viscosity is a function of hematocrit, vessel radius, blood flow rate, and other parameters. In the microcirculation, blood flows through a complex branching network of arterioles, capillaries, and venules. Arterioles are typically 10 to 150 μm in diameter, capillaries are 4 to 8 μm, and venules are 10 to 200 μm. Now we will discuss vascular wall mechanics and blood flow in vessels of different size in more detail.

Mechanics of the Microvascular Wall

The wall of arterioles comprises the intima, which contains a single layer of contiguous endothelial cells; the media, which contains a single layer of smooth muscle cells in terminal and medium-size arterioles or several layers in the larger arterioles; and the adventitia, which contains collagen fibers with occasional fibroblasts and mast cells. Fibers situated between the endothelium and the smooth muscle cells comprise the elastica lamina. The single layer of smooth muscle cells terminates at the capillaries and reappears at the level of small venules; the capillary wall is devoid of smooth muscle cells. Venules typically have a larger diameter and smaller wall thickness-to-diameter ratio than arterioles of the corresponding branching order.

Most of our knowledge of the mechanics of the microvascular wall comes from *in vivo* or *in vitro* measurements of vessel diameter as a function of transmural pressure [Davis, 1993; Shoukas and Bohlen, 1990]. Development of isolated arteriole preparations has made it possible to precisely control the transmural pressure during experiments. In addition, these preparations allow one to separate the effects of metabolic factors and blood flow rate from the effect of pressure by controlling both the chemical environment and the flow rate through the vessel. Arterioles and venules exhibit vascular tone, i.e., their diameter is maximal when smooth muscle is completely relaxed (inactivated). When the vascular smooth muscle is constricted, small arterioles may even completely close their lumen to blood flow, presumably by buckling endothelial cells. Arterioles exhibit a *myogenic response* not observed in other blood vessels, with the exception of cerebral arteries: within a certain physiological pressure range the vessels constrict in response to elevation of transmural pressure and dilate in response to reduction of transmural pressure; in other words, in a certain range of pressures the slope of the pressure–diameter relationship is negative. Arterioles of different size exhibit different degrees of myogenic responsiveness. This effect has been documented in many tissues both *in vivo* and *in vitro* (in isolated arterioles), and it has been shown to play an important role in regulation of blood flow and capillary pressure (see Section 14.4 on Regulation of Blood Flow).

The stress–strain relationship for a thin-walled microvessel can be derived from the experimentally obtained pressure–diameter relationship using the law of Laplace. Stress in the vessel wall can be decomposed into passive and active components. The passive component corresponds to the state of complete vasodilation. The active component determines the vascular tone and the myogenic response. Steady-state stress–strain relationships are, generally, nonlinear. For arterioles, diameter variations of 50 to 100% under physiological conditions are not unusual, so that finite deformations have to be considered in formulating the constitutive relationship for the wall (relationship between stress, strain, and their temporal derivatives). Pertinent to the question of microvascular mechanics is the mechanical interaction of a vessel with its environment that consists of connective tissue, parenchymal cells, and extracellular fluid. There is ultrastructural evidence that blood vessels are tethered to the surrounding tissue, so that mechanical forces can be generated when the vessels constrict or dilate, or when the tissue is moving, e.g., in contracting striated muscle, myocardium, or intestine. Little quantitative information is currently available about the magnitude of these forces, chiefly because of the difficulty of such measurements. A recently reported technique, magnetic tweezers [Guilford and Gore, 1995], opens a new way of investigating the mechanics of the microvascular wall *in vivo* and its interaction with the surrounding tissue.

Under time-dependent conditions microvessels exhibit viscoelastic behavior. In response to a stepwise change in the transmural pressure, arterioles typically respond with a fast "passive" change in diameter followed by a slow "active" response with a characteristic time on the order of tens of seconds. For example, when the pressure is suddenly increased, the vessel diameter will quickly increase, with subsequent vasoconstriction which may result in a lower value of steady-state diameter than that prior to the increase in pressure. Therefore, to accurately describe the time-dependent vessel behavior, the constitutive relationship between stress and strain or pressure and diameter must also contain temporal derivatives of these variables. Theoretical analysis of the resulting nonlinear equations shows that such constitutive equations lead to predictions of spontaneous oscillations of vessel diameter (*vasomotion*) under certain conditions [Ursino and Fabbri, 1992]. Vasomotion has been observed *in vivo* in various tissues and under various physiological conditions [Griffith, 1996]. Whether experimentally observed vasomotion and its effect on blood flow (flow motion) can be quantitatively described by the published theoretical studies remains to be established. It should be noted that there are mechanisms leading to spontaneous flow oscillations that are associated with blood rheology and not with vascular wall mechanics [Kiani et al., 1994].

For most purposes, capillary compliance is not taken into account. However, in some situations, such as analysis of certain capillary water transport experiments or leukocyte flow in a capillary, this view is not adequate and capillary compliance has to be accounted for. Since the capillary wall is devoid of smooth muscle cells, much of this compliance is passive and its magnitude is small. However, the presence of contractile proteins in the cytoskeleton of capillary endothelial cells opens a possibility of active capillary constriction or dilation.

Capillary Blood Flow

Progress in this area is closely related to studies of the mechanics of blood cells described elsewhere in this book. In narrow capillaries RBCs flow in single file, separated by gaps of plasma. They deform and assume a parachute-like shape, generally non-axisymmetric, leaving a submicron plasma sleeve between the RBC and endothelium. In the smallest capillaries, their shape is sausage-like. The hemoglobin solution inside an RBC is a Newtonian fluid. The constitutive relationship for the membrane is expressed by the Evans–Skalak finite-deformations model. The coupled mechanical problem of membrane and fluid motion has been extensively investigated using both analytical and numerical approaches [Skalak et al., 1989]. An important result of the theoretical studies is the prediction of the apparent viscosity of blood. While these predictions are in agreement with *in vitro* studies in glass tubes, they underestimate a few available *in vivo* capillary measurements of apparent viscosity. In addition, *in vivo* capillary hematocrit in some tissues is lower than predicted from *in vitro* studies with tubes of the same size. To explain the low values of hematocrit, RBC interactions with the endothelial glycocalyx have been implicated

[Desjardins and Duling, 1990]. Recent direct measurements of microvascular resistance and theoretical analysis provide further evidence of the role of the glycocalyx [Pries et al., 1997; Secomb et al., 1998].

The motion of leukocytes through blood capillaries has also been studied thoroughly in recent years. Because leukocytes are larger and stiffer than RBCs, under normal flow conditions an increase in capillary resistance caused by a single leukocyte may be 1000 times greater than that caused by a single RBC [Schmid-Schönbein et al., 1989]. Under certain conditions flow stoppage may occur caused by leukocyte plugging. After a period of ischemia, red blood cell and leukocyte plugging may prevent tissue reperfusion (ischemia–reperfusion injury) [Granger and Schmid-Schönbein, 1994]. Chemical bonds between membrane-bound receptors and endothelial adhesion molecules play a crucial role in leukocyte–endothelium interactions. Methods of cell and molecular biology permit manipulation of the receptors and thus make it possible to study leukocyte microcirculatory mechanics at the molecular level. More generally, methods of cell and molecular biology open new and powerful ways to study cell micromechanics and cell–cell interactions [Chien, 1992].

Arteriolar and Venular Blood Flow

The cross-sectional distribution of RBCs in arterioles and venules is nonuniform. A concentrated suspension of RBCs forms a core surrounded by a cell-free or cell-depleted layer of plasma. This "lubrication" layer of lower viscosity fluid near the vessel wall results in lower values of the apparent viscosity of blood compared to its bulk viscosity, producing the Fahraeus–Lindqvist effect [Lipowsky, 1987]. There is experimental evidence that velocity profiles of RBCs are generally symmetric in arterioles, except very close to vascular bifurcations, but are generally asymmetric in venules; the profiles are blunted in vessels of both types [Ellsworth and Pittman, 1986; Tangelder et al., 1986]. Moreover, flow in the venules may be stratified as the result of converging blood streams that do not mix rapidly. The key to understanding the pattern of arteriolar and venular blood flow is the mechanics of flow at vascular bifurcations, diverging for arteriolar flow and converging for venular flow [Das et al., 1997]. One of the main unresolved questions is under what physiological conditions does RBC aggregation affect venular velocity distribution and vascular resistance. Much is known about aggregation *in vitro*, but *in vivo* knowledge is incomplete [Cabel et al., 1997].

Problems of leukocyte distribution in the microcirculation and their interaction with the microvascular endothelium have attracted considerable attention in recent years. Leukocyte rolling along the walls of venules, but not arterioles, has been demonstrated. This effect results from differences in the microvascular endothelium, not the flow conditions [Ley and Gaehtgens, 1991]. Platelet distribution in the lumen is important because of their role in blood coagulation. Detailed studies of platelet distribution in arterioles and venules show that the cross-sectional distribution of these disk-shaped blood elements is dependent on the blood flow rate and vessel hematocrit [Woldhuis et al., 1992].

To conclude, many important features of blood flow through arterioles and venules are qualitatively known and understood; however, a rigorous theoretical description of flow as a suspension of discrete particles is not available. Such description is necessary for a quantitative understanding of the mechanisms of the nonuniform distribution of blood cells in microvessels.

Microvascular Networks: Structure and Hemodynamics

Microvascular networks in different organs and tissues differ in their appearance and structural organization. Methods have been developed to quantitatively describe network architectonics and hemodynamics [Popel, 1987; Hudetz, 1997]. The microvasculature is an adaptable structure capable of changing its structural and functional characteristics in response to various stimuli [Skalak and Price, 1996]. Microvascular *angiogenesis* and rarefaction (anti-angiogenesis) are important examples of this adaptive behavior that play important physiological and pathophysiological roles. Methods of fractal analysis have been applied to interpret experimental data on angioarchitecture and blood flow distribution in the microcirculation; these methods explore the property of geometrical and flow similarity that exists at different scales in the network [Bassingthwaighte et al., 1994]. Microvascular hydraulic pressure varies systematically between consecutive branching orders, decreasing from the systemic values down to 20 to 25 mmHg in the capillaries and decreasing further by 10 to 15 mmHg in the venules. Mean microvascular blood flow rate in arterioles decreases towards the capillaries, in inverse proportion to the number of "parallel" vessels, and increases from capillaries through the venules. In addition to this longitudinal variation of blood flow and pressure among different branching orders, there are significant variations among vessels of the same branching order, referred to as flow heterogeneity. The heterogeneity of blood flow and RBC distribution in microvascular networks has been well documented in a variety of organs and tissues. This phenomenon may have important implications for tissue exchange processes, so significant efforts have been devoted to the quantitative analysis of blood flow in microvascular networks. A mathematical model of blood flow in a network can be formulated as follows: (1) network topology or vessel interconnections have to be specified; (2) the diameter and length of every vascular segment have to be known (alternatively, vessel diameter can be specified as a function of transmural pressure and perhaps some other parameters; these relationships are discussed in the preceding section on wall mechanics); (3) the apparent viscosity of blood has to be specified as a function of vessel diameter, local hematocrit, and shear rate; (4) a relationship between RBC flow rates and bulk blood flow rates at diverging bifurcations has to be specified, a relationship often referred to as the "bifurcation law"; and (5) boundary conditions have to be specified at the inlet vessel branches: bulk flow rate as well as RBC flow rate, or hematocrit (alternatively, pressure can be specified at both inlet and outlet branches). This set of generally nonlinear equations can be solved to yield pressure at each bifurcation, blood flow rate through each segment, and discharge or microvessel hematocrit in each segment. These equations also predict vessel diameters if vessel compliance is taken into account. The calculated variables can then be compared with experimental data. Such a detailed comparison was reported for rat mesentery [Pries et al., 1996]. The authors found a good agreement between theoretical and experimental data when histograms of parameter distributions were compared, but poor agreement, particularly for vessel hematocrit, when comparison was done on a vessel-by-vessel basis. In these calculations, the expression for apparent viscosity was taken from *in vitro* experiments. The agreement was improved when the apparent viscosity was increased substantially from its *in vitro* values, particularly in the smallest vessels. Thus, a working hypothesis was put forward that *in vivo* apparent viscosity in small vessels is higher than the corresponding *in vitro* viscosity in glass tubes. Recent experimental and theoretical studies support this hypothesis [Pries et al., 1997; Secomb et al., 1998].

14.3   Mass Transport in the Microcirculation

Transport of Oxygen and Carbon Dioxide

One of the most important functions of the microcirculation is the delivery of O_2 to tissue and the removal of waste products, particularly of CO_2, from tissue. O_2 is required for aerobic intracellular respiration for the production of adenosine triphosphate (ATP). CO_2 is produced as a by-product of these biochemical reactions. Tissue metabolic rate can change drastically, e.g., in aerobic muscle in the transition from rest to exercise, which necessitates commensurate changes in blood flow and O_2 delivery. One of the major issues studied is how O_2 delivery is matched to O_2 demand under different physiological and pathological conditions. This question arises for short-term or long-term regulation of O_2 delivery in an individual organism, organ, or tissue, as well as in the evolutionary sense, phylogeny. The hypothesis of symmorphosis, a fundamental balance between structure and function, has been formulated for the respiratory and cardiovascular systems and tested in a number of animal species [Weibel et al., 1992; Roy and Popel, 1996].

In the smallest exchange vessels (capillaries, small arterioles, and venules), O_2 molecules (1) are released from hemoglobin inside RBCs, (2) diffuse through the plasma, and (3) cross the endothelium, the extravascular space, and parenchymal cells until they reach the mitochondria where they are utilized in the process of oxidative phosphorylation. The nonlinear relationship between hemoglobin saturation with O_2 and the local O_2 tension (PO_2) is described by the *oxyhemoglobin dissociation curve* (ODC). The theory of O_2 transport from capillaries to tissue was conceptually formulated by August Krogh in 1918 and it has dominated the thinking of physiologists. The model he formulated considered a cylindrical tissue volume supplied by a single central capillary; this element was considered the building block for the entire tissue. A constant metabolic rate was assumed and PO_2 at the capillary–tissue interface was specified. The solution to the corresponding transport equation is the Krogh–Erlang equation describing the radial variation of O_2 tension in tissue. Over the years, the *Krogh tissue cylinder model* has been modified by many investigators to include transport processes in the capillary and PO_2-dependent consumption. However, in the past few years new conceptual models of O_2 transport have emerged. First, it was discovered experimentally and subsequently corroborated by theoretical analysis that capillaries are not the only source of oxygen, but arterioles (*precapillary O_2 transport*) and, to a smaller extent venules (postcapillary O_2 transport) also participate in tissue oxygenation; in fact, a complex pattern of O_2 exchange may exist between arterioles, venules, and adjacent capillary networks [Ellsworth et al., 1994]. Second, theoretical analysis of intracapillary transport suggested that a significant part of the resistance to O_2 transport, on the order of 50%, is located within the capillary, primarily due to poor diffusive conductance of the plasma gaps between the erythrocytes [Roy and Popel, 1996]. Third, the effect of *myoglobin-facilitated O_2 diffusion* in red muscle fibers and cardiac myocytes has been re-evaluated: however, its significance must await additional experimental studies. Fourth, geometric and hemodynamic heterogeneities in O_2 delivery have been quantified experimentally and attempts have been made to model them. Discussion of theoretical issues of O_2 transport can be found in Popel [1989] and Hellums et al. [1996] and experimental findings are reviewed in Pittman [1995] and Sibbald et al. [1998]. One important emerging area of application of this knowledge is artificial oxygen carriers, hemoglobin-based and non-hemoglobin-based [Winslow et al., 1997].

Transport of CO_2 is coupled with O_2 through the Bohr effect (effect of CO_2 tension on the blood O_2 content) and the Haldane effect (effect of PO_2 on the blood CO_2 content). Diffusion of CO_2 is faster than that of O_2 because CO_2 solubility in tissue is higher; theoretical studies predict that countercurrent exchange of CO_2 between arterioles and venules is of major importance so that equilibration of CO_2 tension with surrounding tissue should occur before capillaries are reached. Experiments are needed to test these theoretical predictions.

Transport of Solutes and Water

The movement of solute molecules across the capillary wall occurs primarily by two mechanisms: (1) diffusion and (2) solvent drag. Diffusion is the passive mechanism of transport which rapidly and efficiently transports small solutes over the small distances (tens of microns) between the blood supply (capillaries) and tissue cells. Solvent drag refers to the movement of solute that is entrained in the bulk flow of fluid across the capillary wall and is generally negligible, except in cases of large molecules with small diffusivities and high transcapillary fluid flow.

The capillary wall is composed of a single layer of endothelial cells about 1 μm thick. Lipid soluble substances (e.g., O_2) can diffuse across the entire wall surface, whereas water soluble substances (e.g., glucose) are restricted to small aqueous pathways equivalent to cylindrical pores 8 to 9 nm in diameter. Total pore area is about 0.1% of the surface area of a capillary. The permeability of the capillary wall to a particular substance depends upon the relative size of the substance and the pore ("restricted" diffusion). The efficiency of diffusive exchange can be increased by increasing the number of perfused capillaries (e.g., heart and muscle tissue from rest to exercise), since this increases the surface area available for exchange and decreases the distances across which molecules must diffuse.

The actual pathways through which small solutes traverse the capillary wall appear to be in the form of clefts between adjacent endothelial cells. Rather than being open slits, these porous channels appear to contain a matrix of small cylindrical fibers (perhaps glycosaminoglycans) that occupy about 5% of the volume of these pathways. The permeability properties of the capillary endothelium are modulated by a number of factors, among which are plasma protein concentration and composition, rearrangement of the endothelial cell glycocalyx, calcium influx into the endothelial cell, and endothelial cell membrane potential. Many of the studies that have established our current understanding of the endothelial exchange barrier have been carried out on single perfused capillaries of the frog mesentery; some recent investigations have examined these transport issues in mammalian tissues [Curry, 1994; Michel, 1996; Weinbaum, 1998]. There could be, in addition to the porous pathways, nonporous pathways that involve selective uptake of solutes and subsequent transcellular transport. In order to study such pathways, one must try to minimize the contributions to transcapillary transport from solvent drag.

The processes whereby water passes back and forth across the capillary wall are called filtration and absorption. The flow of water depends upon the relative magnitude of hydraulic and osmotic pressures across the capillary wall and is described quantitatively by the Kedem–Katchalsky equations (the particular form of the equations applied to capillary water transport is referred to as *Starling's Law*). Overall, in the steady-state there is an approximate balance between hydraulic and osmotic pressures which leads to a small net flow of water. Generally, more fluid is filtered than is reabsorbed; the overflow is carried back to the vascular system by the lymphatic circulation. The lymphatic network is composed of a large number of small vessels, the terminal branches of which are closed. Flap valves (similar to those in veins) ensure unidirectional flow of lymph back to the central circulation. The smallest (terminal) vessels are very permeable, even to proteins which occasionally leak from systemic capillaries. Lymph flow is determined by interstitial fluid pressure and the lymphatic "pump" (one-way flap valves and skeletal muscle contraction). Control of interstitial fluid protein concentration is one of the most important functions of the lymphatic system. If more net fluid is filtered than can be removed by the lymphatics, the volume of interstitial fluid increases. This fluid accumulation is called edema. This circumstance is important clinically since solute exchange (e.g., O_2) decreases due to the increased diffusion distances produced when the accumulated fluid pushes the capillaries, tethered to the interstitial matrix, away from each other.

14.4   Regulation of Blood Flow

The cardiovascular system controls blood flow to individual organs (1) by maintaining arterial pressure within narrow limits and (2) by allowing each organ to adjust its vascular resistance to blood flow so that each receives an appropriate fraction of the cardiac output. There are three major mechanisms that control the function of the cardiovascular system: local, neural, and humoral. The sympathetic nervous system and circulating hormones both provide overall vasoregulation, and thus coarse flow control, to all vascular beds. The local mechanisms provide finer regional control within a tissue, usually in response to local changes in tissue activity or local trauma. The three mechanisms can work independently of each other, but there are also interactions among them.

The classical view of blood flow control involved the action of vasomotor influences on a set of vessels called the "resistance vessels," generally arterioles and small arteries smaller than about 100 μm diameter, that controlled flow to and within an organ. The notion of "precapillary sphincters" that control flow in individual capillaries has been abandoned in favor of the current idea that the terminal arterioles control the flow in small capillary networks that branch off of these arterioles. In recent years, it has become clear that the resistance to blood flow is distributed over a wider range of vessel branching orders with diameters up to 500 μm. There are at least two mechanisms to be discussed below that are available for coordinating the actions of local control processes over wider regions.

Neurohumoral Regulation of Blood Flow

The role of neural influences on the vasculature varies greatly from organ to organ. Although all organs receive sympathetic innervation, regulation of blood flow in the cerebral and coronary vascular beds occurs mostly through intrinsic local (metabolic) mechanisms. The circulations in skeletal muscle, skin, and some other organs, however, are significantly affected by the sympathetic nerves. In general, the level of intrinsic myogenic activity and sympathetic discharge sets the state of vascular smooth muscle contraction (basal vascular tone) and hence vascular resistance in organs. This basal tone is modulated by circulating and local vasoactive influences, e.g., endothelium-derived relaxing factor (EDRF), identified as nitric oxide (NO), endothelin, and vasoactive substances released from parenchymal cells.

Local Regulation of Blood Flow

In addition to neural and humoral mechanisms for regulating the function of the cardiovascular system, there are mechanisms intrinsic to the various tissues which can operate independently of neurohumoral influences. The site of local regulation is the microcirculation. Common examples of local control processes are (1) autoregulation of blood flow, (2) reactive hyperemia, and (3) active (or functional) hyperemia. The two major theories of local regulation are (1) the myogenic hypothesis which states, in essence, that the vascular smooth muscle actively contracts in response to stretch (see section on Mechanics of the Microvascular Wall in this chapter) and (2) the metabolic hypothesis which states that there is a link between blood flow and tissue metabolism [Duling, 1991].

Cells have a continuous need for O_2 and also continuously produce metabolic wastes, some of which are vasoactive (usually vasodilators). Under normal conditions there is a balance between O_2 supply and demand, but imbalances give rise to adjustments in blood flow that bring supply back into register with demand. Consider exercising the skeletal muscle as an example. With the onset of exercise, metabolite production and O_2 requirements increase. The metabolites diffuse away from their sites of production and reach the vasculature. Vasodilation ensues, lowering resistance to blood flow. The resulting increase in blood flow increases the O_2 supply and finally a new steady-state is achieved in which O_2 supply and demand are matched. This scenario operates for other tissues in which metabolic activity changes.

The following O_2-linked metabolites have been implicated as potential chemical mediators in the metabolic hypothesis: adenosine (from ATP hydrolysis: ATP→ADP→AMP→adenosine), H^+, and lactate (from lactic acid generated by glycolysis). Their levels are increased when there is a reduction in O_2 supply relative to demand (i.e., tissue hypoxia). The production of more CO_2 as a result of increased tissue activity (leading to increased oxidative metabolism) leads to vasodilation through increased H^+ concentration. Increased potassium ions and interstitial fluid osmolarity (i.e., more osmotically active particles) transiently cause vasodilation under physiological conditions associated with increased tissue activity.

A recent novel idea for oxygen-linked vasoregulation is that the RBC itself could act as a mobile sensor for hypoxia [Ellsworth et al., 1995]. The mechanism works as follows. Under conditions of low oxygen and pH, the RBC releases ATP which binds to purinergic receptors on the endothelial cells. This leads to the production in the endothelial cells of the vasodilator, NO. Since the most likely location for hypoxia would be in or near the venular network, the local vasodilatory response to NO is propagated to upstream vessels causing arteriolar vasodilation (see the section below on Coordination of Vasomotor Responses).

Some Functions of the Endothelium

Endothelial cells form the lining of all blood vessels. They provide a smooth, non-thrombogenic surface with which the blood is in intimate contact. In addition to their passive permeability barrier function, endothelial cells produce a number of important vasoactive substances, among which are: (1) prostacyclin, a potent vascular smooth muscle relaxant and inhibitor of platelet aggregation; (2) endothelin, a potent vasoconstrictor peptide; and (3) NO that mediates the vasodilatory effect of a number of vasodilators (e.g., acetylcholine) [Furchgott and Vanhouttee, 1989; Ignarro, 1989].

It has been observed in arteries and arterioles that increases in blood flow lead to vasodilation (flow-dependent dilation) [Smieško and Johnson, 1993]. This phenomenon appears to be mediated by NO release from the endothelium in those vessels. The sequence of events is: (1) blood flow increases; (2) shear stress at the vessel wall increases (thereby increasing the viscous drag on the endothelial lining of the vessel); (3) NO is released in response to the mechanical stimulus; and (4) vascular smooth muscle relaxation (vasodilation) occurs in response to the elevated level of NO.

Coordination of Vasomotor Responses

Communication via gap junctions between the two active cell types in the blood vessel wall, smooth muscle and endothelial cells, appears to play an important role in coordinating the responses among resistance elements in the vascular network [Duling, 1991; Segal, 1992]. There is chemical and electrical coupling between the cells of the vessel wall, and this signal, in response to locally released vasoactive substances (e.g., from vessel wall, RBCs, or parenchymal cells) can travel along a vessel in either direction with a length constant of about 2 mm. There are two immediate consequences of this communication. A localized vasodilatory stimulus of metabolic origin will be propagated to contiguous vessels, thereby lowering the resistance to blood flow in a larger region. In addition, this more generalized vasodilation should increase the homogeneity of blood flow in response to the localized metabolic event. The increase in blood flow produced as a result of this vasodilation will cause flow also to increase at upstream sites. The increased shear stress on the endothelium as a result of the flow increase will lead to vasodilation of these larger upstream vessels. Thus, the neurohumoral and local responses are linked together in a complex control system that matches regional perfusion to the local metabolic needs [Segal and Kurjiaka, 1995; Muller et al., 1996].

Acknowledgments

This work was supported by National Heart, Lung, and Blood Institute grants HL 18292 and HL 52684.

Defining Terms

Angiogenesis: The growth and development of the vessels.
Apparent viscosity: The viscosity of a Newtonian fluid that would require the same pressure difference to produce the same blood flow rate through a circular vessel as the blood.
Fahraeus effect: Microvessel hematocrit is smaller than hematocrit in the feed or discharge reservoir.
Fahraeus–Lindqvist effect: Apparent viscosity of blood in a microvessel is smaller than the bulk viscosity measured with a rotational viscometer or a large-bore capillary viscometer.
Krogh tissue cylinder model: A cylindrical volume of tissue supplied by a central cylindrical capillary.
Myogenic response: Vasoconstriction in response to elevated transmural pressure and vasodilation in response to reduced transmural pressure.
Myoglobin-facilitated O_2 diffusion: An increase of O_2 diffusive flux as a result of myoglobin molecules acting as a carrier for O_2 molecules.
Oxyhemoglobin dissociation curve: The equilibrium relationship between hemoglobin saturation and O_2 tension.
Poiseuille's law: The relationship between volumetric flow rate and pressure difference for steady flow of a Newtonian fluid in a long circular tube.
Precapillary O_2 transport: O_2 diffusion from arterioles to the surrounding tissue.
Starling's law: The relationship between water flux through the endothelium and the difference between the hydraulic and osmotic transmural pressures.
Vasomotion: Spontaneous rhythmic variation of microvessel diameter.

References

Bassingthwaighte, J.B., Liebovitch, L.S., and West, B.J. 1994. *Fractal Physiology.* Oxford University Press, New York.
Cabel, M., Meiselman, H.J., Popel, A.S., and Johnson, P.C. 1997. Contribution of red cell aggregation to venous vascular resistance. *Am. J. Physiol. (Heart Circ. Physiol.)* 272:H1020–H1032.
Chien, S. 1992. Blood cell deformability and interactions: from molecules to micromechanics and microcirculation. *Microvasc. Res.* 44:243–254.

Curry, F.-R.E. 1994. Regulation of water and solute exchange in microvessel endothelium: studies in single perfused capillaries. *Microcirculation* 1:11–26.

Das, B., Enden, G., and Popel, A.S. 1997. Stratified multiphase model for blood flow in a venular bifurcation. *Ann. Biomed. Eng.* 25:135–153.

Davis, M.J. 1993. Myogenic response gradient in an arteriolar network. *Am. J. Physiol. (Heart Circ. Physiol.)* 264:H2168–H2179.

Desjardins, C. and Duling, B.R. 1990. Heparinase treatment suggests a role for the endothelial cell glycocalyx in regulation of capillary hematocrit. *Am. J. Physiol. (Heart Circ. Physiol.)* 258:H647–H654.

Duling, B.R. 1991. Control of striated muscle blood flow. In *The Lung: Scientific Foundations*, R.G. Crystal and J.B. West, Eds., Raven Press, New York, 1497–1505.

Ellsworth, M.L., and Pittman, R.N. 1986. Evaluation of photometric methods for quantifying convective mass transport in microvessels. *Am. J. Physiol. (Heart Circ. Physiol.)* 251:H869–H879.

Ellsworth, M.L., Ellis, C.G., Popel, A.S., and Pittman, R.N. 1994. Role of microvessels in oxygen supply to tissue. *News Physiol. Sci.* 9:119–123.

Ellsworth, M.L., Forrester, T., Ellis, C.G., and Dietrich, H.H. 1995. The erythrocyte as a regulator of vascular tone. *Am. J. Physiol. (Heart Circ Physiol)* 269:H2155–H2161.

Furchgott, R.F. and Vanhoutte, P.M. 1989. Endothelium-derived relaxing and contracting factors. *FASEB J.* 3:2007–2018.

Griffith, T.M. 1996. Temporal chaos in the microcirculation. *Cardiovasc. Res.* 31:342–358.

Granger, D.N. and Schmid-Schönbein, G.W. 1994. *Physiology and Pathophysiology of Leukocyte Adhesion.* Oxford University Press, New York.

Guilford, W.H. and Gore, R.W. 1995. The mechanics of arteriole-tissue interaction. *Microvasc. Res.* 50:260–287.

Hellums, J.D., Nair, P.K., Huang, N.S., and Ohshima, N. 1996. Simulation of intraluminal gas transport processes in the microcirculation. *Ann. Biomed. Eng.* 24:1–24.

Hudetz, A.G. 1997. Blood flow in the cerebral capillary network: a review emphasizing observations with intravital microscopy. *Microcirculation* 4:233–252.

Ignarro, L.J. 1989. Biological actions and properties of endothelium-derived nitric oxide formed and released from artery and vein. *Circ. Res.* 65:1–21.

Kiani, M.F., Pries, A.R., Hsu, L.L., Sarelius, I.H., and Cokelet, G.R. 1994. Fluctuations in microvascular blood flow parameters caused by hemodynamic mechanisms. *Am. J. Physiol. (Heart Circ. Physiol.)* 266:H1822–H1828.

Ley, K. and Gaehtgens, P. 1991. Endothelial, not hemodynamic, differences are responsible for preferential leukocyte rolling in rat mesenteric venules. *Circ. Res.* 69:1034–1041.

Lipowsky, H.H. 1987. Mechanics of blood flow in the microcirculation. In *Handbook of Bioengineering*, R. Skalak and S. Chien, Eds., McGraw Hill, New York, 18.1–18.25.

Michel, C.C. 1996. Starling: the formulation of his hypothesis of microvascular fluid exchange and its significance after 100 years. *Exp. Physiol.* 82:1–30.

Muller, J.M., Davis, M.J., and Chilian, W.M. 1996. Integrated regulation of pressure and flow in the coronary microcirculation. *Cardiovasc. Res.* 32:668–678.

Pittman, R.N. 1995. Influence of microvascular architecture on oxygen exchange in skeletal muscle. *Microcirculation* 2:1–18.

Popel, A.S. 1987. Network models of peripheral circulation. In *Handbook of Bioengineering*, R. Skalak and S. Chien, Eds., McGraw Hill, New York, 20.1–20.24.

Popel, A.S. 1989. Theory of oxygen transport to tissue. *Crit. Rev. Biomed. Eng.* 17:257-321.

Popel, A.S., Greene, A.S., Ellis, C.G., Ley, K.F., Skalak, T.C., and Tonellato, P.J. 1998. The Microcirculation Physiome project. *Ann. Biomed. Eng.,* 26:911-913.

Pries, A.R., Secomb, T.W., and Gaehtgens, P. 1996. Biophysical aspects of blood flow in the microvasculature. *Cardiovasc. Res.* 32:654–667.

Pries, A.R., Secomb, T.W., Jacobs, H., Sperandio, M., Osterloh, K., and Gaehtgens, P. 1997. Microvascular blood flow resistance: role of endothelial surface layer. *Am. J. Physiol. (Heart Circ. Physiol.)* 273:H2272–H2279.

Roy, T.K. and Popel, A.S. 1996. Theoretical predictions of end-capillary PO_2 in muscles of athletic and non-athletic animals at VO_{2max}. *Am. J. Physiol. (Heart Circ. Physiol.)* 271:H721–H737.

Schmid-Schönbein, G.W., Skalak, T.C., and Sutton, D.W. 1989. Bioengineering analysis of blood flow in resting skeletal muscle. In *Microvascular Mechanics*, J.S. Lee and T.C. Skalak, Eds., Springer-Verlag, New York, 65–99.

Secomb, T.W., Hsu, R., and Pries, A.R. 1998. A model for red blood cell motion in glycocalyx-lined capillaries. *Am. J. Physiol. (Heart Circ. Physiol.)* 274:H1016–H1022.

Segal, S.S. 1992. Communication among endothelial and smooth muscle cells coordinates blood flow control during exercise. *News Physiol. Sci.* 7:152-156.

Segal, S.S. and Kurjiaka, D.T. 1995. Coordination of blood flow control in the resistance vasculature of skeletal muscle. *Med. Sci. Sports Exerc.* 27:1158–1164.

Shoukas, A.A. and Bohlen, H.G. 1990. Rat venular pressure-diameter relationships are regulated by sympathetic activity. *Am. J. Physiol. (Heart Circ. Physiol.)* 259:H674–H680.

Sibbald, W.J., Messmer, K., and Fink, M.P., Eds. 1998. *Tissue Oxygenation in Acute Medicine.* Springer-Verlag, New York.

Skalak, T.C. and Price, R.J. 1996. The role of mechanical stresses in microvascular remodeling. *Microcirculation* 3:143–165.

Skalak, R., Özkaya, N., and Skalak, T.C. 1989. Biofluid mechanics. *Ann. Rev. Fluid Mech.* 21:167–204.

Smieško, V. and Johnson, P.C. 1993. The arterial lumen is controlled by flow-related shear stress. *News Physiol. Sci.* 8:34–38.

Tangelder, G.J., Slaaf, D.W., Muitjens, A.M.M., Arts, T., Oude Egbrink, M.G.A., and Reneman, R.S. 1986. Velocity profiles in blood platelets and red blood cells flowing in arterioles of the rabbit mesentery. *Circ. Res.* 59:505-514.

Ursino, M. and Fabbri, G. 1992. Role of the myogenic mechanism in the genesis of microvascular oscillations (vasomotion): analysis with a mathematical model. *Microvasc. Res.* 43:156–177.

Weinbaum, S. 1998. 1997. Whitaker Distinguished Lecture: models to solve mysteries in biomechanics at the cellular level; a new view of fiber matrix layers. *Ann. Biomed. Eng.* 26:627–43.

Weibel, E.R., Taylor, C.R., and Hoppeler, H. 1992. Variations in function and design: testing symmorphosis in the respiratory system. *Respir. Physiol.* 87:325–348.

Winslow, R.M., Vandergriff, K.D., and Intaglietta, M., Eds. 1997. *Advances in Blood Substitutes. Industrial Opportunities and Medical Challenges.* Birkhauser, Boston.

Woldhuis, B., Tangelder,G.J., Slaaf, D.W., and Reneman, R.S. 1992. Concentration profile of blood platelets differs in arterioles and venules. *Am. J. Physiol. (Heart Circ. Physiol.)* 262:H1217–H1223.

Further Information

The two-volume set, *Handbook of Physiology*, Section 2: *The Cardiovascular System*, Vol. IV, Parts 1 and 2, *Microcirculation*, Renkin, E.M. and Michel, C.C. Eds., 1984, American Physiological Society, remains the most comprehensive overview of the field.

Original research articles on microcirculation can be found in academic journals. *Microcirculation, Microvascular Research, American Journal of Physiology (Heart and Circulatory Physiology), Journal of Vascular Research, Biorheology, Annals of Biomedical Engineering,* and *Journal of Biomechanical Engineering* publish articles emphasizing engineering approaches.

15

Mechanics and Deformability of Hematocytes

Richard E. Waugh
University of Rochester

Robert M. Hochmuth
Duke University

The term *hematocytes* refers to the circulating cells of the blood. These are divided into two main classes: erythrocytes, or red cells, and leukocytes, or white cells. In addition to these are specialized cell-like structures called *platelets.* The mechanical properties of these cells are of special interest because of their physiological role as circulating corpuscles in the flowing blood. The importance of the mechanical properties of these cells and their influence on blood flow is evident in a number of hematological pathologies. The properties of the two main types of hematocytes are distinctly different. The essential character of a red cell is that of an elastic bag enclosing a newtonian fluid of comparatively low viscosity. The essential behavior of white cells is that of a high viscous fluid drop with a more or less constant cortical (surface) tension. Under the action of a given force, red cells deform much more readily than white cells. In this chapter we focus on descriptions of the behavior of the two cell types separately, concentrating on the viscoelastic characteristics of the red cell membrane and the fluid of characteristics of the white cell cytosol.

15.1 Fundamentals

Stresses and Strains in Two Dimensions

The description of the mechanical deformation of the membrane is cast in terms of principal *force resultants* and principle *extension ratios* of the surface. The force resultants, like conventional three-dimensional strain, are generally expressed in terms of a tensorial quantity, the components of which depend on coordinate rotation. For the purposes of describing the constitutive behavior of the surface, it is convenient to express the surface resultants in terms of rotationally invariant quantities. These can be either the principal force resultants N_1 and N_2, or the isotropic resultant $\overline{N}$ and the maximum shear resultant N_s. The surface strain is also a tensorial quantity but may be expressed in terms of the principal extension ratios of the surface λ_1 and λ_2. The rate of surface shear deformation [Evans and Skalak, 1979] is given by:

$$V_s = \left(\frac{\lambda_2}{\lambda_1}\right)^{1/2} \frac{d}{dt}\left(\frac{\lambda_1}{\lambda_2}\right)^{1/2} \tag{15.1}$$

The membrane deformation is calculated from observed macroscopic changes in cell geometry, usually with the use of simple geometric shapes to approximate the cell shape. The membrane force resultants are calculated from force balance relationships. For example, in the determination of the *area expansivity modulus* of the red cell membrane or the *cortical tension* in neutrophils, the force resultants in the plane of the membrane of the red cell or the white cell are isotropic. In this case, as long as the membrane surface of the cell does not stick to the pipette, the membrane force resultant can be calculated from the law of Laplace:

$$\Delta P = 2\overline{N}\left(\frac{1}{R_p} - \frac{1}{R_c}\right) \tag{15.2}$$

where R_p is the radius of the pipette, R_c is the radius of the spherical portion of the cell outside the pipette, $\overline{N}$ is the isotropic force resultant (tension) in the membrane, and ΔP is the aspiration pressure in the pipette.

Basic Equations for Newtonian Fluid Flow

The constitutive relations for fluid flow in a sphere undergoing axisymmetric deformation can be written:

$$\sigma_{rr} = -p + 2\eta \frac{\partial V_r}{\partial r} \tag{15.3}$$

$$\sigma_{r\theta} = \eta\left[\frac{1}{r}\frac{\partial V_r}{\partial \theta} + r\frac{\partial}{\partial r}\left(\frac{V_\theta}{r}\right)\right] \tag{15.4}$$

where σ_{rr} and $\sigma_{r\theta}$ are components of the stress tensor, p is the hydrostatic pressure, r is the radial coordinate, θ is the angular coordinate in the direction of the axis of symmetry in spherical coordinates, and V_r and V_θ are components of the fluid velocity vector. These equations effectively define the material viscosity, η. In general, η may be a function of the strain rate. The second term in Eq. (15.3) contains the radial strain rate $\dot{\varepsilon}_{rr}$ and the bracketed term in Eq. (15.4) corresponds to $\dot{\varepsilon}_{r\theta}$. For the purposes of evaluating this dependence, it is convenient to define the mean shear rate $\dot{\gamma}_m$ averaged over the cell volume and duration of the deformation process t_e:

$$\dot{\gamma}_m = \left(\frac{3}{4}\frac{1}{t_e}\int_0^{t_e}\int_0^{R(t)}\int_0^{\pi}\frac{r^2}{R^3}\left(\dot{\varepsilon}_{ij}\dot{\varepsilon}_{ij}\right)\sin\theta\,d\theta\,dr\,dt\right)^{\frac{1}{2}} \tag{15.5}$$

where repeated indices indicate summation.

TABLE 15.1   Parameter Values for a Typical
Red Blood Cell (37°C)

Area	132 μm²
Volume	96 μm²
Sphericity	0.77
Membrane area modulus	480 mN/m
Membrane shear modulus	0.006 mN/m
Membrane viscosity	0.00036 mN·s/m
Membrane bending stiffness	0.2×10^{-18} J
Thermal area expansivity	0.12%/°C
$\dfrac{1}{V}\dfrac{dV}{dT}$	−0.14%/°C

15.2   Red Cells

Size and Shape

The normal red cell is a biconcave disk at rest. The average human cell is approximately 7.7 μm in diameter and varies in thickness from ~2.8 μm at the rim to ~1.4 μm at the center [Fung et al., 1981]. However, red cells vary considerably in size even within a single individual. The mean surface area ~130 μm² and the mean volume is 96 μm³ (Table 15.1), but the range of sizes within a population is gaussian-distributed with standard deviations of ~15.8 μm² for the area and ~16.1 μm³ for the volume [Fung et al., 1981]. Cells from different species vary enormously in size, and tables for different species have been tabulated elsewhere [Hawkey et al., 1991].

Red cell deformation takes place under two important constraints: fixed surface area and fixed volume. The constraint of fixed volume arises from the impermeability of the membrane to cations. Even though the membrane is highly permeable to water, the inability of salts to cross the membrane prevents significant water loss because of the requirement for colloidal osmotic equilibrium [Lew and Bookchin, 1986]. The constraint of fixed surface area arises from the large resistance of bilayer membranes to changes in area per molecule [Needham and Nunn, 1990]. These two constraints place strict limits on the kinds of deformations that the cell can undergo and the size of the aperture that the cell can negotiate. Thus, a major determinant of red cell deformability is its ratio of surface area to volume. One measure of this parameter is the *sphericity*, defined as the dimensionless ratio of the two-thirds power of the cell volume to the cell area times a constant that makes its maximum value 1.0:

$$S = \frac{4\pi}{\left(4\pi/3\right)^{2/3}} \cdot \frac{V^{2/3}}{A}. \tag{15.6}$$

The mean value of sphericity of a normal population of cells was measured by interference microscopy to be 0.79 with a standard deviation (SD) of 0.05 at room temperature [Fung et al., 1981]. Similar values were obtained using micropipettes: mean = 0.81, SD = 0.02 [Waugh and Agre, 1988]. The membrane area increases with temperature, and the membrane volume decreases with temperature, so the sphericity at physiological temperature, is expected to be somewhat smaller. Based on measurements of the thermal area expansivity of 0.12%/°C [Waugh and Evans, 1979] and a change in volume of −0.14%/°C [Waugh and Evans, 1979], the mean sphericity at 37°C is estimated to be 0.76–0.78. (See Table 15.1.)

Red Cell Cytosol

The interior of a red cell is a concentrated solution of hemoglobin, the oxygen-carrying protein, and it behaves as a newtonian fluid [Cokelet and Meiselman, 1968]. In a normal population of cells there is a

TABLE 15.2   Viscosity of Red Cell Cytosol (37°C)

Hemoglobin Concentration (g/l)	Measured Viscosity[a] (mPa·s)	Best Fit Viscosity[b] (mPa·s)
290	4.1–5.0	4.2
310	5.2–6.6	5.3
330	6.6–9.2	6.7
350	8.5–13.0	8.9
370	10.8–17.1	12.1
390	15.0–23.9	17.2

[a] Data taken from Cokelet and Meiselman [1968] and Chien and coworkers [1970].
[b] Fitted curve from Ross and Minton [1977].

distribution of hemoglobin concentrations in the range 29 to 39 g/dl. The viscosity of the cytosol depends on the hemoglobin concentration as well as temperature. (See Table 15.2.) Based on theoretical models [Ross and Minton, 1977], the temperature dependence of the cytosolic viscosity is expected to be the same as that of water; that is, the ratio of cytosolic viscosity at 37°C to the viscosity at 20°C is the same as the ratio of water viscosity at those same temperatures. In most cases, even in the most dense cells, the resistance to flow of the cytosol is small compared with the viscoelastic resistance of the membrane when membrane deformations are appreciable.

Membrane Area Dilation

The large resistance of the membrane to area dilation has been characterized in micromechanical experiments. The changes in surface area that can be produced in the membrane are small, and so they can be characterized in terms of a simple hookean elastic relationship between the isotropic force resultant $\overline{N}$ and the fractional change in surface area $\alpha = A/A_o - 1$:

$$\overline{N} = K\alpha. \tag{15.7}$$

The proportionality constant K is called the *area compressibility modulus* or the *area expansivity modulus*. Early estimates placed its value at room temperature at ~450 mN/m [Evans and Waugh, 1977] and showed a dependence of the modulus on temperature, its value changing from ~300 mN/m at 45°C to a value of ~600 mN/m at 5°C [Waugh and Evans, 1979]. Subsequently it was shown that the measurement of this parameter using micropipettes is affected by extraneous electric fields, and the value at room temperature was corrected upward to ~500 mN/m [Katnik and Waugh, 1990]. The values in Table 15.3 are based on this measurement, and the fractional change in the modulus with temperature is based on the original micropipette measurements [Waugh and Evans, 1979].

TABLE 15.3   Temperature Dependence of Viscoelastic Coefficients of the Red Cell Membrane

Temperature (°C)	K (mN/m)[a]	μ_m (mN/m)[b]	η_m (mN·s/m)[c]
5	670	0.0078	0.0021
15	580	0.0072	0.0014
25	500	0.0065	0.00074
37	400	0.0058	0.00036
45	330	0.0053	—

[a] Based on a value of the modulus at 35°C of 500 mN/m and the fractional change in modulus with temperature measured by Waugh and Evans [1979].
[b] Based on linear regression to the data of Waugh and Evans [1979].
[c] Data from Hochmuth and colleagues [1980].

Membrane Shear Deformation

The shear deformations of the red cell surface can be large, and so a simple linear relationship between force and extension is not adequate for describing the membrane behavior. The large resistance of the membrane composite to area dilation led early investigators to postulate that the membrane maintained constant surface density during shear deformation, that is, that the surface was two-dimensionally incompressible. Most of what exists in the literature about the shear deformation of the red cell membrane is based on this assumption. Only very recently has experimental evidence emerged that this assumption is an oversimplification of the true cellular behavior, and that deformation produces changes in the local surface density of the membrane elastic network [Discher et al., 1994]. Nevertheless, the older simpler relationships provide a description of the cell behavior that can be used for many applications, and so the properties of the cell defined under that assumption are summarized here.

For a simple, two-dimensional, incompressible, hyperelastic material, the relationship between the membrane shear force resultant N_s and the material deformation [Evans and Skalak, 1979] is:

$$N_s = \frac{\mu_m}{2}\left(\frac{\lambda_1}{\lambda_2} - \frac{\lambda_2}{\lambda_1}\right) + 2\eta_m V_s \qquad (15.8)$$

where λ_1 and λ_2 are the principal extension ratios for the deformation and V_s is the rate of surface shear deformation [Eq. (15.1)]. The *membrane shear modulus* μ_m and the *membrane viscosity* η_m are defined by this relationship. Values for these coefficients at different temperatures are given in Table 15.3.

Stress Relaxation and Strain Hardening

Subsequent to these original formulations a number of refinements to these relationships have been proposed. Observations of persistent deformations after micropipette aspiration for extended periods of time formed the basis for the development of a model for long-term stress relaxation [Markle et al., 1983]. The characteristic times for these relaxations were on the order of 1–2 h, and these times were thought to correlate with permanent rearrangements of the membrane elastic network.

Another type of stress relaxation is thought to occur over very short times (~0.1 s) after rapid deformation of the membrane either by micropipette [Chien et al., 1978] or in cell extension experiments (Waugh and Bisgrove, unpublished observations). This phenomenon is thought to be due to transient entanglements within the deforming network. Whether the phenomenon actually occurs remains controversial. The stresses relax rapidly, and it is difficult to account for inertial effects of the measuring system and to reliably assess the intrinsic cellular response.

Finally, there has been some evidence that the coefficient for shear elasticity may be a function of the surface extension, increasing with increasing deformation. This was first proposed by Fischer in an effort to resolve discrepancies between theoretical predictions and observed behavior of red cells undergoing dynamic deformations in fluid shear [Fischer et al., 1981]. Increasing elastic resistance with extension has also been proposed as an explanation for discrepancies between theoretical predictions based on a constant modulus and measurements of the length of a cell projection into a micropipette [Waugh and Marchesi, 1990]. However, due to the approximate nature of the mechanical analysis of cell deformation in shear flow, and the limits of optical resolution in micropipette experiments, the evidence for a dependence of the modulus on extension is not clear-cut, and this issue remains unresolved.

New Constitutive Relations for the Red Cell Membrane

The most modern picture of membrane deformation recognizes that the membrane is a composite of two layers with distinct mechanical behavior. The membrane bilayer, composed of phospholipids and integral membrane proteins, exhibits a large elastic resistance to area dilation but is fluid in surface shear. The membrane skeleton, composed of a network of structural proteins at the cytoplasmic surface of the bilayer, is locally compressible and exhibits an elastic resistance to surface shear. The assumption that the membrane skeleton is locally incompressible is no longer applied. This assumption had been challenged over the years on the basis of theoretical considerations, but only very recently has experimental evidence emerged that shows definitively that the membrane skeleton is compressible. [Discher et al., 1994]. This has led to a new constitutive model for membrane behavior [Mohandas and Evans, 1994]. The principal stress resultants in the membrane skeleton are related to the membrane deformation by:

$$N_1 = \mu_N \left(\frac{\lambda_1}{\lambda_2} - 1 \right) + K_N \left(\lambda_1 \lambda_2 - \frac{1}{(\lambda_1 \lambda_2)n} \right) \tag{15.9}$$

and

$$N_2 = \mu_N \left(\frac{\lambda_2}{\lambda_1} - 1 \right) + K_N \left(\lambda_1 \lambda_2 - \frac{1}{(\lambda_1 \lambda_2)n} \right) \tag{15.10}$$

where μ_N and K_N are the shear and isotropic moduli of the membrane skeleton, respectively, and n is a parameter to account for molecular crowding of the skeleton in compression. The original modulus based on the two-dimensionally incompressible case is related to these moduli by:

$$\mu_m \approx \frac{\mu_N K_N}{\mu_N + K_N} \tag{15.11}$$

Values for the coefficients determined from fluorescence measurements of skeletal density distributions during micropipette aspiration studies are $\mu_N \approx 6 \times 10^{-3}$ mN/m, and $K_N/\mu_N \approx 2$. The value for n is estimated to be ≥ 2 [Mohandas and Evans, 1994; Discher, et al., 1994].

These new concepts for membrane constitutive behavior have only recently been introduced and have yet to be thoroughly explored. The temperature dependence of these moduli is unknown, and the implications such a model will have on interpretation of dynamic deformations of the membrane remain to be resolved.

Bending Elasticity

Even though the membrane is very thin, it has a high resistance to surface dilation. This property, coupled with the finite thickness of the membrane, gives the membrane a small but finite resistance to bending. This resistance is characterized in terms of the *membrane-bending modulus*. The bending resistance of biological membranes is inherently complex because of their lamellar structure. There is a local resistance to bending due to the inherent stiffness of the individual leaflets of the membrane bilayer. (Because the membrane skeleton is compressible, it is thought to contribute little if anything to the membrane-bending stiffness.) In addition to this local stiffness, there is a *nonlocal bending resistance* due to the net compression and expansion of the adjacent leaflets resulting from the curvature change. The nonlocal contribution is complicated by the fact that the leaflets may be redistributed laterally within the membrane capsule to equalize the area per molecule within each leaflet. The situation is further complicated by the likely possibility that molecules may exchange between leaflets to alleviate curvature-induced dilation/compression. Thus, the bending stiffness measured by different approaches probably reflects contributions from both local and nonlocal mechanisms, and the measured values may differ because of different contributions from the two mechanisms. Estimates based on buckling instabilities during micropipette aspiration give a value of $\sim 0.18 \times 10^{-18}$ J [Evans, 1983]. Measurements based on the mechanical formation of lipid tubes from the cell surface give a value of $\sim 0.20 \times 10^{-18}$ J [Hwang and Waugh, 1997].

15.3 White Cells

Whereas red cells account for approximately 40% of the blood volume, white cells occupy less than 1% of the blood volume. Yet because white cells are less deformable, they can have a significant influence on blood flow, especially in the microvasculature. Unlike red cells, which are very similar to each other, as are platelets, there are several different kinds of white cells or *leukocytes*. In general, the white cells are classified into groups according to their appearance when viewed with the light microscope. Thus, there are *granulocytes, monocytes,* and *lymphocytes* [Alberts et al., 1994]. The granulocytes with their many internal granules are separated into *neutrophils, basophils,* and *eosinophils* according to the way each cell stains. The neutrophil, also called a *polymorphonuclear leukocyte* because of its segmented or multilobed nucleus, is the most common white cell in the blood. (See Table 15.4). The lymphocytes, which constitute 20–40% of the white cells and which are further subdivided into *B lymphocytes* and *killer* and *helper T lymphocytes,* are the smallest of the white cells. The other types of leukocytes are found with much less frequency. Most of the geometric and mechanical studies of white cells reported below have focused on the neutrophil because it is the most common cell in the circulation, although the lymphocyte is now starting to receive more attention.

Size and Shape

White cells at rest are spherical. The surfaces of white cells contain many folds, projections, and "microvilli" to provide the cells with sufficient membrane area to deform as they enter capillaries with diameters much smaller than the resting diameter of the cell. (Without the reservoir of membrane area in these folds, the constraints of constant volume and membrane area would make a spherical cell essentially undeformable.) The excess surface area of the neutrophil, when measured in a wet preparation, is slightly more than twice the apparent surface area of a smooth sphere with the same diameter [Evans and Yeung, 1989; Ting-Beall et al., 1993]. It is interesting to note that each type of white cell has its own unique surface topography, which allows one to readily determine if a cell is, for example, either a neutrophil or monocyte or lymphocyte [Hochmuth et al., 1995].

The cell volumes listed in Table 15.4 were obtained with the light microscope, either by measuring the diameter of the spherical cell or by aspirating the cell into a small glass pipette with a known diameter and then measuring the resulting length of the cylindrically shaped cell. Other values for cell volume obtained using transmission electron microscopy are somewhat smaller, probably because of cell shrinkage due to fixation and drying prior to measurement [Schmid-Schönbein et al., 1980; Ting-Beall et al., 1995]. Although the absolute magnitude of the cell volume measured with the electron microscope may be erroneous, if it is assumed that all parts of the cell dehydrate equally when they are dried in preparation for viewing, then this approach can be used to determine the volume occupied by the nucleus (Table 15.4)

TABLE 15.4 Size and Appearance of White Cells in the Circulation

	Occurrence[a] (% of WBCs)	Cell Volume[b] (μm^3)	Cell Diameter[b] (μm)	Nucleus[c] (% Cell Volume)	Cortical Tension (mN/m)
Granulocytes					
Neutrophils	50–70	300–310	8.2–8.4	21	0.024–0.035[d]
Basophils	0–1				
Eosinophils	1–3			18	
Monocytes	1–5	400	9.1	26	0.06[e]
Lymphocytes	20–40	220	7.5	44	0.035[e]

[a] Diggs et al. [1985].

[b] Ting-Beall et al. [1993]. (Diameter calculated from the volume of a sphere).

[c] Schmid-Schönbein et al. [1980].

[d] Evans and Yeung [1989], Needham and Hochmuth [1992], Tsai et al. [1993, 1994].

[e] Preliminary data, Hochmuth, Zhelev, and Ting-Beall.

and other organelles of various white cells. The volume occupied by the granules in the neutrophil and eosinophil (recall that both are granulocytes) is 15 and 23%, respectively, whereas the granular volume in monocytes and lymphocytes is less than a few percent.

Mechanical Behavior

The early observations of Bagge and colleagues [1977] led them to suggest that the neutrophil behaves as a simple viscoelastic solid with a Maxwell element (an elastic and viscous element in series) in parallel with an elastic element. This elastic element in the model was thought to pull the unstressed cell into its spherical shape. Subsequently, Evans and Kukan [1984] and Evans and Yeung [1989] showed that the cells flow continuously into a pipette, with no apparent approach to a static limit, when a constant suction pressure was applied. Thus, the cytoplasm of the neutrophil should be treated as a liquid rather than a solid, and its surface has a persistent *cortical tension* that causes the cell to assume a spherical shape.

Cortical Tension

Using a micropipet and a small suction pressure to aspirate a hemispherical projection from a cell body into the pipette, Evans and Yeung measured a value for the cortical tension of 0.035 mN/m. Needham and Hochmuth [1992] measured for cortical tension of individual cells that were driven down a tapered pipette in a series of equilibrium positions. In many cases the cortical tension increased as the cell moved further into the pipette, which means that the cell has an apparent area expansion modulus [Eq. (15.7)]. They obtained an average value of 0.04 mN/m for the expansion modulus and an extrapolated value for the cortical tension (at zero area dilation) in the resting state of 0.024 mN/m. The importance of the actin cytoskeleton in maintaining cortical tension was demonstrated by Tsai et al. [1994]. Treatment of the cells with a drug that disrupts actin filament structure (CTB = cytochalasin B) resulted in a decrease in cortical tension from 0.027 to 0.022 mN/m at a CTB concentration of 3 μM and to 0.014 mN/m at 30 μM.

Preliminary measurements in one of the authors' laboratories indicate that the value for the cortical tension of a monocyte is about double that for a granulocyte, that is, 0.06 mN/m, and the value for a lymphocyte is about 0.035 mN/m.

Bending Rigidity

The existence of a cortical tension suggests that there is a cortex—a relatively thick layer of F-actin filaments and myosin—that is capable of exerting a finite tension at the surface. If such a layer exists, it would have a finite thickness and bending rigidity. Zhelev and colleagues [1994] aspirated the surface of neutrophils into pipettes with increasingly smaller diameters and determined that the surface had a bending modulus of about 1 to 2×10^{-18} J, which is 5 to 50 times the bending moduli for erythrocyte or lipid bilayer membranes. The thickness of the cortex should be smaller than the radius of smallest pipette used in this study, which was 0.24 μm.

Apparent Viscosity

Using their model of the neutrophil as a newtonian liquid drop with a constant cortical tension and (as they showed) a negligible surface viscosity, Yeung and Evans [1989] analyzed the flow of neutrophils into a micropipette and obtained a value for the *cytoplasmic viscosity* of about 200 Pa·s. In their experiments, the aspiration pressures were on the order of 10 to 1000 Pa. Similar experiments by Needham and Hochmuth [1990] using the same newtonian model (with a negligible surface viscosity) but using higher aspiration pressures (ranging from 500 to 2000 Pa) gave an average value for the cytoplasmic viscosity of 135 Pa·s for 151 cells from five individuals. The apparent discrepancy between these two sets of experiments was resolved to a large extent by Tsai et al. [1993], who demonstrated that the neutrophil viscosity decreases with increasing rate of deformation. They proposed a model of the cytosol as a *power law fluid:*

$$\eta = \eta_c \left(\frac{\dot{\gamma}_m}{\dot{\gamma}_c} \right)^{-b} \tag{15.12}$$

where $b = 0.52$, $\dot{\gamma}_m$ is defined by Eq. (15.5), and η_c is a characteristic viscosity of 130 Pa·s when the characteristic mean shear rate, $\dot{\gamma}_c$ is 1 s^{-1}. These values are based on an approximate method for calculating the viscosity from measurements of the total time it takes for a cell to enter a micropipette. Because of different approximations used in the calculations, the values of viscosity reported by Tsai et al. [1993] tend to be somewhat smaller than those reported by Evans and coworkers or Hochmuth and coworkers. Nevertheless, the shear rate dependence of the viscosity is the same, regardless of the method of calculation. Values for the viscosity are given in Table 15.5.

In addition to the dependence of the viscosity on shear rate, there is also evidence that it depends on the extent of deformation. In micropipette experiments the initial rate at which the cell enters the pipette is significantly faster than predicted, even when the shear rate dependence of the viscosity is taken into account. In a separate approach, the cytosolic viscosity was estimated from observation of the time course of the cell's return to a spherical geometry after expulsion from a micropipette. When the cellular deformations were large, a viscosity of 150 Pa·s was estimated [Tran-Son-Tay et al., 1991], but when the deformation was small, the estimated viscosity was only 60 Pa·s [Hochmuth et al., 1993]. Thus, it appears that the viscosity is smaller when the magnitude of the deformation is small and increases as deformations become large.

An alternative attempt to account for the initial rapid entry of the cell into micropipettes involved the application of a *Maxwell fluid* model with a constant cortical tension. Dong and coworkers [1988] used this model to analyze both the shape recovery of neutrophils following small, complete deformations in pipettes and the small-deformation aspiration of neutrophils into pipettes. However, in another study by Dong and coworkers [1991], they used a finite-element, numerical approach and a Maxwell model with constant cortical tension to describe the continuous, finite-deformation flow of a neutrophil into a pipette. But in order to fit the theory to the data for the increase in length of the cell in the pipette with time, Dong and colleagues [1991] had to steadily increase both the elastic and viscous coefficients in their finite-deformation Maxwell model. This shows that even a Maxwell model is not adequate for describing the rheological properties of the neutrophil.

Although it is clear that the essential behavior of the cell is fluid, the simple fluid drop model with a constant and uniform viscosity does not match the observed time course of cell deformation in detail [Drury and Dembo, 1998]. Better agreement between theory and experiment is obtained if approximate account is taken for the shear rate dependence of the viscosity, but some discrepancies remain.

TABLE 15.5 Viscous Parameters of White Blood Cells

Cell Type	Range of Viscosities (Pa·s)[a]		Characteristic Viscosity (Pa·s)	Shear Rate Dependence (*b*)
	Min.	Max.		
Neutrophil	50	500	130[b]	0.52[b]
(30-μM CTB)	41	52	54[b]	0.26[b]
Monocyte	70	1000		
LC60 (G1)			220[c]	0.53[c]
HL60 (S)			330[c]	0.56[c]

[a] Evans and Yeung [1989], Needham and Hochmuth [1992], Tsai et al. [1993, 1994].
[b] Tsai et al. [1993, 1994].
[c] Tsai and Waugh [1996a].

A quantitative model that accounts for the dependence of viscosity on the magnitude of deformation has not yet been formulated, and a complete analysis in which the viscosity is made to vary in space and time according to the local shear rate and extent of deformation has not yet been performed. Nevertheless, the fluid drop model captures the essential behavior of the cell, and when it is applied consistently (that is, for similar rate and extent of deformation) it provides a sound basis for predicting cell behavior and comparing the behaviors of different types of cells.

Although the mechanical properties of the neutrophil have been studied extensively as discussed above, the other white cells have not been studied in depth. Preliminary unpublished results from Robert Hochmuth's laboratory indicate that monocytes are somewhat more viscous (from roughly 30% to a factor of 2) than neutrophils under similar conditions in both recovery experiments and experiments in which the monocyte flows into a pipette. A lymphocyte, when aspirated into a small pipette so that its relatively large nucleus is deformed, behaves as an elastic body in that the projection length into the pipette increases linearly with the suction pressure. This elastic behavior appears to be due to the deformation of the nucleus, which has an apparent area elastic modulus of 2 mN/m. A lymphocyte recovers its shape somewhat more quickly than the neutrophil does, although this recovery process is driven both by the cortical tension and the elastic nucleus. These preliminary results are discussed by Tran-Son-Tay and coworkers [1994]. Finally, the properties of a human myeloid leukemic cell line (HL60) thought to resemble immature neutrophils of the bone marrow have also been characterized, as shown in Table 15.5. The apparent cytoplasmic viscosity varies both as a function of the cell cycle and during maturation toward a more neutrophil-like cell. The characteristic viscosity ($\dot{\gamma}_c = 1\,s^{-1}$) is 220 Pa·s for HL60 cells in the G1 stage of the cell cycle. This value increases to 400 Pa·s for cells in the S phase, but decreases with maturation, so that 7 days after induction the properties approach those of neutrophils (130 Pa·s) [Tsai and Waugh, 1996b].

It is important to note in closing that the characteristics described above apply to passive leukocytes. It is the nature of these cells to respond to environmental stimulation and engage in active movements and shape transformations. *White cell activation* produces significant heterogeneous changes in cell properties. The cell projections that form as a result of stimulation (called *pseudopodia*) are extremely rigid, whereas other regions of the cell may retain the characteristics of a passive cell. In addition, the cell can produce protrusive or contractile forces ranging from 2.0–10.7 nN [Guilford, et al., 1995]. The changes in cellular mechanical properties that result from cellular activation are complex and only beginning to be formulated in terms of mechanical models. This is expected to be a major focus of research in the coming years.

Summary

Constitutive equations that capture the essential features of the responses of red blood cells and passive leukocytes have been formulated, and material parameters characterizing the cellular behavior have been measured. The red cell response is dominated by the cell membrane which can be described as a hyper-viscoelastic, two-dimensional continuum. The passive white cell behaves like a highly viscous fluid drop, and its response to external forces is dominated by the large viscosity of the cytosol. Refinements of these constitutive models and extension of mechanical analysis to activated white cells is anticipated as the ultrastructural events that occur during cellular deformation are delineated in increasing detail.

Defining Terms

Area expansivity modulus: A measure of the resistance of a membrane to area dilation. It is the proportionality between the isotropic force resultant in the membrane and the corresponding fractional change in membrane area (units: 1 mN/m = 1 dyn/cm).

Cortical tension: Analogous to surface tension of a liquid drop, it is a persistent contractile force per unit length at the surface of a white blood cell (units: 1 mN/m = 1 dyn/cm).

Cytoplasmic viscosity: A measure of the resistance of the cytosol to flow (units: 1 Pa·s = 10 poise).

Force resultant: The stress in a membrane integrated over the membrane thickness. It is the two-dimensional analog of stress with units of force/length (units: 1 mN/m = 1 dyn/cm).

Maxwell fluid: A constitutive model in which the response of the material to applied stress includes both an elastic and viscous response in series. In response to a constant applied force, the material will respond elastically at first, then flow. At fixed deformation, the stresses in the material will relax to zero.

Membrane-bending modulus: The intrinsic resistance of the membrane to changes in curvature. It is usually construed to exclude nonlocal contributions. It relates the moment resultants (force times length per unit length) in the membrane to the corresponding change in curvature (inverse length) (units: 1 N·m = 1 joule = 10^7 erg).

Membrane shear modulus: A measure of the elastic resistance of the membrane to surface shear deformation; that is, changes in the shape of the surface at constant surface area [Eq. (15.8)] (units: 1 mN/·m = 1 dyn/cm).

Membrane viscosity: A measure of the resistance of the membrane to surface shear flow, that is, to the rate of surface shear deformation [Eq. (15.8)] (units: 1 mN·s/m = 1 $m^{Pa·s}$ m = 1 dyn · s/cm = 1 surface poise).

Nonlocal bending resistance: A resistance to bending resulting from the differential expansion and compression of the two adjacent leaflets of a lipid bilayer. It is termed *nonlocal* because the leaflets can move laterally relative to one another to relieve local strains, such that the net resistance to bending depends on the integral of the change in curvature of the entire membrane capsule.

Power law fluid: A model to describe the dependence of the cytoplasmic viscosity on rate of deformation [Eq. (15.12)].

Principal extension ratios: The ratios of the deformed length and width of a rectangular material element (in principal coordinates) to the undeformed length and width.

Sphericity: A dimensionless ratio of the cell volume (to the 2/3 power) to the cell area. Its value ranges from near zero to one, the maximum value corresponding to a perfect sphere [Eq. (15.6)].

White cell activation: The response of a leukocyte to external stimuli that involves reorganization and polymerization of the cellular structures and is typically accompanied by changes in cell shape and cell movement.

References

Alberts B, Bray D, Lewis J et al. 1994. *Molecular Biology of the Cell*, 3rd ed. New York, Garland Publishing.

Bagge U, Skalak R, Attefors R. 1977. Granulocyte rheology. *Adv Microcirc* 7:29.

Chien S, Sung KLP, Skalak R et al. 1978. Theoretical and experimental studies on viscoelastic properties of erythrocyte membrane. *Biophys J* 24:463.

Chien S, Usami S, Bertles JF. 1970. Abnormal rheology of oxygenated blood in sickle cell anemia. *J Clin Invest* 49:623.

Cokelet GR, Meiselman HJ. 1968. Rheological comparison of hemoglobin solutions and erythrocyte suspensions. *Science* 162:275.

Diggs LW, Sturm D, Bell A. 1985. *The Morphology of Human Blood Cells*, 5th ed. Abbott Park, IL, Abbott Laboratories.

Discher DE, Mohandas N, and Evans EA. 1994. Molecular maps of red cell deformation: hidden elasticity and in situ connectivity. *Science* 266:1032.

Dong C, Skalak R, Sung K-LP. 1991. Cytoplasmic rheology of passive neutrophils. *Biorheology* 28:557.

Dong C, Skalak R, Sung K-LP et al. 1988. Passive deformation analysis of human leukocytes. *J Biomech Eng* 110:27.

Drury JL, Dembo M. 1999. Hydrodynamics of micropipette aspiration. *Biophys J* 76:110.

Evans EA. 1983. Bending elastic modulus of red blood cell membrane derived from buckling instability in micropipet aspiration tests. *Biophys J* 43:27.

Evans E, Kukan B. 1984. Passive material behavior of granulocytes based on large deformation and recovery after deformation tests. *Blood* 64:1028.

Evans EA, Skalak R. 1979. Mechanics and thermodynamics of biomembrane. *Crit Rev Bioeng* 3:181.

Evans EA, Waugh R. 1977. Osmotic correction to elastic area compressibility measurements on red cell membrane. *Biophys J* 20:307.

Evans E, Yeung A. 1989. Apparent viscosity and cortical tension of blood granulocytes determined by micropipet aspiration. *Biophys J* 56:151.

Fischer TM, Haest CWM, Stohr-Liesen M, et al. The stress-free shape of the red blood cell membrane. *Biophys J* 34:409.

Fung YC, Tsang WCO, Patitucci P. 1981. High-resolution data on the geometry of red blood cells. *Biorheology* 18:369.

Guilford WH, Lantz RC, Gore RW. 1995. Locomotive forces produced by single leukocytes in vivo and *in vitro*. *Am J Physiol* 268:C1308.

Hawkey CM, Bennett PM, Gascoyne SC et al. 1991. Erythrocyte size, number and haemoglobin content in vertebrates. *Br J Haematol* 77:392.

Hochmuth RM, Buxbaum KL, Evans EA. 1980. Temperature dependence of the viscoelastic recovery of red cell membrane. *Biophys J* 29:177.

Hochmuth RM, Ting-Beall HP, Beaty BB et al. 1993. Viscosity of passive human neutrophils undergoing small deformations. *Biophys J* 64:1596.

Hochmuth RM, Ting-Beall HP, Zhelev DV. 1995. The mechanical properties of individual passive neutrophils *in vitro*. In DN Granger, GW Schmid-Schönbein (Eds.), *Physiology and Pathophysiology of Leukocyte Adhesion*, pp. 83–96. London, Oxford University Press.

Hwang WC, Waugh RE. 1997. Energy of dissociation of lipid bilayer from the membrane skeleton of red cells. *Biophys J* 72:2669.

Katnik C, Waugh R. 1990. Alterations of the apparent area expansivity modulus of red blood cell membrane by electric fields. *Biophys J* 57:877.

Lew VL, Bookchin RM. 1986. Volume, pH and ion content regulation human red cells: analysis of transient behavior with an integrated model. *J Membr Biol* 10:311.

Markle DR, Evans EA, Hochmuth RM. 1983. Force relaxation and permanent deformation of erythrocyte membrane. *Biophys J* 42:91.

Mohandas N, Evans E. 1994. Mechanical properties of the red cell membrane in relation to molecular structure and genetic defects. *Annu Rev Biophys Biomol Struct* 23:787.

Needham D, Hochmuth RM. 1992. A sensitive measure of surface stress in the resting neutrophil. *Biophys J* 61:1664.

Needham D, Hochmuth RM. 1990. Rapid flow of passive neutrophils into a 4 μm pipet and measurement of cytoplasmic viscosity. *J Biomech Eng* 112:269.

Needham D, Nunn RS. 1990. Elastic deformation and failure of lipid bilayer membranes containing cholesterol. *Biophys J* 58:997.

Ross PD, Minton AP. 1977. Hard quasispherical model for the viscosity of hemoglobin solutions. *Biochem Biophys Res Commun* 76:971.

Schmid-Schönbein GW, Shih YY, Chien S. 1980. Morphometry of human leukocytes. *Blood* 56(5):866.

Ting-Beall HP, Needham D, Hochmuth RM. 1993. Volume and osmotic properties of human neutrophils. *Blood* 81(10):2774.

Ting-Beall HP, Zhelev DV, Hochmuth RM. 1995. A comparison of different drying procedures for scanning electron microscopy using human leukocytes. *Microsc Res Tech* 32:357.

Tran-Son-Tay R, Kirk TF III, Zhelev DV et al. 1994. Numerical simulation of the flow of highly viscous drops down a tapered tube. *J Biomech Eng* 116:172.

Tran-Son-Tay R, Needham D, Yeung A et al. 1991. Time-dependent recovery of passive neutrophils after large deformation. *Biophys J* 60:856.

Tsai MA, Frank RS, Waugh RE. 1993. Passive mechanical behavior of human neutrophils: power-law fluid. *Biophys J* 65:2078.

Tsai MA, Frank RS, Waugh RE. 1994. Passive mechanical behavior of human neutrophils: effect of Cytochalasin B. *Biophys J* 66:2166.

Tsai MA, Waugh RE, Keng PC. 1996a. Cell cycle dependence of HL-60 cell deformability. *Biophys J* 70:2023.

Tsai MA, Waugh RE, Keng PC. 1996b. Changes in HL-60 cell deformability during differentiation induced by DMSO. *Biorheology* 33:1.

Waugh RE, Agre P. 1988. Reductions of erythrocyte membrane viscoelastic coefficients reflect spectrin deficiencies in hereditary spherocytosis. *J Clin Invest* 81:133.

Waugh RE, Bauserman RG. 1995. Physical measurements of bilayer-skeletal separation forces. *Ann Biomed Eng* 23:308.

Waugh R, Evans EA. 1979. Thermoelasticity of red blood cell membrane. *Biophys J* 26:115.

Waugh RE, Marchesi SL. 1990. Consequences of structural abnormalities on the mechanical properties of red blood cell membrane. In CM Cohen and J Palek (Eds.), *Cellular and Molecular Biology of Normal and Abnormal Erythrocyte Membranes*, pp. 185–199. New York, Alan R. Liss.

Yeung A, Evans E. 1989. Cortical shell-liquid core model for passive flow of liquid-like spherical cells into micropipets. *Biophys J* 56:139.

Zhelev DV, Needham D, Hochmuth RM. 1994. Role of the membrane cortex in neutrophil deformation in small pipets. *Biophys J* 67:696.

Further Information

Basic information on the mechanical analysis of biomembrane deformation can be found in Evans and Skalak [1979], which also appeared as a book under the same title (CRC Press, Boca Raton, FL, 1980). A more recent work that focuses more closely on the structural basis of the membrane properties is Berk and colleagues, Chapter 15, pp. 423–454, in the book *Red Blood Cell Membranes: Structure, Function, Clinical Implications*, edited by Peter Agre and John Parker (Marcel Dekker, New York, 1989). More detail about the membrane structure can be found in other chapters of that book.

Basic information about white blood cell biology can be found in the book by Alberts and coworkers [1994]. A more thorough review of white blood cell structure and response to stimulus can be found in two reviews by T. P. Stossel, one entitled, "The mechanics response of white blood cells," pp. 325–342, in the book *Inflammation: Basic Principles and Clinical Correlates*, edited by J. I. Galin and colleagues (Raven Press, New York, 1988), and the second entitled, "The molecular basis of white blood cell motility," pp. 541–562, in the book *The Molecular Basis of Blood Diseases*, edited by G. Stamatoyannopoulos and colleagues (W. B. Saunders, Philadelphia, 1994). Recent advances in white cell rheology can be found in the book *Cell Mechanics and Cellular Engineering*, edited by Van C. Mow and coworkers (Springer-Verlag, New York, 1994).

16

The Venous System

Artin A. Shoukas
The Johns Hopkins University

Carl F. Rothe
Indiana University

The venous system not only serves as a conduit for the return of blood from the capillaries to the heart but also provides a dynamic, variable blood storage compartment that influences cardiac output. The systemic (noncardiopulmonary) venous system contains more than 75% of the blood volume of the entire systemic circulation. Although the heart is the source of energy for propelling blood throughout the circulation, filling of the right heart before the subsequent beat is primarily passive. The subsequent amount of blood ejected is exquisitely sensitive to the transmural filling pressure. (For example, a change of right heart filling pressure of 1 cm water can cause the cardiac output to change by about 50%).

Because the blood vessels are elastic and have smooth muscle in their walls, contraction or relaxation of the smooth muscle can quickly redistribute blood between the periphery and the heart to influence cardiac filling and thus cardiac output. Even though the right ventricle is not essential for life, its functioning acts to reduce the central venous pressure to facilitate venous return [1]. It largely determines the magnitude of the cardiac output by influencing the degree of filling of the left heart. Dynamic changes in venous tone, by redistributing blood volume, can thus, at rest, change cardiac output over a range of more than ±20%. The dimensions of the vasculature influence both blood flow—by way of their resistive properties—and contained blood volume—by way of their capacitive properties. The arteries have about 10 times the resistance of the veins, and the veins are more than 10 times as compliant as the arteries.

The conduit characteristics of the venous system primarily depend on the anatomy of the system. Valves in the veins of the limbs are crucial for reducing the pressure in dependent parts of the body. Even small movements from skeletal muscle activity tend to compress the veins and move blood toward the heart. A competent valve then blocks back flow, thus relieving the pressure when the movement stops. Even a few steps can reduce the transmural venous pressure in the ankle from as much as 100 mmHg to about 20 mmHg. Without this mechanism, transcapillary movement of fluid into the extravascular spaces results in edema. Varicose (swollen) veins and peripheral pooling of blood can result from damage to the venous valves. During exercise, the rhythmic contraction of the skeletal muscles, in conjunction with venous valves, provides an important mechanism—the skeletal muscle pump—aiding the large increases in blood flow through the muscles without excessive increases in capillary pressure and blood pooling in the veins of the muscles. Without this mechanism, the increase in venous return leading to the dramatic increases in cardiac output would be greatly limited.

16.1   Definitions

Capacitance

Capacitance is a general term that relates the magnitude of contained volume to the transmural pressure across the vessel walls and is defined by the pressure–volume relationship. In living blood vessels, the pressure–volume relationship is complex and nonlinear. At transmural pressure near zero, there is a finite volume within the vessels (see definition of *unstressed volume*). If this volume is then removed from the vessels, there is only a small decrease in transmural pressure as the vessel collapses from a circular cross-section to an elliptical one. This is especially true for superficial or isolated venous vessels. However, for vessels which are tethered or embedded in tissue a negative pressure may result without appreciably changing the shape of the vessels. With increases in contained volume, the vessel becomes distended, and there is a concomitant increase in transmural pressure. The incremental change in volume to incremental change in transmural pressure is often relatively constant. At very high transmural pressures vessels become stiffer, and the incremental volume change to transmural pressure change is small. Because all blood vessels exhibit these nonlinearities, no single parameter can describe capacitance; instead, the entire pressure–volume relationship must be considered.

Compliance

Vascular compliance (C) is defined as the slope of the pressure-volume relationship. It is the ratio of the change in incremental volume (ΔV) to a change in incremental transmural pressure (ΔP). Thus $C = \Delta V/\Delta P$. Because the pressure–volume relationship is nonlinear, the slope of the relationship is not constant over its full range of pressures, and so the compliance should be specified at a given pressure. Units of compliance are those of volume divided by pressure, usually reported in ml/mmHg. Values are typically normalized to wet tissue weight or to total body weight. When the compliance is normalized by the total contained blood volume, it is termed the *vascular distensibility* and represents the fractional change in volume ($\Delta V/V$) per change in transmural pressure; $D = (\Delta V/V)\Delta P$, where V is the volume at control or at zero transmural pressure.

Unstressed Volume

Unstressed volume (V_0) is the volume in the vascular system when the transmural pressure is zero. It is a calculated volume obtained by extrapolating the relatively linear segment of the pressure–volume relationship over the normal operating range to zero transmural pressure. Many studies have shown that reflexes and drugs have quantitatively more influence on V_0 than on the compliance.

Stressed Volume

The *stressed volume* (V_s) is the volume of blood in the vascular system that must be removed to change the computed transmural pressure from its prevailing value to zero transmural pressure. It is computed as the product of the vascular compliance and transmural distending pressure: $V_s = C \times P$. The total contained blood volume at a specific pressure (P) is the sum of stressed and unstressed volume. The unstressed volume is then computed as the total blood volume minus the stressed volume. Because of the marked nonlinearity around zero transmural pressure and the required extrapolation, both V_0 and V_s are virtual volumes.

Capacity

Capacity refers to the amount of blood volume contained in the blood vessels at a specific distending pressure. It is the sum of the unstressed volume and the stressed volume, $V = V_0 + V_s$.

Mean Filling Pressure

If the inflow and outflow of an organ are suddenly stopped, and blood volume is redistributed so that all pressures within the vasculature are the same, this pressure is the *mean filling pressure* [5]. This pressure can be measured for the systemic or pulmonary circuits or the body as a whole. The arterial pressure often does not equal the venous pressure as flow is reduced to zero, because blood must move from the distended arterial vessels to the venous beds during the measurement maneuver, and flow may stop before equilibrium occurs. This is because smooth-muscle activity in the arterial vessels, rheological properties of blood, or high interstitial pressures act to impede the flow, thus corrections must often be made [4,5]. The experimentally measured mean filling pressure provides a good estimate of P_v, (the pressure in the minute venules), for estimating venous stressed volume.

Venous Resistance

Venous resistance (R) refers to the hindrance to blood flow through the venous vasculature caused by friction of the moving blood along the venous vascular wall. By definition it is the ratio of the pressure gradient between the entrance of the venous circulation, namely the capillaries, and the venous outflow divided by the venous flow rate. Thus,

$$R = \frac{\left(P_c = P_{ra}\right)}{F} \tag{16.1}$$

where R is the venous resistance, P_c is the capillary pressure, P_{ra} is the right atrial pressure, and F is the venous flow. As flow is decreased to zero, arterial closure may occur, leading to a positive perfusion pressure at zero flow. With partial collapse of veins, a Starling resistor-like condition is present in which an increase in outlet pressure has no influence on flow until the outlet pressure is greater than the "waterfall" pressure.

Venous Inertance

Venous inertance (I_v) is the opposition to a change in flow rate related to the mass of the bolus of blood that is accelerated or decelerated. The inertance I_v for a cylindrical tube with constant cross-sectional area is $I_v = L\rho/A$, where L is the length of the vessel, ρ is the density of the blood, and A is the cross-sectional area [9].

16.2   Methods To Measure Venous Characteristics

Our knowledge of the nature and role of the capacitance characteristics of the venous system has been limited by the difficulty of measuring the various variables needed to compute parameter values. State-of-the-art equipment is often needed because of the low pressures and many disturbing factors present. Many of the techniques that have been used to measure venous capacitance require numerous assumptions that may not be correct or are currently impossible to evaluate [4].

Resistance

For the estimate of vascular resistance, the upstream to outflow pressure gradient across the tissues must be estimated along with a measure of flow. Pressures in large vessels are measured with a catheter connected to a pressure transducer, which typically involves measurement of minute changes in resistance elements attached to a stiff diaphragm which flexes proportionally to the pressure. For the veins in tissue, the upstream pressure, just downstream from the capillaries, is much more difficult to measure because of the minute size (ca. 15 µm) of the vessels. For this a servo-null micropipette technique may be used.

A glass micropipette with a tip diameter of about 2 μm is filled with a 1–2 mol saline solution. When the pipette is inserted into a vein, the pressure tends to drive the lower conductance blood plasma into the pipette. The conductance is measured using an AC-driven bridge. A servosystem, driven by the imbalance signal, is used to develop a counter pressure to maintain the interface between the low-conductance filling solution and the plasma near the tip of the pipette. This counter pressure, which equals the intravascular pressure, is measured with a pressure transducer. Careful calibration is essential.

Another approach for estimating the upstream pressure in the veins is to measure the mean filling pressure of the organ (see above) and assume that this pressure is the upstream venous pressure. Because this venous pressure must be less than the capillary pressure and because most of the blood in an organ is in the small veins and venules, this assumption, though tenuous, is not unreasonable. To measure flow many approaches are available including electromagnetic, transit-time ultrasonic, or Doppler ultrasonic flowmeters. Usually the arterial inflow is measured with the assumption that the outflow is the same. Indicator dilution techniques are also used to estimate average flow. They are based on the principle that the reduction in concentration of infused indicator is inversely proportional to the rate of flow. Either a bolus injection or a continuous infusion may be used. Adequacy of mixing of indicator across the flow stream, lack of collateral flows, and adequately representative sampling must be considered [2].

Capacitance

For estimating the capacitance parameters of the veins, contained volume, rather than flow, and transmural pressure, rather than the longitudinal pressure gradient, must be measured. Pressures are measured as described above. For the desired pressure–volume relationship the total contained volume must be known.

Techniques used to measure total blood volume include *indicator dilution*. The ratio of the integral of indicator concentration time to that of concentration is used to compute the mean transit time (MIT) following the sudden injection of a bolus of indicator [2,4]. The active volume is the product of MTT and flow, with flow measured as outlined above. Scintigraphy provides an image of the distribution of radioactivity in tissues. A radioisotope, such as technicium 99 that is bound to red blood cells which in turn are contained within the vasculature, is injected and allowed to equilibrate. A camera, with many collimating channels sensitive to the emitted radiation, is placed over the tissue. The activity recorded is proportional to the volume of blood. Currently it is not possible to accurately calibrate the systems to provide measures of blood volume because of uncertain attenuation of radiation by the tissue and distance. Furthermore, delimiting a particular organ within the body and separating arterial and venous segments of the circulation are difficult.

Compliance

To estimate compliance, changes in volume are needed. This is generally easier than measuring the total blood volume. Using *plethysmography*, a rigid container is placed around the organ, and a servo system functions to change the fluid volume in the chamber to maintain the chamber pressure-constant. The consequent volume change is measured and assumed to be primarily venous, because most of the vascular volume is venous. With a tight system and careful technique, at the end of the experiment both inflow and outflow blood vessels can be occluded and then the contained blood washed out and measured to provide a measure of the total blood volume [12].

Gravimetric Techniques

Gravimetric techniques can be used to measure changes in blood volume. If the organ can be isolated and weighed continuously with the blood vessels intact, changes in volume can be measured in response to drugs or reflexes. With an important modification, this approach can be applied to an organ or the systemic circulation; the tissues are perfused at a constant rate, and the outflow is emptied at a constant pressure into a reservoir. Because the reservoir is emptied at a constant rate for the perfusion, changes in reservoir volume reflect an opposite change in the perfused tissue blood volume [8]. To measure compliance, the outflow pressure is changed (2–5 mmHg) and the corresponding change in reservoir volume noted. With the inflow and outflow pressure held constant, the pressure gradients are assumed to be constant so that 100% of an outflow pressure change can be assumed to be transmitted to the primary capacitance vessels. Any reflex or drug-induced change in reservoir volume may be assumed to be inversely related to an active change in vascular volume [7,8,10]. If resistances are also changed by the reflex or drug, then corrections are needed and the interpretations are more complex.

Outflow Occlusion

If the outflow downstream from the venous catheter is suddenly occluded, the venous pressure increases, and its rate of increase is measured. The rate of inflow is also measured so that the compliance can be estimated as the ratio flow to rate of pressure rise: Compliance in ml/mmHg = (flow in ml/min)/(rate of venous pressure rise in mmHg/min). The method is predicated on the assumption that the inflow continues at a constant rate and that there is no pressure gradient between the pressure measuring point and the site of compliance for the first few seconds of occlusion when the rate of pressure rise is measured.

Integral of Inflow Minus Outflow

With this technique both inflow and outflow are measured and the difference integrated to provide the volume change during an experimental forcing. If there is a decrease in contained volume, the outflow will be transiently greater than the inflow. The volume change gives a measure of the response to drugs or reflexes. Following a change in venous pressure, the technique can be used to measure compliance. Accurate measures of flow are needed. Serious errors can result if the inflow is not measured but is only assumed to be constant during the experimental protocol. With all methods dependent on measured or controlled flow, small changes in zero offset, which is directly or indirectly integrated, leads to serious error after about 10 minutes, and so the methods are not useful for long-term or slow responses.

16.3 Typical Values

Cardiac output, the *sine qua non* of the cardiovascular system, averages about 100 ml/(min-kg). It is about 90 in humans, is over 110 ml/(min-kg) in dogs and cats, and is even higher on a body weight basis in small animals such as rats and mice. The mean arterial blood pressure in relaxed, resting, conscious mammals averages about 90 mmHg. The mean circulatory filling pressure averages about 7 mmHg, and the central venous pressure just outside the right heart about 2 mmHg. The blood volume of the body is about 75 ml/kg, but in humans it is about 10% less, and it is larger in small animals. It is difficult to measure accurately because the volume of distribution of the plasma is about 10% higher than that of the red blood cells.

Vascular compliance averages about 2 ml (mmHg-kg body weight). The majority is in the venules and veins. Arterial compliance is only about 0.05 ml/mmHg-kg). Skeletal muscle compliance is less than that of the body as a whole, whereas the vascular compliance of the liver is about 10 times that of other organs. The stressed volume is the product of compliance and mean filling pressure and so is about 15 ml/kg. By difference, the unstressed volume is about 60 ml/kg.

As flow is increased through a tissue, the contained volume increases even if the outflow pressure is held constant, because there is a finite pressure drop across the veins which is increased as flow increases. This increase in upstream distending pressure acts to increase the contained blood volume. The volume sensitivity to flow averages about 0.1 ml per 1 ml/min change in flow [6]. For the body as a whole, the sensitivity is about 0.25 ml per 1 ml/min with reflexes blocked, and with reflexes intact it averages about 0.4 ml/min^3. Using similar techniques, it appears that the passive compensatory volume redistribution from the peripheral toward the heart during serious left heart failure is similar in magnitude to a reflex-engendered redistribution from activation of venous smooth muscle [6].

The high-pressure carotid sinus baroreceptor reflex system is capable of changing the venous capacitance [10]. Over the full operating range of the reflex it is capable of mobilizing up to 7.5 ml/kg of blood by primarily changing the unstressed vascular volume with little or no changes in venous compliance [7,8]. Although this represents only a 10% change in blood volume, it can cause nearly a 100% change in cardiac output. It is difficult to say with confidence what particular organ and/or tissue is contributing to this blood volume mobilization. Current evidence suggests that the splanchnic vascular bed contributes significantly to the capacitance change, but this also may vary between species [11].

Acknowledgments

This work was supported by National Heart Lung and Blood Institute grants HL 19039 and HL 07723.

References

1. Furey SAI, Zieske H, Levy MN. 1984. The essential function of the right heart. *Am Heart J* 107:404.
2. Lassen NA, Perl W. 1979. *Tracer Kinetic Methods in Medical Physiology.* New York, Raven Press.
3. Numao Y, Iriuchijima J. 1977. Effect of cardiac output on circulatory blood volume. *Jpn J Physiol* 27:145.
4. Rothe CF. 1983. Venous system: physiology of the capacitance vessels. In JT Shepherd, FM Abboud (Eds.), *Handbook of Physiology: The Cardiovascular System,* Sec. 2, Vol. 3, Pt. 1, pp 397–452. Bethesda, MD, American Physiology Society.
5. Rothe CF. 1993. Mean circulatory filling pressure: its meaning and measurement. *J Appl Physiol* 74:499.
6. Rothe CF, Gaddis ML. 1990. Autoregulation of cardiac output by passive elastic characteristics of the vascular capacitance system. *Circulation* 81:360.
7. Shoukas AA, MacAnespie CL, Brunner MJ et al. 1981. The importance of the spleen in blood volume shifts of the systemic vascular bed caused by the carotid sinus baroreceptor reflex in the dog. *Circ Res* 49:759.
8. Shoukas AA, Sagawa K. 1973. Control of total systemic vascular capacity by the carotid sinus baroreceptor reflex. *Circ Res* 33:22.
9. Rose W, Shoukas AA. 1993. Two-port analysis of systemic venous and arterial impedances. *Am J Physiol* 265(Heart Circ Physiol 34):H1577.
10. Shoukas AA. 1993. Overall systems analysis of the carotid sinus baroreceptor reflex control of the circulation. *Anesthesiology* 79:1402.
11. Haase E, Shoukas AA. 1991. The role of the carotid sinus baroreceptor reflex on pressure and diameter relations of the microvasculature of the rat intestine. *Am J Physiol* 260:H752.
12. Zink J, Delaive J, Mazerall E, Greenway CV. 1976. An improved plethsmograph with servo control of hydrostatic pressure. *J Appl Physiol* 41(1):107.

17

Mechanics of Tissue and Lymphatic Transport

Alan R. Hargen
University of California, San Diego and NASA Ames Research Center

Geert W. Schmid-Schönbein
University of California, San Diego

17.1 Introduction

Transport of fluid and metabolites from blood to tissue is critically important for maintaining the viability and function of cells within the body. Similarly, transport of fluid and waste products from tissue to the *lymphatic system* of vessels and nodes is also crucial to maintain tissue and organ health. Therefore, it is important to understand the mechanisms for transporting fluid containing micro- and macromolecules from blood to tissue and the drainage of this fluid into the lymphatic system. Because of the succinct nature of this chapter, readers are encouraged to consult more complete reviews of blood, tissue, and lymphatic transport by Aukland and Reed [1993], Bert and Pearce [1984], Casley-Smith [1982], Curry [1984], Hargens [1986], Jain [1987], Lai-Fook [1986], Levick [1984], Reddy [1986], Schmid-Schönbein [1990], Schmid-Schönbein and Zweifach [1994], Staub [1988], Staub, Hogg, and Hargens [1987], Taylor and Granger [1984], Zweifach and Lipowsky [1984], and Zweifach and Silverberg [1985].

Most previous studies of blood, tissue, and lymphatic transport have used isolated organs or whole animals under general anesthesia. Under these conditions, transport of fluid and metabolites is artificially low in comparison to animals which are actively moving. In some cases, investigators employed passive motion by connecting an animal's limb to a motor in order to facilitate studies of blood to lymph transport and lymphatic flow. However, new methods and technology allow studies of physiologically active animals so that a better understanding of the importance of transport phenomena in moving tissues is now apparent. Therefore, the major focus of this chapter emphasizes recent developments in the understanding of the mechanics of tissue and lymphatic transport.

The majority of the fluid that is filtered from the microcirculation into the interstitial space is carried out of the tissue via the lymphatic network. This unidirectional transport system originates with a set of blind channels in distal regions of the microcirculation. It carries a variety of interstitial molecules, proteins, metabolites, and even cells along channels deeply embedded in the tissue parenchyma towards a set of sequential lymph nodes and eventually back into the venous system via the right and left thoracic ducts. The lymphatics are the pathways for immune surveillance by the lymphocytes and thus are one of the important highways of the immune system.

17.2 Basic Concepts of Tissue and Lymphatic Transport

Transcapillary Filtration

Because lymph is formed from fluid filtered from the blood, an understanding of transcapillary exchange must be considered first. Usually pressure parameters favor filtration of fluid across the *capillary* wall to the *interstitium* (J_c) according to the Starling–Landis equation:

$$J_c = L_p A\left[\left(P_c - P_t\right) - \sigma_p\left(\pi_c - \pi_t\right)\right] \tag{17.1}$$

where: J_c = net transcapillary fluid transport
L_p = hydraulic conductivity of capillary wall
A = capillary surface area
P_c = capillary blood pressure
P_t = interstitial fluid pressure
σ_p = reflection coefficient for protein
π_c = capillary blood colloid osmotic pressure
π_t = interstitial fluid colloid osmotic pressure

In many tissues, fluid transported out of the capillaries is passively drained by the initial lymphatic vessels so that:

$$J_c = J_1 \tag{17.2}$$

where J_1 = lymph flow. Pressure within the initial lymphatic vessels, P_1, depends on higher interstitial fluid pressure P_t for establishing lymph flow:

$$P_t \geq P_1 \tag{17.3}$$

Starling Pressures and Edema Prevention

Hydrostatic and *colloid osmotic pressures* within the blood and interstitial fluid primarily govern transcapillary fluid shifts (Fig. 17.1). Although input arterial pressure averages about 100 mmHg at heart level, capillary blood pressure P_c is significantly reduced due to resistance R, according to the Poiseuille equation (Eq. 17.4):

$$R = \frac{8\eta l}{\pi r^4} \tag{17.4}$$

where: η = blood viscosity
l = vessel length between feed artery and capillary
r = radius

Therefore, normally at heart level, P_c is approximately 30 mmHg. However, during upright posture, P_c at foot level is about 90 mmHg and only about 25 mmHg at head level [Parazynski et al., 1991]. Differences in P_c between capillaries of the head and feet are due to gravitational components of blood pressure according to $\rho g h$. For this reason, volumes of transcapillary filtration and lymph flows are generally higher in tissues of the lower body as compared to those of the upper body. Moreover, one might expect much more sparse distribution of lymphatic vessels in upper body tissues. In fact, tissues of the lower body of humans and other tall animals have efficient skeletal muscle pumps, prominent

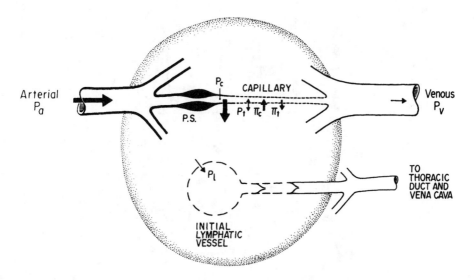

FIGURE 17.1 Starling pressures which regulate transcapillary fluid balance. Pressure parameters which determine direction and magnitude of transcapillary exchange include capillary blood pressure P_c, interstitial fluid pressure P_t (directed into capillary when positive or directed into tissue when negative), plasma colloidal osmotic pressure π_c, and interstitial fluid colloidal osmotic pressure π_t. Precapillary sphincters (PS) regulate P_c, capillary flow, and capillary surface area A. It is generally agreed that a hydrostatic pressure gradient (P_t > lymph pressure P_l) drains off excess interstitial fluid under conditions of net filtration. Relative magnitudes of pressures are depicted by the size of arrows. (*Source:* Hargens AR. 1986. Interstitial fluid pressure and lymph flow. In: R Skalak, S Chien (Eds.), *Handbook of Bioengineering*, Vol. 19, pp. 1–35. New York, McGraw-Hill. With permission.)

lymphatic systems, and noncompliant skin and fascial boundaries to prevent dependent *edema* [Hargens et al., 1987].

Other pressure parameters in the Starling–Landis Eq. (17.1) such as P_t, π_c, and π_t are not as sensitive to changes in body posture as is P_c. Typical values for P_t range from –2 mmHg to 10 mmHg depending on the tissue or organ under investigation [Wiig, 1990]. However, during movement, P_t in skeletal muscle increases to 150 mmHg or higher [Murthy et al., 1994], providing a mechanism to promote lymphatic flow and venous return via the skeletal pump (Fig. 17.2). Blood colloid osmotic pressure π_c usually ranges between 25 and 35 mmHg and is the other major force for retaining plasma within the vascular system and preventing edema. Interstitial π_t depends on the reflection coefficient of the capillary wall (σp ranges from 0.5 to 0.9 for different tissues) as well as washout of interstitial proteins during high filtration rates [Aukland and Reed, 1993]. Typically π_t ranges between 8 and 15 mmHg with higher values in upper body tissues compared to those in the lower body [Aukland and Reed, 1993; Parazynski et al., 1991]. Precapillary sphincter activity (Fig. 17.1) also decreases blood flow, decreases capillary filtration area A, and reduces P_c in dependent tissues of the body to help prevent edema during upright posture [Aratow et al., 1991].

Interstitial Flow and Lymph Formation

As stated in the Introduction, many previous investigators were convinced that interstitial flow of proteins was limited by simple diffusion according to Fick's equation (Eq. 17.5):

$$J_p = -D \, \partial cp/\partial x \tag{17.5}$$

where: J_p = one dimensional protein flux
       D = diffusion coefficient
      $\partial cp/\partial x$ = concentration gradient of protein through interstitium

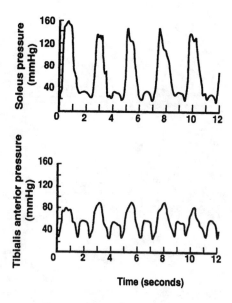

FIGURE 17.2 Simultaneous intramuscular pressure oscillations in the soleus (top panel) and the tibialis anterior (bottom panel) muscles during plantar- and dorsiflexion exercise. Soleus muscle is an integral part of the calf muscle pump. (*Source:* Murthy G, DE Watenpaugh, RE Ballard, et al., 1994. Supine exercise during lower body negative pressure effectively simulates upright exercise in normal gravity. *J. Appl. Physiol.* 76:2742. Modified with permission.)

However, recent experimental and theoretical understandings of the dependence of volume and solute flows on hydrostatic and osmotic pressures [Hammel, 1994; Hargens and Akeson, 1986] strongly suggest that convective flow plays an important role in interstitial flow and tissue nutrition. For example, in the presence of osmotic or hydrostatic pressure gradients, protein transport J_p is coupled to fluid transport according to Eq. (17.6):

$$J_p = \bar{c}_p J_v \tag{17.6}$$

where: $\bar{c}_p$ = average protein concentration
J_v = volume flow of fluid

Most investigators note that lymph formation and flow greatly depend upon tissue movement or activity related to muscle contraction. It is also generally agreed that formation of initial lymph depends solely on the composition of nearby interstitial fluid and pressure gradients across the interstitial–lymphatic boundary [Hargens, 1986; Zweifach and Lipowsky, 1984]. For this reason, lymph formation and flow can be quantified by measuring disappearance of isotope-labeled albumin from subcutis or skeletal muscle [Reed et al., 1985].

Lymphatic Architecture

To understand lymph transport in engineering terms it is paramount that we develop a detailed picture of the lymphatic network topology and vessel morphology. This task is facilitated by a number of morphological and ultrastructural studies from past decades which give a general picture of the morphology and location of lymphatic vessels in different tissues. Lymphatics have been studied by injections of macroscopic and microscopic contrast media and by light and electron microscopic sections. The display of the lymphatics is organ specific; there are many variations in lymphatic architecture [Schmid-Schönbein, 1990]. We will focus this discussion predominantly on skeletal muscle, the intestines, and skin. But, the mechanisms may also, in part, be relevant with respect to other organs.

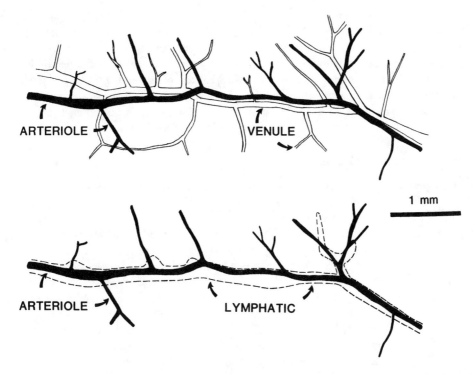

FIGURE 17.3 Tracing of a typical lymphatic channel (bottom panel) in rat spinotrapezius muscle after injection with a micropipette of a carbon contrast suspension. All lymphatics are of the initial type and are closely associated with the arcade arterioles. Few lymphatics follow the path of the arcade venules, or their side branches, the collecting venules or the transverse arterioles. [Skalak et al., 1986.]

In skeletal muscle, lymphatics are positioned in *immediate* proximity of the arterioles [Skalak et al., 1984]. The majority of feeder arteries in skeletal muscle and most, but not all of the arcade arterioles, are closely accompanied by a lymphatic vessel (Fig. 17.3). Lymphatics can be traced along the entire length of the arcade arterioles, but they can be traced only over relatively short distances (less than 50 μm) into the side branches of the arcades, the transverse (terminal) arterioles which supply the blood into the capillary network. Systematic reconstructions of the lymphatics in skeletal muscle have yielded little evidence for lymphatic channels that enter into the capillary network per se [Skalak et al., 1984]. Thus, the network density of lymphatics is quite low compared to the high density of the capillary network in the same tissue, a characteristic feature of lymphatics in most organs [Skalak et al., 1986]. The close association between lymphatics and the vasculature is also present in the skin [Ikomi and Schmid-Schönbein, 1995], and in other organs and may extend into the central vasculature.

Lymphatic Morphology

Histological sections of the lymphatics permit the classification into distinctly different subsets, *initial* lymphatics and *collecting* lymphatics. The initial lymphatics (denoted also as terminal or capillary lymphatics) form a set of blind endings in the tissue, which feed into the collecting lymphatics, and which in turn are the conduits into the lymph nodes. While both initial and collecting lymphatics are lined by a highly attenuated endothelium, only the collecting lymphatics have a smooth muscle in their media. In accordance, collecting lymphatics exhibit spontaneous narrowing of their lumen, while there is no evidence for contractility in the initial lymphatics. Contractile lymphatics are capable of peristaltic smooth muscle contractions, which in conjunction with periodic opening and closing of the intraluminal valves permits unidirectional fluid transport. The lymphatic smooth muscle has an adrenergic innervation [Ohhashi et al., 1982], exhibits myogenic contraction [Hargens and Zweifach, 1977; Mizuno et al., 1997], and reacts to a variety of vasoactive stimuli [Ohhashi et al., 1978; Benoit, 1997], including signals which involve nitric oxide [Ohhashi and Takahashi, 1991; Bohlen and Lash, 1992; Yokoyama and Ohhashi, 1993]. None of these contractile features has been documented in the initial lymphatics.

The lymphatic endothelium has a number of similarities with the vascular endothelium. It forms a continuous lining. There are numerous caveolae, Weibel Palade bodies, but lymphatic endothelium has fewer interendothelial adhesion complexes and there is a discontinuous basement membrane. The residues of the basement membrane are attached to interstitial collagen via anchoring filaments [Leak and Burke, 1968] which provide relatively firm attachment of the endothelium to the interstitial structures.

Lymphatic Network Display

One of the interesting aspects regarding lymphatic transport in skeletal muscle is the fact that all lymphatics *inside* the muscle parenchyma are of the non-contractile *initial* type [Skalak et al., 1984]. *Collecting* lymphatics can only be observed outside the muscle fibers as a conduit to adjacent lymph nodes. The fact that all lymphatics inside the tissue parenchyma are of the initial type is not unique to skeletal muscle but has been demonstrated in other organs [Unthank and Bohlen, 1988; Yamanaka et al., 1995]. The initial lymphatics are positioned in the adventitia of the arcade arterioles surrounded by collagen fibers (Fig. 17.4). In this position they are in immediate proximity to the arteriolar smooth muscle and adjacent to myelinated nerve fibers and a set of mast cells that accompany the arterioles. The initial lymphatics are frequently sandwiched between arteriolar smooth muscles and their paired venules, and they in turn are embedded between the skeletal muscle fibers [Skalak et al., 1984]. The initial lymphatics are firmly attached to the adjacent basement membrane and collagen fibers via anchoring filaments [Leak and Burke, 1968]. The basement membrane of the lymphatic endothelium is discontinuous, especially at the interendothelial junctions, so that macromolecular material and even large colloidal particles enter the initial lymphatics [Casley-Smith, 1962; Bach and Lewis, 1973; Strand and Persson, 1979; Bollinger et al., 1981; Ikomi et al., 1996].

The lumen cross-section of the initial lymphatics is highly irregular in contrast to the overall circular cross-section of the collecting lymphatics (Fig. 17.4). The lumen cross-section of the initial lymphatics is partially or completely collapsed and may frequently span around the arcade arteriole. In fact, we have documented cases in which the arcade arteriole is completely surrounded by an initial lymphatic channel, highlighting the fact that the activity of the lymphatics is closely linked to that of the arterioles [Ikomi and Schmid-Schönbein, 1995]. Initial lymphatics in skeletal muscle have intra-luminal valves which consist of bileaflets and a funnel structure [Mazzoni et al., 1987]. The leaflets are flexible structures and are opened and closed by a viscous pressure drop along the valve funnel. In a closed position the leaflets are able to support considerable pressures [Eisenhoffer et al., 1995; Ikomi et al., 1997]. This arrangement serves to preserve normal valve function even in initial lymphatics with irregularly shaped lumen cross-sections.

The lymphatic endothelial cells are attenuated and have many of the morphological characteristics of vascular endothelium, including expression of P-selectin, von Willebrand factor [Di Nucci et al., 1996], and factor VIII [Schmid-Schönbein, 1990]. An important difference between vascular and lymphatic endothelium lies in the arrangement of the endothelial junctions. In the initial lymphatics, the endothelial cells lack tight junctions [Schneeberger and Lynch, 1984] and are frequently encountered in an overlapping but open position, so that proteins, colloid particles, and even chylomicron particles can readily pass through the junctions [Casley-Smith, 1962; Casley-Smith, 1964; Leak, 1970]. Examination of the junctions with scanning electron microscopy shows that there exists a periodic *interdigitating* arrangement of endothelial extensions. Individual extensions are attached via anchoring filaments to the underlying basement membrane and connective tissue, but the two extensions of adjacent endothelial cell resting on top of each other are not attached by interendothelial adhesion complexes. Mild mechanical stretching of the initial lymphatics shows that the endothelial extensions can be separated in part from each other, indicating that the membranes of two neighboring lymphatic endothelial cells are not attached to each other, but are firmly attached to the underlying basement membrane [Castenholz, 1984]. Lymphatic endothelium does not exhibit continuous junctional complexes, and instead has a "streak and dot"

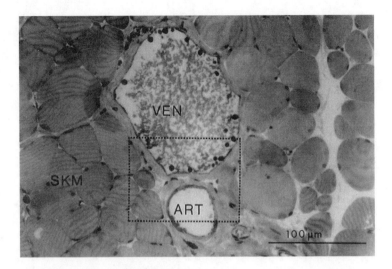

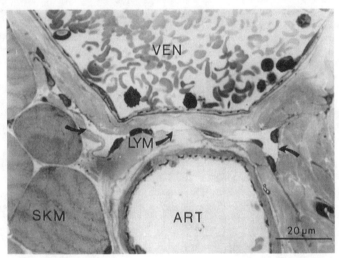

FIGURE 17.4   Histological cross-sections of lymphatics (LYM) in rat skeletal muscle before (p. 253) and after (p. 254) contraction of the paired arcade arterioles (ART). The lymphatic channel is of the initial type with a single attenuated endothelial layer (curved arrows). Note, that in the dilated arteriole, the lymphatic is essentially compressed (p. 253) while the lymphatic is expanded after arteriolar contraction (p. 254) which is noticeable by the folded endothelial cells in the arteriolar lumen. In both cases the lumen cross-sectional shape of the initial lymphatic channels is highly irregular. All lymphatics within skeletal muscle (SKM) have these characteristic features. [Skalak et al., 1984.]

immunostaining pattern of VE-cadherin and associated intracellular proteins—desmoplakin and plako-globulin [Schmelz et al., 1994]. But the staining pattern is not uniform for all lymphatics, and in larger lymphatics a more continuous pattern is present. This highly specialized arrangement has been referred to in the following as the *lymphatic endothelial microvalves* [Schmid-Schönbein, 1990].

Mechanics of Lymphatic Valves

In contrast to the central large valves in the heart, which are closed by inertial fluid forces, the lymphatic valves are small and the fluid Reynolds number is almost zero. Thus, since no inertial forces are available to open and close these valves, a different valve morphology has evolved. They form long funnel-shaped channels which are inserted into the lymph conduits and attached at their base and the funnel is prevented from inversion by attachment via a buttress to the lymphatic wall. The valve wall structure consists of a

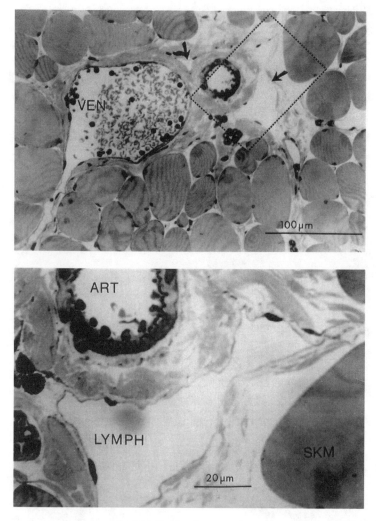

FIGURE 17.4 (continued)

collagen layer sandwiched between two endothelial layers, and the entire structure is quite deformable under mild physiological fluid pressures. The funnel structure serves to create a *viscous pressure gradient* which is sufficient to generate, during forward fluid motion, a pressure drop to open the values and upon flow reversal to close the valves [Mazzoni et al., 1987].

Lymph Pump Mechanisms

One of the important questions which is fundamental in lymphology is: How do fluid particles in the interstitium find their way into the initial lymphatics? In light of the relative sparse display of the initial lymphatics, a directed convective transport is required—provided by either a hydrostatic or a colloid osmotic pressure drop [Zweifach and Silberberg, 1979]. The documentation of the exact mechanism has remained an elusive target. Several proposals have been advanced (which are discussed in detail in Schmid-Schönbein, 1990). Briefly, a number of authors have postulated that there exists a constant pressure drop from the interstitium into the initial lymph which may support a steady fluid motion into the lymphatics. But, repeated measurements with different techniques have uniformly failed to provide supporting evidence for a *steady* pressure drop to transport fluid into the initial lymphatics [Zweifach and Prather, 1975; Clough and Smaje, 1978]. Under steady-state conditions, a steady pressure drop does not exist in the vicinity of the initial lymphatics in skeletal muscle within the resolution of the measurement (about 0.2 cm H_2O) [Skalak et al., 1984]. An order of magnitude estimate of the pressure drop to be expected at the relatively slow flow rates of the lymphatics shows, however, that the pressure drop from the interstitium may be significantly lower [Schmid-Schönbein, 1990]. Furthermore, the assumption of a *steady* pressure drop is not in agreement with the substantial evidence that lymph flow rate is enhanced under unsteady conditions (see below). Some investigators have postulated an osmotic pressure in the lymphatics to aspirate fluid into the initial lymphatics [Casley-Smith, 1972] due to ultrafiltration across the lymphatic endothelium, a mechanism that has been referred to as "bootstrap effect" [Perl, 1975]. Critical tests of this hypothesis, such as the microinjection of hyperosmotic protein solutions, have not led to a uniformly accepted hypothesis for lymph formation involving an osmotic pressure. Others have suggested a retrograde aspiration mechanism, such that the recoil in the collecting lymphatics serves to lower the pressure in the initial lymphatics upstream of the collecting lymphatics [Reddy, 1986; Reddy and Patel, 1995], or an electric charge difference across lymphatic endothelium [O'Morchoe et al., 1984].

Tissue Mechanical Motion and Lymphatic Pumping

An intriguing feature of the lymphatic pressure is that lymphatic flow rates depend on tissue motion. In a resting tissue, the lymph flow rate is relatively small, but different forms of tissue motion serve to enhance the lymph flow. This was originally shown for pulsatile pressure in the rabbit ear. Perfusion of the ear with steady pressure (even at the same mean pressure) serves to stop most lymph transport, while pulsatile pressure promotes lymph transport [Parsons and McMaster, 1938]. In light of the paired arrangement of the arterioles and lymphatics, periodic expansion of the arterioles leads to compression of the adjacent lymphatics, and vice versa: a reduction of the arteriolar diameter during the pressure reduction phase leads to expansion of the adjacent lymphatics [Skalak et al., 1984] (Fig. 17.4). Vasomotion which is associated with a slower contraction of the arterioles, but with a larger amplitude than pulsatile pressure, serves to increase lymph formation [Intaglietta and Gross, 1982; Colantuoni et al., 1984]. In addition, muscle contractions, simple walking [Olszewski and Engeset, 1980], respiration, intestinal peristalsis, skin compression [Ohhashi et al., 1991], and other tissue motions are associated with an increase in the lymph flow rates. Periodic tissue motions are significantly more effective to enhance the lymph flow than, for example, elevation of the venous pressure [Ikomi et al., 1996], which is also associated with enhanced fluid filtration [Renkin et al., 1977].

A requirement for lymph fluid flow is the periodic expansion and compression of the initial lymphatics. Since initial lymphatics do not have their own smooth muscle, the expansion and compression of the initial lymphatics depends on the motion of tissue in which they are embedded. In skeletal muscle, the strategic location of the initial lymphatics in the adventitia of the arterioles provides the opportunity for the expansion and compression to be achieved via several mechanisms: (1) arteriolar pressure pulsations or vasomotion, (2) active or passive skeletal muscle contractions, or (3) external muscle compression. Direct measurements of the cross-sectional area of the initial lymphatics during arteriolar contractions or during skeletal muscle shortening support this hypothesis [Skalak et al., 1984; Mazzoni et al., 1990] (Fig. 17.5). The different lymph pump mechanisms are additive. Resting skeletal muscle has much lower lymph flow rates (provided largely by the arteriolar pressure pulsation and vasomotion) than skeletal muscle during exercise (produced by a combination of intramuscular pressure pulsations and skeletal muscle shortening).

Measurements of lymph flow rates in an afferent lymph vessel (diameter about 300 to 500 μm, proximal to the popliteal node) in the hind leg [Ikomi and Schmid-Schönbein, 1996] serve to demonstrate that lymph fluid formation can be influenced by passive or active motion of the surrounding tissue. The lymphatics in this tissue region drain muscle and skin of the hind leg, and the majority are of the *initial* type, whereas collecting lymphatics are detected outside the tissue parenchyma in the fascia proximal to the node. Without whole leg rotation, lymph flow remains at low non-zero values. If the pulse pressure is stopped, the lymph flow falls to values below the detection limit (less than about 10% of the values during pulse pressure). Introduction of whole leg passive rotation causes a strong, frequency-dependent

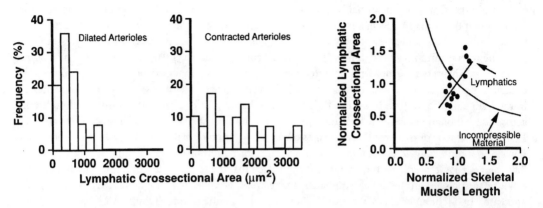

FIGURE 17.5 Histograms of initial lymphatic cross-sectional area in rat spinotrapezius m. before (left) and after (middle) contraction of the paired arteriole with norepinephrine; lymphatic cross-sectional area as a function of muscle length during active contraction or passive stretch (right). Cross-sectional area and muscle length are normalized with respect to the values *in vivo* in resting muscle. Note the expansion of the initial lymphatics with contraction of the arterioles or muscle stretch. [Skalak et al., 1984; Mazzoni et al., 1990.]

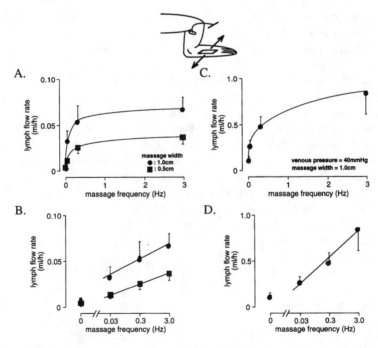

FIGURE 17.6 Lymph flow rates in a prenodal afferent lymphatic draining the hindleg as a function of the frequency of a periodic surface shear motion (massage) without (panels A, B) and with (panels C, D) elevation of the venous pressure by placement of a cuff. Zero frequency refers to a resting leg with a lymph flow rate which depends on pulse pressure. The amplitudes of the tangential skin shear motion were 1 cm and 0.5 cm (panels A, B) and 1 cm in the presence of the elevated venous pressure (panels C, D). Note that the ordinates in panels C and D are larger than those in panels A and B. [Ikomi and Schmid-Schönbein, 1996.]

lymph flow rate which increases linearly with the logarithm of frequency between 0.03 Hz and 1.0 Hz (Fig. 17.6). Elevation of the venous pressure, which serves to enhance fluid filtration from the vasculature and elevates the flow rates, does not significantly alter the dependency on periodic tissue motion [Ikomi et al., 1996].

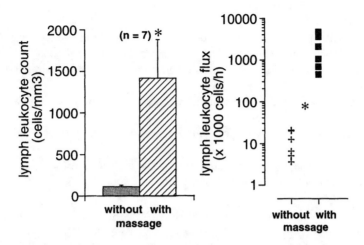

FIGURE 17.7   Lymph leukocyte count (left) and leukocyte flux (right) before and after application of periodic hind leg skin shear motion (massage) at a frequency of about 1 Hz and amplitude of 1 cm. The flux rates were computed from the product of lymph flow rates and the lymphocyte counts. *Statistically significant differences from case without massage. (Adapted from Ikomi et al. [1996].)

Similarly, application of passive tissue compression on the skin serves to elevate the lymph flow rate in a frequency-dependent manner. The lymph flow rates are determined to a significant degree by the *local* action of the lymph pump, since arrest of the heart beat and reduction of the central blood pressure to zero do not stop lymph flow, and instead reduces the lymph flow rates only about 50% during continued leg motion or application of periodic shear stress to the skin for several hours [Ikomi and Schmid-Schönbein, 1996]. Periodic compression of the initial lymphatics also serves to enhance proteins and even lymphocyte counts in the lymphatics [Ikomi et al., 1996] (Fig. 17.7). Thus either arteriolar smooth muscle or parenchymal skeletal muscle activity serve to expand and compress the initial lymphatics in skeletal muscle. The mechanisms serve to adjust lymph flow rates according to organ activity, such that a resting skeletal muscle has low lymph flow rates. During normal daily activity or mild or strenuous exercise, the lymph flow rates as well as the protein and cell transport in the lymphatics increases [Olszewski et al., 1977].

In conclusion, the lymphatics are a unique transport system that is even present in primitive physiological systems. It likely carries out a multitude of functions, many of which have yet to be discovered. Details of its operation and its growth kinetics must await a more detailed bioengineering analysis, especially at the molecular level [Jeltsch et al., 1997].

Acknowledgments

We thank Karen Hutchinson for expert manuscript assistance. This work was supported by NASA grants 199-14-12-04 and 199-26-12-38 and NSF grant IBN-9512778.

Defining Terms

Capillary: The smallest blood vessel of the body which provides oxygen and other nutrients to nearby cells and tissues.

Colloid osmotic pressure: A negative pressure which depends on protein concentration (mainly of albumin and globulins) and prevents excess filtration across the capillary wall.

Edema: Excess fluid or swelling within a given tissue.

Interstitium: The space between cells of various tissues of the body. Normally fluid and proteins within this space are transported from the capillary to the initial lymphatic vessel.

Lymphatic system: The clear network of vessels which return excess fluid and proteins to the blood via the thoracic duct.

References

Aratow M, Hargens AR, Meyer JU et al. 1991. Postural responses of head and foot cutaneous microvascular flow and their sensitivity to bed rest. *Aviat. Space Environ. Med.* 62:246.

Aukland K, Reed RK. 1993. Interstitial-lymphatic mechanisms in the control of extracellular fluid volume. *Physiol. Rev.* 73:1.

Bach C, Lewis GP. 1973. Lymph flow and lymph protein concentration in the skin and muscle of the rabbit hind limb. *J. Physiol. (Lond.)* 235:477.

Benoit JN. 1997. Effects of alpha-adrenergic stimuli on mesenteric collecting lymphatics in the rat. *Am. J. Physiol.* 273:R331.

Bert JL, Pearce RH. 1984. The interstitium and microvascular exchange. In *Handbook of Physiology: The Cardiovascular System: Microcirculation,* Sec. 2, Vol. 4, Pt. 1, pp. 521–547, Bethesda, MD, American Physiological Society.

Bohlen HG, Lash JM. 1992. Intestinal lymphatic vessels release endothelial-dependent vasodilators. *Am. J. Physiol.* 262:H813.

Bollinger A, Jäger K, Sgier F et al. 1981. Fluorescence microlymphography. *Circulation* 64:1195.

Casley-Smith JR. 1962. The identification of chylomicra and lipoproteins in tissue sections and their passage into jejunal lacteals. *J. Cell Biol.* 15:259.

Casley-Smith JR. 1964. Endothelial permeability—the passage of particles into and out of diaphragmatic lymphatics. *Quart. J. Exp. Physiol.* 49:365.

Casley-Smith JR. 1972. The role of the endothelial intercellular junctions in the functioning of the initial lymphatics. *Angiologica* 9:106.

Casley-Smith JR. 1982. Mechanisms in the formation of lymph. *Cardiovascular Physiology IV, International Review of Physiology,* pp. 147–187, Baltimore, MD, University Park Press.

Castenholz A. 1984. Morphological characteristics of initial lymphatics in the tongue as shown by scanning electron microscopy. *Scan. Electr. Microsc.* 1984:1343.

Clough G, Smaje LH. 1978. Simultaneous measurement of pressure in the interstitium and the terminal lymphatics of the cat mesentery. *J. Physiol. (Lond.)* 283:457.

Colantuoni A, Bertuglia S, Intaglietta M. 1984. A quantitation of rhythmic diameter changes in arterial microcirculation. *Am. J. Physiol.* 246:H508.

Curry F-RE. 1984. Mechanics and thermodynamics of transcapillary exchange. In *Handbook of Physiology: The Cardiovascular System: Microcirculation,* Sec. 2, Vol. 4, Pt. 1, pp. 309–374, Bethesda, MD, American Physiological Society.

Di Nucci A, Marchetti C, Serafini S et al. 1996. P-selectin and von Willebrand factor in bovine mesenteric lymphatics: an immunofluorescent study. *Lymphology* 29:25.

Eisenhoffer J, Kagal A, Klein T et al. 1995. Importance of valves and lymphangion contractions in determining pressure gradients in isolated lymphatics exposed to elevations in outflow pressure. *Microvasc. Res.* 49:97.

Hammel HT. 1994. How solutes alter water in aqueous solutions. *J. Phys. Chem.* 98:4196.

Hargens AR. 1986. Interstitial fluid pressure and lymph flow. *Handbook of Bioengineering,* 19:1–35, New York, McGraw-Hill.

Hargens AR, Akeson WH. 1986. Stress effects on tissue nutrition and viability. In *Tissue Nutrition and Viability,* pp. 1–24, New York, Springer-Verlag.

Hargens AR, Zweifach BW. 1977. Contractile stimuli in collecting lymph vessels. *Am. J. Physiol.* 233:H57.

Hargens AR, Millard RW, Pettersson K et al. 1987. Gravitational haemodynamics and oedema prevention in the giraffe. *Nature* 329:59.

Ikomi F, Schmid-Schönbein GW. 1995. Lymph transport in the skin. *Clin. Dermatol.* 13(5):419, Elsevier Science Inc.

Ikomi F, Schmid-Schönbein GW. 1996. Lymph pump mechanics in the rabbit hind leg. *Am. J. Physiol.* 271:H173.

Ikomi F, Hunt J, Hanna G et al. 1996. Interstitial fluid, protein, colloid and leukocyte uptake into interstitial lymphatics. *J. Appl. Physiol.* 81:2060.

Ikomi F, Zweifach BW, Schmid-Schönbein GW. 1997. Fluid pressures in the rabbit popliteal afferent lymphatics during passive tissue motion. *Lymphology* 30:13.

Intaglietta M, Gross JF. 1982. Vasomotion, tissue fluid flow and the formation of lymph. *Int. J. Microcirc. Clin. Exp.* 1:55.

Jain RK. 1987. Transport of molecules in the tumor interstitium: a review. *Cancer Res.* 47:3039.

Jeltsch M, Kaipainen A, Joukov V et al. 1997. Hyperplasia of lymphatic vessels in VEGF-C transgenic mice. *Science* 276:1423.

Lai-Fook SJ. 1986. Mechanics of lung fluid balance. *Crit. Rev. Biomed. Eng.* 13:171.

Leak LV. 1970. Electron microscopic observations on lymphatic capillaries and the structural components of the connective tissue-lymph interface. *Microvasc. Res.* 2:361.

Leak LV, Burke JF. 1968. Ultrastructural studies on the lymphatic anchoring filaments. *J. Cell Biol.* 36:129.

Levick JR. 1984. *Handbook of Physiology: The Cardiovascular System: Microcirculation,* Sec. 2, Vol. 4, Pt. 1, pp. 917–947, Bethesda, MD, American Physiological Society.

Mazzoni MC, Skalak TC, Schmid-Schönbein GW. 1987. The structure of lymphatic valves in the spinotrapezius muscle of the rat. *Blood Vessels* 24:304.

Mazzoni MC, Skalak TC, Schmid-Schönbein GW. 1990. The effect of skeletal muscle fiber deformation on lymphatic volume. *Am. J. Physiol.* 259:H1860.

Mizuno R, Dornyei G, Koller A et al. 1997. Myogenic responses of isolated lymphatics: Modulation by endothelium. *Microcirculation* 4:413.

Murthy G, Watenpaugh DE, Ballard RE et al. 1994. Supine exercise during lower body negative pressure effectively simulates upright exercise in normal gravity. *J. Appl. Physiol.* 76:2742.

O'Morchoe CCC, Jones WRI, Jarosz HM et al. 1984. Temperature dependence of protein transport across lymphatic endothelium in vitro. *J. Cell Biol.* 98:629.

Ohhashi T, Takahashi N. 1991. Acetylcholine-induced release of endothelium-derived relaxing factor from lymphatic endothelial cells. *Am. J. Physiol.* 260:H1172.

Ohhashi T, Kawai Y, Azuma T. 1978. The response of lymphatic smooth muscles to vasoactive substances. *Plügers Arch.* 375:183.

Ohhashi T, Kobayashi S, Tsukahara S et al. 1982. Innervation of bovine mesenteric lymphatics: from the histochemical point of view. *Microvasc. Res.* 24:377.

Ohhashi T, Yokoyama S, Ikomi F. 1991. Effects of vibratory stimulation and mechanical massage on micro- and lymph-circulation in the acupuncture points between the paw pads of anesthetized dogs. In *Recent Advances in Cardiovascular Diseases,* pp. 125–133, Osaka, National Cardiovascular Center.

Olszewski WL., Engeset A. 1980. Intrinsic contractility of prenodal lymph vessels and lymph flow in human leg. *Am. J. Physiol.* 239:H775.

Olszewski WL, Engeset A, Jaeger PM et al. 1977. Flow and composition of leg lymph in normal men during venous stasis, muscular activity and local hyperthermia. *Acta. Physiol. Scand.* 99:149.

Parazynski SE, Hargens AR, Tucker B et al. 1991. Transcapillary fluid shifts in tissues of the head and neck during and after simulated microgravity. *J. Appl. Physiol.* 71:2469.

Parsons RJ, McMaster PD. 1938. The effect of the pulse upon the formation and flow of lymph. *J. Exp. Med.* 68:353.

Perl W. 1975. Convection and permeation of albumin between plasma and interstitium. *Microvasc. Res.* 10:83

Reddy NP. 1986. Lymph circulation: physiology, pharmacology, and biomechanics. *Crit. Rev. Biomed. Sci.* 14:45.

Reddy NP, Patel K. 1995. A mathematical model of flow through the terminal lymphatics. *Med. Eng. Phy.* 17:134.

Reed RK, Johansen S, Noddeland H. 1985. Turnover rate of interstitial albumin in rat skin and skeletal muscle. Effects of limb movements and motor activity. *Acta Physiol. Scand.* 125:711.

Renkin EM, Joyner WL, Sloop CH et al. 1977. Influence of venous pressure on plasma-lymph transport in the dog's paw: convective and dissipative mechanisms *Microvasc. Res.* 14:191.

Schmelz M, Moll R, Kuhn C et al. 1994. Complex adherentes, a new group of desmoplakin-containing junctions in endothelial cells: II. Different types of lymphatic vessels. *Differentiation* 57:97.

Schmid-Schönbein GW. 1990. Microlymphatics and lymph flow. *Physiol. Rev.* 70:987.

Schmid-Schönbein GW. Zweifach BW. 1994. Fluid pump mechanisms in initial lymphatics. *News Physiol. Sci.* 9:67.

Schneeberger EE, Lynch RD. 1984. Tight junctions: their structure, composition and function. *Circ. Res.* 5:723.

Skalak TC, Schmid-Schönbein GW, Zweifach BW. 1984. New morphological evidence for a mechanism of lymph formation in skeletal muscle. *Microvasc. Res.* 28:95.

Skalak TC, Schmid-Schönbein GW, Zweifach BW. 1986. Lymph transport in skeletal muscle. In *Tissue Nutrition and Viability,* pp. 243–262, New York, Springer-Verlag.

Staub NC. 1988. New concepts about the pathophysiology of pulmonary edema. *J. Thorac. Imaging* 3:8.

Staub NC, Hogg JC, Hargens AR. 1987. *Interstitial-Lymphatic Liquid and Solute Movement,* pp. 1–290, Basel, Karger.

Strand S-E, Persson BRR. 1979. Quantitative lymphoscintigraphy I: Basic concepts for optimal uptake of radiocolloids in the parasternal lymph nodes of rabbits. *J. Nucl. Med.* 20:1038.

Taylor AE, Granger DN. 1984. Exchange of macromolecules across the microcirculation. *Handbook of Physiology: The Cardiovascular System: Microcirculation,* Sec. 2, Vol. 4, Pt. 1, pp. 467–520, Bethesda, MD, American Physiological Society.

Unthank JL, Bohlen HG. 1988. Lymphatic pathways and role of valves in lymph propulsion from small intestine. *Am. J. Physiol.* 254:G389.

Wiig H. 1990. Evaluation of methodologies for measurement of interstitial fluid pressure (Pi): physiological implications of recent Pi data. *Crit. Rev. Biomed. Eng.* 18:27.

Yamanaka Y, Araki K, Ogata T. 1995. Three-dimensional organization of lymphatics in the dog small intestine: a scanning electron microscopic study on corrosion casts. *Arch. Hist. Cyt.* 58:465.

Yokoyama S, Ohhashi T. 1993. Effects of acetylcholine on spontaneous contractions in isolated bovine mesenteric lymphatics. *Am. J. Physiol.* 264:H1460.

Zweifach BW, Silberberg A. 1979. The interstitial-lymphatic flow system. In *International Review of Physiology—Cardiovascular Physiology III,* pp. 215–260, Baltimore, MD, University Park Press.

Zweifach BW, Silverberg A. 1985. The interstitial-lymphatic flow system. In *Experimental Biology of the Lymphatic Circulation,* pp. 45–79, Amsterdam, Elsevier.

Zweifach BW, Lipowsky HH. 1984. Pressure-flow relations in blood and lymph microcirculation. In *Handbook of Physiology: The Cardiovascular System: Microcirculation,* Sec. 2, Vol. 4, Pt. 1, pp. 251–307, Bethesda, MD, American Physiological Society.

Zweifach BW, Prather JW. 1975. Micromanipulation of pressure in terminal lymphatics of the mesentary. *J. Appl. Physiol.* 228:1326.

Further Information

Drinker, C.K. and J.M. Yoffey. 1941. *Lymphatics, Lymph and Lymphoid Tissue: Their Physiological and Clinical Significance.* Harvard University Press, Cambridge, MA. This is a classic treatment of the lymphatic circulation by two pioneers in the field of lymphatic physiology.

Yoffey, J.M. and F.C. Courtice. 1970. *Lymphatics, Lymph and the Lymphomyeloid Complex.* Academic Press, New York. This is a classic book in the field of lymphatic physiology. The book contains a comprehensive review of pertinent literature and experimental physiology on the lymphatic system.

18

Cochlear Mechanics

Charles R. Steele
Stanford University

Gary J. Baker
Stanford University

Jason A. Tolomeo
Stanford University

Deborah E. Zetes-Tolomeo
Stanford University

18.1 Introduction

The inner ear is a transducer of mechanical force to appropriate neural excitation. The key element is the receptor cell, or hair cell, which has cilia on the apical surface and afferent (and sometimes efferent) neural synapses on the lateral walls and base. Generally for hair cells, mechanical displacement of the cilia in the forward direction toward the tallest cilia causes the generation of electrical impulses in the nerves, while backward displacement causes inhibition of spontaneous neural activity. Displacement in the lateral direction has no effect. For moderate frequencies of sinusoidal ciliary displacement (20 to 200 Hz), the neural impulses are in synchrony with the mechanical displacement, one impulse for each cycle of excitation. Such impulses are transmitted to the higher centers of the brain and can be perceived as sound. For lower frequencies, however, neural impulses in synchrony with the excitation are apparently confused with the spontaneous, random firing of the nerves. Consequently, there are three mechanical devices in the inner ear of vertebrates that provide perception in the different frequency ranges. At zero frequency, i.e., linear acceleration, the otolith membrane provides a constant force acting on the cilia of hair cells. For low frequencies associated with rotation of the head, the semicircular canals provide the proper force on cilia. For frequencies in the hearing range, the cochlea provides the correct forcing of hair cell cilia. In nonmammalian vertebrates, the equivalent of the cochlea is a bent tube, and the upper frequency of hearing is around 7 kHz. For mammals, the upper frequency is considerably higher, 20 kHz for man but extending to almost 200 kHz for toothed whales and some bats. Other creatures, such as certain insects, can perceive high frequencies, but do not have a cochlea nor the frequency discrimination of vertebrates.

Auditory research is a broad field [Keidel and Neff, 1976]. The present article provides a brief guide of a restricted view, focusing on the transfer of the input sound pressure into correct stimulation of hair cell cilia in the cochlea. In a general sense, the mechanical functions of the semicircular canals and the otoliths are clear, as are the functions of the outer ear and middle ear; however, the cochlea continues to elude a reasonably complete explanation. Substantial progress in cochlear research has been made in the past decade, triggered by several key discoveries, and there is a high level of excitement among workers in the area. It is evident that the normal function of the cochlea requires a full integration of mechanical, electrical, and chemical effects on the milli-, micro-, and nanometer scales. Recent texts, which include details of the anatomy, are by Pickles [1988] and Gulick and coworkers [1989]. A summary of analysis and data related to the macromechanical aspect up to 1982 is given by Steele [1987], and more recent surveys specifically on the cochlea are by de Boer [1991], Dallos [1992], Hudspeth [1989], Ruggero [1993], and Nobili and colleagues [1998].

18.2 Anatomy

The cochlea is a coiled tube in the shape of a snail shell (cochlea = schnecke = snail), with length about 35 mm and radius about 1 mm in humans. There is not a large size difference across species: the length is 60 mm in elephants and 7 mm in mice. There are two and one-half turns of the coil in humans and dolphins, and five turns in guinea pigs. Despite the correlation of coiling with hearing capability of land animals [West, 1985], no significant effect of the coiling on the mechanical response has yet been identified.

Components

The cochlea is filled with fluid and divided along its length by two partitions. The main partition is at the center of the cross-section and consists of three segments: (1) on one side—the *bony shelf* (or *primary spiral osseous lamina*), (2) in the middle, an elastic segment (*basilar membrane*) (shown in Fig. 18.1), and (3) on the other side, a thick support (*spiral ligament*). The second partition is *Reissner's membrane*, attached at one side above the edge of the bony shelf and attached at the other side to the wall of the cochlea. *Scala media* is the region between Reissner's membrane and the basilar membrane, and is filled with *endolymphatic fluid*. This fluid has an ionic content similar to intracellular fluid, high in potassium and low in sodium, but with a resting positive electrical potential of around +80 mV. The electrical potential is supplied by the *stria vascularis* on the wall in scala media. The region above Reissner's membrane is *scala vestibuli*, and the region below the main partition is *scala tympani*. Scala vestibuli and scala tympani are connected at the apical end of the cochlea by an opening in the bony shelf, the *helicotrema*, and are filled with *perilymphatic fluid*. This fluid is similar to extracellular fluid, low in potassium and high in sodium with zero electrical potential. Distributed along the scala media side of

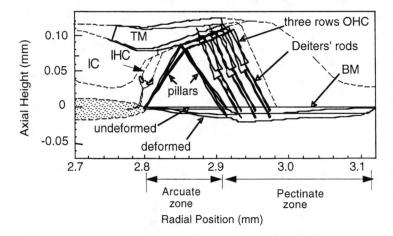

FIGURE 18.1 Finite element calculation for the deformation of the cochlear partition due to pressure on the basilar membrane (BM). Outer hair cell (OHC) stereocilia are sheared by the motion of the pillars of Corti and reticular lamina relative to the tectorial membrane (TM). The basilar membrane is supported on the left by the bony shelf and on the right by the spiral ligament. The inner hair cells (IHC) are the primary receptors, each with about 20 afferent synapses. The inner sulcus (IC) is a fluid region in contact with the cilia of the inner hair cells.

the basilar membrane is the sensory epithelium, the *organ of Corti*. This contains one row of *inner hair cells* and three rows of *outer hair cells*. In humans, each row contains about 4000 cells. Each of the inner hair cells has about twenty afferent synapses; these are considered to be the primary receptors. In comparison, the outer hair cells are sparsely innervated but have both afferent (5%) and efferent (95%) synapses.

The basilar membrane is divided into two sections. Connected to the edge of the bony shelf, on the left in Fig. 18.1, is the *arcuate zone*, consisting of a single layer of transverse fibers. Connected to the edge of the spiral ligament, on the right in Fig. 18.1, is the *pectinate zone*, consisting of a double layer of transverse fibers in an amorphous ground substance. The *arches of Corti* form a truss over the arcuate zone, which consist of two rows of *pillar cells*. The foot of the inner pillar is attached at the point of connection of the bony shelf to the arcuate zone, while the foot of the outer pillar cell is attached at the common border of the arcuate zone and pectinate zone. The heads of the inner and outer pillars are connected and form the support point for the *recticular lamina*. The other edge of the recticular lamina is attached to the top of *Henson cells*, which have bases connected to the basilar membrane. The inner hair cells are attached on the bony shelf side of the inner pillars, while the three rows of outer hair cells are attached to the recticular lamina. The region bounded by the inner pillar cells, the recticular lamina, the Henson cells, and the basilar membrane forms another fluid region. This fluid is considered to be perilymph, since it appears that ions can flow freely through the arcuate zone of the basilar membrane. The cilia of the hair cells protrude into the endolymph. Thus the outer hair cells are immersed in perilymph at 0 mV, have an intracellular potential of −70 mV, and have cilia at the upper surface immersed in endolymph at a potential of +80 mV. In some regions of the ears of some vertebrates [Freeman and Weiss, 1990], the cilia are free standing. However, mammals always have a *tectorial membrane*, originating near the edge of the bony shelf and overlying the rows of hair cells parallel to the recticular lamina. The tallest rows of cilia of the outer hair cells are attached to the tectorial membrane. Under the tectorial membrane and inside the inner hair cells is a fluid space, the *inner sulcus,* filled with endolymph. The cilia of the inner hair cells are not attached to the overlying tectorial membrane, so the motion of the fluid in the inner sulcus must provide the mechanical input to these primary receptor cells. Since the inner sulcus is found only in mammals, the fluid motion in this region generated by acoustic input may be crucial to high frequency discrimination capability.

With a few exceptions of specialization, the dimensions of all the components in the cross section of the mammalian cochlea change smoothly and slowly along the length, in a manner consistent with high stiffness at the base, or input end, and low stiffness at the apical end. For example, in the cat the basilar membrane width increases from 0.1 to 0.4 mm while the thickness decreases from 13 to 5 μm. The density of transverse fibers decreases more than the thickness, from about 6000 fibers per μm at the base to 500 per μm at the apex [Cabezudo 1978].

Material Properties

Both perilymph and endolymph have the viscosity and density of water. The bone of the wall and the bony shelf appear to be similar to compact bone, with density approximately twice that of water. The remaining components of the cochlea are soft tissue with density near that of water. The stiffnesses of the components vary over a wide range, as indicated by the values of Young's modulus listed in Table 18.1. These values are taken directly or estimated from many sources, including the stiffness measurements in the cochlea by Békésy [1960], Gummer and coworkers [1981], Strelioff and Flock [1984], Miller [1985], Zwislocki and Cefaratti [1989], and Olson and Mountain [1994].

18.3 Passive Models

The anatomy of the cochlea is complex. By modeling, one attempts to isolate and understand the essential features. Following is an indication of propositions and the controversy associated with a few such models.

TABLE 18.1   Typical Values and Estimates for Young's Modulus E

Compact bone	20	GPa
Keratin	3	GPa
Basilar membrane fibers	1.9	GPa
Microtubules	1.2	GPa
Collagen	1	GPa
Reissner's membrane	60	MPa
Actin	50	MPa
Red blood cell, extended (assuming thickness = 10 nm)	45	MPa
Rubber, elastin	4	MPa
Basilar membrane ground substance	200	kPa
Tectorial membrane	30	kPa
Jell-O	3	kPa
Henson's cells	1	kPa

Resonators

The ancient Greeks suggested that the ear consisted of a set of tuned resonant cavities. As each component in the cochlea was discovered subsequently, it was proposed to be the tuned resonator. The most well known resonance theory is Helmholtz's. According to this theory, the transverse fibers of the basilar membrane are under tension and respond like the strings of a piano. The short strings at the base respond to high frequencies and the long strings toward the apex respond to low frequencies. The important feature of the Helmholz theory is the *place principle*, according to which the receptor cells at a certain *place* along the cochlea are stimulated by a certain frequency. Thus, the cochlea provides a real-time frequency separation (Fourier analysis) of any complex sound input. This aspect of the Helmholtz theory has since been validated, as each of the some 30,000 fibers exiting the cochlea in the auditory nerve is sharply tuned to a particular frequency. A basic difficulty with such a resonance theory is that sharp tuning requires small damping, which is associated with a long ringing after the excitation ceases. Yet, the cochlea is remarkable for combining sharp tuning with short time delay for the onset of reception and the same short time delay for the cessation of reception.

A particular problem with the Helmholtz theory arises from the equation for the resonant frequency for a string under tension:

$$f = \frac{1}{2b}\sqrt{\frac{T}{\rho h}} \tag{18.1}$$

in which T is the tensile force per unit width, ρ is the density, b is the length, and h is the thickness of the string. In humans, the frequency range over which the cochlea operates is $f = 200$ to $20,000$ Hz, a factor of 100, while the change in length b is only a factor of 5 and the thickness of the basilar membrane h varies the wrong way by a factor of 2 or so. Thus, to produce the necessary range of frequency, the tension T would have to vary by a factor of about 800. In fact, the spiral ligament, which would supply such tension, varies in area by a factor of only 10.

Traveling Waves

No theory anticipated the actual behavior found in the cochlea in 1928 by Békésy [1960]. He observed *traveling waves* moving along the cochlea from base toward apex which have a maximum amplitude at a certain place. The place depends on the frequency, as in the Helmholz theory, but the amplitude envelope is not very localized. In Békésy's experimental models, and in subsequent mathematical and experimental models, the anatomy of the cochlea is greatly simplified. The coiling, Reissner's membrane, and the organ of Corti are all ignored, so the cochlea is treated as a straight tube with a single partition.

(An exception is in Fuhrmann and colleagues [1986]). A gradient in the partition stiffness similar to that in the cochlea, gives beautiful traveling waves in both experimental and mathematical models.

One-Dimensional Model

A majority of work has been based on the assumption that the fluid motion is one dimensional. With this simplification the governing equations are similar to those for an electrical transmission line and for the long wavelength response of an elastic tube containing fluid. The equation for the pressure p in a tube with constant cross-sectional area A and with constant frequency of excitation is:

$$\frac{d^2 p}{dx^2} + \frac{2\rho\omega^2}{AK} p = 0 \tag{18.2}$$

in which x is the distance along the tube, ρ is the density of the fluid, ω is the frequency in radians per second, and K is the generalized partition stiffness, equal to the net pressure divided by the displaced area of the cross-section. The factor of 2 accounts for fluid on both sides of the elastic partition. Often K is represented in the form of a single degree-of-freedom oscillator:

$$K = k + i\omega d - m\omega^2 \tag{18.3}$$

in which k is the static stiffness, d is the damping, and m is the mass density:

$$m = \rho_P \frac{h}{b} \tag{18.4}$$

in which ρ_P is the density of the plate, h is the thickness, and b is the width. Often the mass is increased substantially to provide better curve fits. A good approximation is to treat the pectinate zone of the basilar membrane as transverse beams with simply supported edges, for which

$$k = \frac{10 E h^3 c_f}{b^5} \tag{18.5}$$

in which E is the Young's modulus, and c_f is the volume fraction of fibers. Thus, for the moderate changes in the geometry along the cochlea as in the cat, h decreases by a factor of 2, c_f decreases by a factor of 12, b increases by a factor of 5, and the stiffness k from Eq. (18.4) decreases by five orders of magnitude, which is ample for the required frequency range. Thus, it is the bending stiffness of the basilar membrane pectinate zone and not the tension which governs the frequency response of the cochlea. The solution of Eq. (18.2) can be obtained by numerical or asymptotic (called WKB or CLG) methods. The result is traveling waves for which the amplitude of the basilar membrane displacement builds to a maximum and then rapidly diminishes. The parameters of K are adjusted to obtain agreement with measurements of the dynamic response in the cochlea. Often all the material of the organ of Corti is assumed to be rigidly attached to the basilar membrane so that h is relatively large and the effect of mass m is large. Then the maximum response is near the *in vacua* resonance of the partition given by:

$$\omega^2 = \frac{b}{h} \frac{k}{\rho} \tag{18.6}$$

The following are objections to the one-dimensional model [e.g., Siebert, 1974]: (1) The solutions of Eq. (18.2) show wavelengths of response in the region of maximum amplitude that are small in comparison with the size of the cross section, violating the basic assumption of one-dimensional fluid flow. (2) In the drained cochlea, Békésy [1960] observed no resonance of the partition, so there is no significant partition mass. The significant mass is entirely from the fluid and therefore Eq. (18.6) is not correct. This is consistent with the observations of experimental models. (3) In model studies by Békésy [1960] and others, the localization of response is independent of the area A of the cross-section. Thus Eq. (18.2) cannot govern the most interesting part of the response, the region near the maximum amplitude for a given frequency. (4) Mechanical and neural measurements in the cochlea show dispersion which is incompatible with the one-dimensional model [Lighthill, 1991]. (5) The one-dimensional model fails badly in comparison with experimental measurements in models for which the parameters of geometry, stiffness, viscosity, and density are known.

Nevertheless, the simplicity of Eq. (18.2) and the analogy with the transmission line have made the one-dimensional model popular. We note that there is interest in utilizing the principles in an analog model built on a silicon chip, because of the high performance of the actual cochlea. Watts [1993] reports on the first model with an electrical analog of two-dimensional fluid in the scali. An interesting observation is that the transmission line hardware models are sensitive to failure of one component, while the two-dimensional model is not. In experimental models, Békésy found that a hole at one point in the membrane had little effect on the response at other points.

Two-Dimensional Model

The pioneering work with two-dimensional (2-D) fluid motion was begun in 1931 by Ranke, as reported in Ranke [1950] and discussed by Siebert [1974]. Analysis of 2-D and three-dimensional (3-D) fluid motion without the *a priori* assumption of long or short wavelengths and for physical values of all parameters is discussed by Steele [1987]. The first of two major benefits derived from the 2-D model is the allowance of short wavelength behavior, i.e, the variation in fluid displacement and pressure in the duct height direction. Localized fluid motion near the elastic partition generally occurs near the point of maximum amplitude and the exact value of A becomes immaterial. The second major benefit of a 2-D model is the admission of a stiffness-dominated elastic partition (i.e., massless) which better approximates the physiological properties of the basilar membrane. The two benefits together address all the objections the one-dimensional model discussed previously.

Two-dimensional models start with the Navier–Stokes and continuity equations governing the fluid motion, and an anisotropic plate equation governing the elastic partition motion. The displacement potential φ for the incompressible and inviscid fluid must satisfy Laplace's equation:

$$\varphi_{,xx} + \varphi_{,zz} = 0 \tag{18.7}$$

where x is the distance along the partition and z the distance perpendicular to the partition, and the subscripts with commas denote partial derivatives. The averaged potential and the displacement of the partition are:

$$\overline{\varphi} = \frac{1}{H}\int_0^H \varphi\, dz \quad w = \varphi_z(x, 0) \tag{18.8}$$

so Eq. (18.7) yields the "macro" continuity condition (for constant H):

$$\overline{\varphi}_{,xx} - w = 0 \,. \tag{18.9}$$

An approximate solution is

$$\tilde{\varphi}(x, z, t) = F(x)\cosh\left[n(x)(z - H)\right]e^{i\omega t} \tag{18.10}$$

where F is an unknown amplitude function, n is the local wave number, H is the height of the duct, t is time, and ω is the frequency. This is an exact solution for constant properties and is a good approximation when the properties vary slowly along the partition. The conditions at the plate fluid interface yield the dispersion relation:

$$n\tanh(nH) = \frac{2\rho\omega^2}{AK} H \qquad (18.11)$$

The averaged value Eq. (18.8) of the approximate potential Eq. (18.10) is:

$$\bar{\tilde{\varphi}} = \frac{F\sinh nH}{nH} \qquad (18.12)$$

so the continuity condition Eq. (18.9) yields the equation:

$$\bar{\tilde{\varphi}}_{,xx} + n^2\,\bar{\tilde{\varphi}} = 0 \qquad (18.13)$$

For small values of the wave number n, the system Eq. (18.11) and Eq. (18.13) reduce to the one-dimensional problem Eq. (18.2).

For physiological values of the parameters, the wave number for a given frequency is small at the stapes and becomes large (i.e., short wave lengths) toward the end of the duct. With this formulation, the form of the wave is not assumed. It is clear that for large n the WKB solution will give excellent approximation to the solution of Eq. (18.13) in exponential form. So it is possible to integrate Eq. (18.13) numerically for small n for which the solution is not exponential and match the solution with the WKB approximation. This provides a uniformly valid solution for the entire region of interest without an *a priori* assumption of the wave form.

For a physically realistic model, the mass of the membrane can be neglected and K written as:

$$K = k(1 + i\varepsilon) \qquad (18.14)$$

in which k is the static stiffness. For many polymers, the material damping ε is nearly constant. If the damping comes from the viscous boundary layer of the fluid, then ε is approximated by

$$\varepsilon \approx n\sqrt{\frac{\mu}{2\rho\omega}} \qquad (18.15)$$

in which μ is the viscosity. For water, ε is small with a value near 0.05 at the point of maximum amplitude. The actual duct is tapered, so $H = H(x)$ and additional terms must be added to Eq. (18.13).

The best verification of the mathematical model and calculation procedure comes from comparison with measurements in experimental models for which the parameters are known. Zhou and coworkers [1994] provide the first life-sized experimental model, designed to be similar to the human cochlea, but with fluid viscosity 28 times that of water to facilitate optical imaging. Results are shown in Figs. 18.2 and 18.3. Eq. (18.13) gives rough agreement with the measurements.

Three-Dimensional Model

A further improvement in the agreement with experimental models can be obtained by adding the component of fluid motion in the direction across the membrane for a full 3-D model. The solution by direct numerical means is computationally intensive, and was first carried out by Raftenberg [1990],

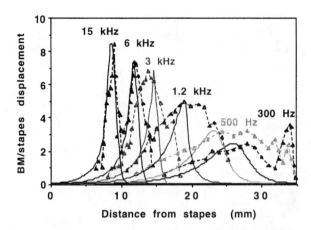

FIGURE 18.2 Comparison of 3-D model calculations (solid curves) with experimental results of Zhou and co-workers [1994] (dashed curves) for the amplitude envelopes for different frequencies. This is a life-sized model, but with an isotropic BM and fluid viscosity 28 times that of water. The agreement is reasonable, except for the lower frequencies.

who reports a portion of his results for the fluid motion around the organ of Corti. Böhnke and colleagues [1996] use the finite element code ANSYS for the most accurate description to date of the structure of the organ of Corti. However, the fluid is not included and only a restricted segment of the cochlea considered. The fluid is also not included in the finite element calculations of Zhang and colleagues [1996]. A "large finite element method", which combines asymptotic and numerical methods for shell analysis, can be used for an efficient computation of all the structural detail as shown in Fig. 18.1. Both the fluid and the details of the structure are considered with a simplified element description and simplified geometry by Kolston and Ashmore [1996], requiring some 10^5 degrees of freedom and hours of computing time (on a 66-MHz PC) for the linear solution for a single frequency. The asymptotic WKB solution, however, provides the basis for computing the 3-D fluid motion [Steele, 1987] and yields excellent agreement with older measurements of the basilar membrane motion in the real cochlea (with a computing time of 1 second per frequency). The 2-D analysis Eq. (18.13) can easily be extended to 3-D. The 3-D WKB calculations are compared with the measurements of the basilar membrane displacement in the experimental model of Zhou and coworkers [1994] in Figs. 18.2 and 18.3. As shown by Taber and Steele [1979], the 3-D fluid motion has a significant effect on the pressure distribution. This is confirmed by the measurements by Olson [1998] for the pressure at different depths in the cochlea, that show a substantial increase near the partition.

18.4   The Active Process

Before around 1980, it was thought that the processing may have two levels. First the basilar membrane and fluid provide the correct place for a given frequency (a purely mechanical "first filter"). Subsequently, the micromechanics and electrochemistry in the organ of Corti, with possible neural interactions, perform a further sharpening (a physiologically vulnerable "second filter").

A hint that the two-filter concept had difficulties was in the measurements of Rhode [1971], who found significant nonlinear behavior of the basilar membrane in the region of the maximum amplitude at moderate amplitudes of tone intensity. Passive models cannot explain this, since the usual mechanical nonlinearities are significant only at very high intensities, i.e., at the threshold of pain. Russell and Sellick [1977] made the first *in vivo* mammalian intracellular hair cell recordings and found that the cells are as sharply tuned as the nerve fibers. Subsequently, improved measurement techniques in several laboratories revealed that the basilar membrane is actually as sharply tuned as the hair cells and the nerve fibers. Thus, the sharp tuning occurs at the basilar membrane. No passive cochlear model, even with

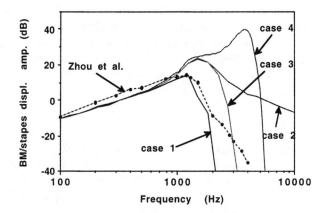

FIGURE 18.3 Comparison of 3-D model calculations with experimental results of Zhou and coworkers [1994] for amplitude at the place $x = 19$ mm as a function of frequency. The scales are logarithmic (20 dB is a factor of 10 in amplitude). Case 1 shows a direct comparison with the physical parameters of the experiment, with isotropic BM and viscosity 28 times that of water. Case 2 is computed for the viscosity reduced to that of water. Case 3 is computed for the BM made of transverse fibers. Case 4 shows the effect of active OHC feed-forward, with the pressure gain $\alpha = 0.21$ and feed-forward distance $\Delta x = 25$ microns. Thus, lower viscosity, BM orthotropy, and active feed-forward all contribute to higher amplitude and increased localization of the response.

physically unreasonable parameters, has yielded amplitude and phase response similar to such measurements. Measurements in a damaged or dead cochlea show a response similar to that of a passive model. Further evidence for an active process comes from Kemp [1978], who discovered that sound pulses into the ear caused echoes coming from the cochlea at delay times corresponding to the travel time to the place for the frequency and back. Spontaneous emission of sound energy from the cochlea has now been measured in the external ear canal in all vertebrates [Probst, 1990]. Some of the emissions can be related to the hearing disability of tinnitus (ringing in the ear). The conclusion drawn from these discoveries is that normal hearing involves an active process in which the energy of the input sound is greatly enhanced. A widely accepted concept is that spontaneous emission of sound energy occurs when the local amplifiers are not functioning properly and enter some sort of limit cycle (Zweig and Shera, 1995). However, there remains doubt about the nature of this process (Allen and Neely [1992], Hudspeth [1989], Nobili and coworkers [1998]).

Outer Hair Cell Electromotility

Since the outer hair cells have sparse afferent innervation, they have long been suspected of serving a basic motor function, perhaps beating and driving the subtectorial membrane fluid. Nevertheless, it was surprising when Brownell and colleagues [1985] found that the outer hair cells have *electromotility*: the cell expands and contracts in an oscillating electric field, either extra- or intracellular. The electromotility exists at frequencies far higher than possible for normal contractile mechanisms [Ashmore, 1987]. The sensitivity is about 20 nm/mV (about 10^5 better than PZT-2, a widely used piezoelectric ceramic). It has not been determined if the electromotility can operate to the 200 kHz used by high frequency mammals. However, a calculation of the cell as a pressure vessel with a fixed charge in the wall [Jen and Steele, 1987] indicates that, despite the small diameter (10 μm), the viscosity of the intra- and extracellular fluid is not a limitation to the frequency response. In a continuation of the work reported by Hemmert and coworkers [1996], the force generation is found to continue to 80 kHz in the constrained cell. In contradiction, however, the same laboratory (Preyer and coworkers [1996]) finds that the intracellular voltage change due to displacement of the cilia drops off at a low frequency.

The motility appears to be due to a passive piezoelectric behavior of the cell plasma membrane [Kalinec and colleagues, 1992]. Iwasa and Chadwick [1992] measured the deformation of a cell under pressure loading and voltage clamping and computed the elastic properties of the wall, assuming isotropy. It appears that for agreement with both the pressure and axial stiffness measurements, the cell wall must be orthotropic, similar to a filament reinforced pressure vessel with close to the optimum filament angle of 38° (Tolomeo and Steele, 1995). Holley [1990] finds circumferential filaments of the cytoskeleton with an average nonzero angle of about 26°. Mechanical measurements of the cell wall by Tolomeo and coworkers [1996] directly confirm the orthotropic stiffness.

Hair Cell Gating Channels

In 1984, Pickles and colleagues discovered tip links connecting the cilia of the hair cell, as shown in Fig. 18.4, that are necessary for the normal function of the cochlea. These links are about 6 nm in diameter and 200 nm long [Pickles, 1988]. Subsequent work by Hudspeth [1989] and Assad and Corey [1992] shows convincingly that there is a resting tension in the links. A displacement of the ciliary bundle in the excitatory direction causes an opening of ion channels in the cilia, which in turn decreases the intracellular potential. This depolarization causes neural excitation and, in the piezoelectric outer hair cells, a decrease of the cell length.

A purely mechanical analog model of the gating is in Steele [1992], in which the ion flow is replaced by viscous fluid flow and the intracellular pressure is analogous to the voltage. A constant flow-rate pump and leak channel at the base of the cell establish the steady-state condition of negative intracellular pressure, tension in the tip links, and a partially opened gate at the cilia through which there is an average magnitude of flow. The pressure drop of the flow through the gate has a nonlinear negative spring effect on the system. If the cilia are given a static displacement, the stiffness for small perturbation displacement is dependent on the amplitude of the initial displacement, as observed by Hudspeth [1989]. For oscillatory

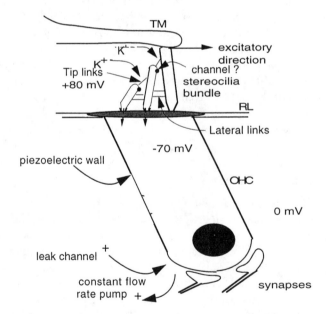

FIGURE 18.4 Model of outer hair cell. The normal pumping of ions produces negative intracellular electrical potential. Displacement of the cilia in the excitatory direction causes an opening of the ion channels in the cilia. There is evidence that the channels are located at the point of closest proximity of the two cilia. The tip and lateral links are important to maintain the correct stiffness and position. The opening of the channels decreases the intracellular potential, causing a piezoelectric contraction of the cell and excitation of the neural synapses. This can be modeled by a constant flow rate pump, leak channel, and spring controlled gate. The mechanical effect of the flow on the gate is important. The inner hair cell also has cilia, no piezoelectric property, but about 20 afferent synapses.

forcing of the cilia, the fluid analog shows that a gain in power is possible, as in an electrical or fluidic amplifier, and that a modest change in the parameters can lead to instability.

Thus, it appears that amplification in the cochlea resides in the gating of the outer hair cell cilia, while the motility is due to passive piezoelectric properties of the cell wall. The flow through the gate has significant nonlinearity at small amplitudes of displacement of the cilia (10 nm). Sufficiently high amplitudes of displacement of the cilia will cause the tip links to buckle. We estimate that this will occur at around 70 dB sound pressure level, thereby turning off the active process for higher sound intensity.

There is evidence reported by Hackney and colleagues [1996] that the channels do not occur at either end of the tip links, but at the region of closest proximity of the cilia, as shown in Fig. 18.4. Mechanical models such as in Furness and colleagues [1997] indicate that the channels at such a location can also be opened by force on the cilia. More elaborate models show that the stiffness of the tip links remains important to the mechanism.

There may be a connection between the gating channels and the discovery by Canlon and coworkers [1988]. They found that acoustic stimulation of the wall of the isolated outer hair cell caused a tonic (DC) expansion of the cell over a narrow frequency band, which is related to the place for the cell along the cochlea. Khanna and coworkers [1989] observe a similar tonic displacement of the whole organ of Corti.

18.5 Active Models

De Boer [1991], Geisler [1993], and Hubbard [1993] discuss models in which the electromotility of the outer hair cells feeds energy into the basilar membrane. The partition stiffness K is expanded from Eq. (18.3) into a transfer function, containing a number of parameters and delay times. These are classed as phenomenological models, for which the physiological basis of the parameters is not of primary concern. The displacement gain may be defined as the ratio of ciliary shearing displacement to cell expansion. For these models, the gain used is larger by orders of magnitude than the maximum found in laboratory measurements of isolated hair cells.

Another approach [Steele and colleagues, 1993], which is physiologically based, appears promising. The outer hair cells are inclined in the propagation direction. Thus, the shearing of the cilia at the distance x causes a force from the hair cells acting on the basilar membrane at the distance $x + \Delta x$. This "feed-forward" law can be expressed in terms of the pressure as:

$$p_{ohc}\left(x + \Delta x\right) = \alpha p\left(x\right) = \alpha\left[2\,p_f\left(x\right) + p_{ohc}\left(x\right)\right] \tag{18.16}$$

where p, the total pressure acting on the basilar membrane, consists of the effective pressure acting on the basilar membrane from the hair cells p_{ohc} and the pressure from the fluid p_f. The coefficient α is the force gain supplied by the outer hair cells. With this law, Eq. (18.11) is replaced by:

$$\left(1 - e^{in\Delta x}\right) n \tanh nH = 2\frac{\rho\omega^2}{AK}H \tag{18.17}$$

from which the local wave number n must be computed numerically. Only two new parameters are needed, the gain α and the spacing Δx. With physiologically reasonable gain $\alpha = 0.18$ and spacing $\Delta x = 20$ μm, the result is an increase of the response of the basilar membrane for higher frequencies by a factor of 10^2 in a narrow sector, apical to the passive peak. The simple feed-forward addition in Eq. (18.17) enhances a narrow band of wave lengths without a closed control loop. At this time, it appears that much of the elaborate structure of the organ of Corti is for the purpose of such a "feed-forward". This approach is also found to work well for the one-dimensional model by Geisler and Sang [1995].The results from the 3-D model for the effect of adding some feed-forward are shown in Fig. 18.3. One defect of the current feed-forward results is that the shift of the maximum response point is about one octave, as seen in Fig. 18.3, rather than one-half octave consistently shown in normal cochlear measurements of mechanical and neural response.

The significant nonlinear effect is the saturation of the active process at high amplitudes. This can be computed by letting the gain α be a function of the amplitude.

In the normal, active cochlea, it was first observed by Khanna and colleagues [1989] and subsequently by Gummer and colleagues [1996], that the tectorial membrane (TM in Fig. 18.1) has a substantially higher amplitude than the basilar membrane. Presumably, the electromotile expansion of the outer hair cells encounters less resistance from the tectorial membrane than the basilar membrane. This shows the importance of getting the correct stiffness and geometry into a model.

18.6   Fluid Streaming?

Békésy [1960] and many others have observed significant fluid streaming in the actual cochlea and in experimental models. Particularly for the high frequencies, it is tempting to seek a component of steady streaming as the significant mechanical stimulation of the inner hair cells [Lighthill, 1992]. Passive models indicated that such streaming occurs only at high sound intensity. Among the many open questions is whether or not the enhancement of amplitude provided by the feed-forward of energy by the outer hair cells and the mechanical nonlinearity at low amplitudes of displacement provided by the ciliary gating can trigger significant streaming. It is clear from the anatomy Fig. 18.1 that the motion and corresponding pressure in the fluid of the inner sulcus is the primary source of excitation for the inner hair cells. The DC pressure associated with DC streaming would be an important effect.

18.7   Clinical Possibilities

A better understanding of the cochlear mechanisms would be of clinical value. Auditory pathology related to the inner ear is discussed by Pickles [1988] and Gulick and colleagues [1989]. The spontaneous and stimulated emissions from the cochlea raise the possibility of diagnosing local inner ear problems, which is being pursued at many centers around the world. The capability for more accurate, physically realistic modeling of the cochlea should assist in this process. A significant step is provided by Zweig and Shera [1995], who find that a random distribution of irregularities in the properties along the cochlea explains much of the emissions.

A patient with a completely nonfunctioning cochlea is referred to as having "nerve deafness". In fact, there is evidence that in many cases the nerves may be intact, while the receptor cells and organ of Corti are defective. For such patients, a goal is to restore hearing with cochlear electrode implants to stimulate the nerve endings directly. Significant progress has been made. However, despite electrode stimulation of nerves at the correct place along the cochlea for a high frequency, the perception of high frequency has not been achieved. So, although substantial advance in cochlear physiology has been made in the recent past, several such waves of progress may be needed to adequately understand the functioning of the cochlea.

References

Allen JB and Neely ST. 1992. Micromechanical models of the cochlea. *Physics Today*, 45:40–47.

Ashmore JF. 1987. A fast motile response in guinea-pig outer hair cells: the cellular basis of the cochlear amplifier. *J. Physiol.* 388:323–347.

Assad JA and Corey DP. 1992. An active motor model for adaptation by vertebrate hair cells. *J. Neurosci.* 12(9):3291–3309.

Békésy G von. 1960. *Experiments in Hearing*. McGraw-Hill, New York.

Böhnke F, von Mikusch-Buchberg J, and Arnold W. 1996. 3-D Finite Elemente Modell des cochleären Verstärkers. *Biomedizinische Technik*. 42:311–312.

Brownell WE, Bader CR, Bertrand D, and de Ribaupierre Y. 1985. Evoked mechanical responses of isolated cochlear outer hair cells. *Science* 227:194–196.

Canlon B, Brundlin L, and Flock Å. 1988. Acoustic stimulation causes tonotopic alterations in the length of isolated outer hair cells from the guinea pig hearing organ. *Proc. Natl. Acad. Sci. USA* 85:7033–7035.

Cabezudo LM. 1978. The ultrastructure of the basilar membrane in the cat. *Acta Otolaryngol.* 86:160–175.

Dallos P. 1992. The active cochlea. *J. Neurosci.* 12(12):4575–4585.

De Boer E. 1991. Auditory physics. Physical principles in hearing theory. III. *Phys. Rep.* 203(3):126–231.

Freeman DM and Weiss TF. 1990. Hydrodynamic analysis of a two-dimensional model for micromechanical resonance of free-standing hair bundles. *Hearing Res.* 48:37–68.

Fuhrmann E, Schneider W, and Schultz M. 1987. Wave propagation in the cochlea (inner ear): effects of Reissner's membrane and non-rectangular cross-section. *Acta Mechanica* 70:15–30.

Furness DN, Zetes DE, Hackney CM, and Steele CR. 1997. Kinematic analysis of shear displacement as a means for operating mechanotransduction channels in the contact region between adjacent stereocilia of mammalian cochlear hair cells. *Proc. R. Soc. Lond. B* 264:45–51.

Geisler CD. 1993. A realizable cochlear model using feedback from motile outer hair cells. *Hearing Res.,* 68:253–262.

Geisler CD and Sang C. 1995. A cochlear model using feed-forword outer-hair-cell forces, *Hearing Res.* 85:132–146.

Gulick WL, Gescheider GA, and Fresina RD 1989. *Hearing: Physiological Acoustics, Neural Coding, and Psychoacoustics.* Oxford University Press, London.

Gummer AW, Johnston BM, and Armstrong NJ. 1981. Direct measurements of basilar membrane stiffness in the guinea pig. *J. Acoust. Soc. Am.* 70:1298–1309.

Gummer AW Hemmert W, and Zenner HP. 1996. Resonant tectorial membrane motion in the inner ear: its crucial role in frequency tuning. *Proc. Natl. Acad. Sci. USA*, 93:8727–8732.

Hackney CM, Furness DN, and Katori Y. 1996. Stereociliary ultrastructure in relation to mechanotransduction: tip links and the contact region. In *Diversity in Auditory Mechanics.* University of California, Berkeley, 173–180.

Hemmert W, Schauz C, Zenner HP, and Gummer AW. 1996. Force generation and mechanical impedance of outer hair cells. In *Diversity in Auditory Mechanics.* University of California, Berkeley, 189–196.

Holley MD. 1990. Cell biology of hair cells. *Seminars in the Neurosciences* 2:41–47.

Hubbard AE. 1993. A traveling wave-amplifier model of the cochlea. *Science* 259:68–71.

Hudspeth AJ. 1989. How the ears work. *Nature* 34:397–404.

Iwasa KH, and Chadwick RS. 1992. Elasticity and active force generation of cochlear outer hair cells. *J. Acoust. Soc. Am.* 92:3169–3173.

Jen DH and Steele CR. 1987. Electrokinetic model of cochlear hair cell motility. *J. Acoust. Soc. Am.* 82:1667–1678.

Kalinec F, Holley MC, Iwasa KH, Lim D, Kachar B. 1992. A membrane-based force generation mechanism in auditory sensory cells. *Proc. Natl. Acad. Sci. USA* 89:8671–8675.

Keidel WD and Neff WD Eds. 1976. *Handbook of Sensory Physiology,* Vol. V: *Auditory System.* Springer-Verlag, Berlin.

Kemp DT 1978. Stimulated acoustic emissions from within the human auditory system. *J. Acoust. Soc. Am.* 64:1386–1391.

Khanna SM, Flock Å, and Ulfendahl M. 1989. Comparison of the tuning of outer hair cells and the basilar membrane in the isolated cochlea. *Acta Otolaryngol. [Suppl.] Stockholm* 467:141–156.

Kolston PJ and Ashmore JF 1996. Finite element micromechanical modeling of the cochlea in three dimensions. *J. Acoust. Soc. Am.* 99:455–467.

Lighthill J. 1991. Biomechanics of hearing sensitivity. *J. Vibration Acoust.* 113:1–13.

Lighthill J. 1992. Acoustic streaming in the ear itself. *J. Fluid Mech.* 239:551–606.

Miller CE. 1985. Structural implications of basilar membrane compliance measurements. *J. Acoust. Soc. Am.* 77:1465–1474.

Nobili R, Mommano F, and Ashmore J. 1998. How well do we understand the cochlea? *TINS* 21(4): 159–166.

Olson ES, and Mountain DC. 1994. Mapping the cochlear partition's stiffness to its cellular architecture. *J. Acoust. Soc. Am.* 95(1):395–400.

Olson ES. 1998. Observing middle and inner ear mechanics with novel intracochlear pressure sensors. *J. Acoust. Soc. Am.* 103(6): 3445–3463.

Pickles JO. 1988. *An Introduction to the Physiology of Hearing*, 2nd ed. Academic Press, London.

Preyer S, Renz S, Hemmert W, Zenner H, and Gummer A. 1996. Receptor potential of outer hair cells isolated from base to apex of the adult guinea-pig cochlea: implications for cochlear tuning mechanisms. *Auditory Neurosci.* 2:145–157.

Probst R. 1990. Otoacoustic emissions: an overview. *Adv. Oto-Rhino-Laryngol.* 44:1–9.

Raftenberg MN. 1990. Flow of endolymph in the inner spiral sulcus and the subtectorial space. *J. Acoust. Soc. Am.* 87(6):2606–2620.

Ranke OF. 1950. Theory of operation of the cochlea: a contribution to the hydrodynamics of the cochlea. *J. Acoust. Soc. Am.* 22:772–777.

Rhode WS. 1971. Observations of the vibration of the basilar membrane in squirrel monkeys using the Mössbauer technique. *J. Acoust. Soc. Am.* 49:1218–1231.

Ruggero MA. 1993. Distortion in those good vibrations. *Curr. Biol.* 3(11):755–758.

Russell IJ and Sellick PM. 1977. Tuning properties of cochlear hair cells. *Nature,* 267:858–860.

Siebert WM. 1974. Ranke revisited—a simple short-wave cochlear model. *J. Acoust. Soc. Am.* 56(2):594–600.

Steele CR. 1987. Cochlear Mechanics. In *Handbook of Bioengineering*, R. Skalak and S. Chien, Eds., pp. 30.11–30.22. McGraw-Hill, New York.

Steele CR. 1992. Electroelastic behavior of auditory receptor cells. *Biomimetics* 1(1):3–22.

Steele CR, Baker G, Tolomeo JA, and Zetes DE. 1993. Electro-mechanical models of the outer hair cell. In *Biophysics of Hair Cell Sensory Systems*, H. Duifhuis, J.W. Horst, P. van Dijk, and S.M. van Netten, Eds., World Scientific, Singapore.

Strelioff D and Flock Å. 1984. Stiffness of sensory-cell hair bundles in the isolated guinea pig cochlea. *Hearing Res.* 15:19–28.

Taber LA and Steele CR. 1979. Comparison of 'WKB' and experimental results for three-dimensional cochlear models. *J. Acous. Soc. Am.* 65:1007–1018.

Tolomeo JA and Steele CR. 1995. Orthotropic piezoelectric propeties of the cochlear outer hair cell wall. *J. Acous. Soc. Am.* 95 (5):3006–3011.

Tolomeo JA, Steele CR, and Holley MC. 1996. Mechanical properties of the lateral cortex of mammalian auditory outer hair cells. *Biophys. J.* 71:421–429.

Watts L. 1993. *Cochlear Mechanics: Analysis and Analog VLSI*, Ph.D. thesis, California Institute of Technology.

West CD. 1985. The relationship of the spiral turns of the cochlea and the length of the basilar membrane to the range of audible frequencies in ground dwelling mammals. *J. Acoust. Soc. Am.* 77(3): 1091–1101.

Zhang L, Mountain DC and Hubbard AE. 1996. Shape changes from base to apex cannot predict characteristic frequency changes. *Diversity in Auditory Mechanics.* University of California, Berkeley, 611–618.

Zhou G, Bintz L, Anderson DZ, and Bright KE. 1994. A life-sized physical model of the human cochlea with optical holographic readout. *J. Acoust. Soc. Am.* 93(3):1516–1523.

Zweig G and Shera CA. 1995. The origin of periodicity in the spectrum of evoked otoacoustic emissions. *J. Acoust. Soc. Am.* 98(4):2018–2047.

Zwislocki JJ, and Cefaratti LK. 1989. Tectorial membrane II: Stiffness measurements *in vivo. Hearing Res.,* 42:211–227.

Further Information

The following are workshop proceedings that document many of the developments:

De Boer E, and Viergever MA, Eds. 1983. *Mechanics of Hearing*. Nijhoff, The Hague.

Allen JB, Hall JL, Hubbard A, Neely ST, and Tubis A, Eds. 1985. *Peripheral Auditory Mechanisms*. Springer, Berlin.

Wilson JP and Kemp DT, Eds. 1988. *Cochlear Mechanisms: Structure, Function, and Models*. Plenum Press, New York.

Dallos P, Geisler CD, Matthews JW, Ruggero MA, and Steele CR, Eds. 1990. *The Mechanics and Biophysics of Hearing*. Springer-Verlag, Berlin.

Duifhuis H, Horst JW, van Kijk P, and van Netten SM, Eds. 1993. *Biophysics of Hair Cell Sensory Systems*. World Scientific, Singapore.

Lewis ER, Long GR, Lyon RF, Narins PM, Steele CR, and Hecht-Poinar E, Eds. 1997. *Diversity in Auditory Mechanics*. World Scientific, Singapore.

19

Vestibular Mechanics

Wallace Grant
*Virginia Polytechnic Institute and
State University*

The vestibular system is responsible for sensing motion and gravity and using this information for control of postural and body motion. This sense is also used to control eyes position during head movement, allowing for a clear visual image. Vestibular function is rather inconspicuous and for this reason is frequently not recognized for its vital roll in maintaining balance and equilibrium and in controlling eye movements. Vestibular function is truly a sixth sense, different from the five originally defined by Greek physicians.

The vestibular system is named for its position within the vestibule of the temporal bone of the skull. It is located in the inner ear along with the auditory sense. The vestibular system has both central and peripheral components. This chapter deals with the mechanical sensory function of the peripheral end organ and its ability to measure linear and angular inertial motion of the skull over the frequency ranges encountered in normal activities.

19.1   Structure and Function

The vestibular system in each ear consists of the *utricle* and *saccule* (collectively called the *otolithic organs*) which are the linear motion sensors, and the three *semicircular canals* (SCCs) which sense rotational motion. The SCCs are oriented in three nearly mutually perpendicular planes so that angular motion about any axis may be sensed. The otoliths and SCCs consist of membranous structures which are situated in hollowed out sections of the temporal bone. This hollowed out section of the temporal bone is called the *bony labyrinth,* and the membranous labyrinth lies within this bony structure. The membranous labyrinth is filled with a fluid called *endolymph,* which is high in potassium, and the volume between the membranous and bony labyrinths is filled with a fluid called *perilymph,* which is similar to blood plasma.

The otoliths sit within the utricle and saccule. Each of these organs is rigidly attached to the temporal bone of the skull with connective tissue. The three semicircular canals terminate on the utricle forming a complete circular fluid path, and the membranous canals are also rigidly attached to the bony skull. This rigid attachment is vital to the roll of measuring inertial motion of the skull.

Each SCC has a bulge called the *ampulla* near the one end, and inside the ampulla is the cupula, which is formed of saccharide gel. The capula forms a complete hermetic seal with the ampulla, and the cupula sits on top of the crista, which contains the sensory receptor cells called *hair cells.* These hair cells have small stereocilia (hairs) which extend into the cupula and sense its deformation. When the head is rotated the endolymph fluid, which fills the canal, tends to remain at rest due to its inertia, the relative flow of fluid in the canal deforms the cupula like a diaphragm, and the hair cells transduce the deformation into nerve signals.

The otolithic organs are flat layered structures covered above with endolymph. The top layer consists of calcium carbonate crystals with *otoconia* which are bound together by a saccharide gel. The middle layer consists of pure saccharide gel, and the bottom layer consists of receptor hair cells which have stereocilia that extend into the gel layer. When the head is accelerated, the dense otoconial crystals tend to remain at rest due to their inertia as the sensory layer tends to move away from the otoconial layer. This relative motion between the octoconial layer deforms the gel layer. The hair cell stereocilia sense this deformation, and the receptor cells transduce this deformation into nerve signals. When the head is tilted, weight acting on the otoconial layer also will deform the gel layer. The hair cell stereocilia also have directional sensitivity which allows them to determine the direction of the acceleration acting in the plane of the otolith and saccule. The planes of the two organs are arranged perpendicular to each other so that linear acceleration in any direction can be sensed. The vestibular nerve, which forms half of the VIII cranial nerve, innervates all the receptor cells of the vestibular apparatus.

19.2 Otolith Distributed Parameter Model

The otoliths are an overdamped second-order system whose structure is shown in Fig. 19.1. In this model the otoconial layer is assumed to be a rigid and nondeformable, the gel layer is a deformable layer of isotropic viscoelastic material, and the fluid endolymph is assumed to be a newtonian fluid. A small element of the layered structure with surface area *dA* is cut from the surface, and a vertical view of this surface element, of width *dx,* is shown in Fig. 19.2. To evaluate the forces that are present, free-body diagrams are constructed of each elemental layer of the small differential strip. See the nomenclature table for a description of all variables used in the following formulas, and for derivation details see Grant and colleagues [1984, 1991].

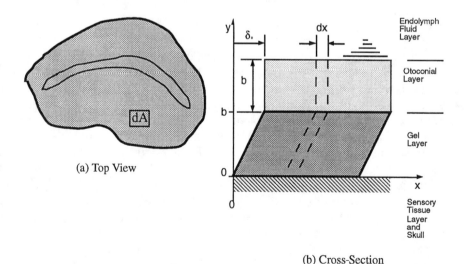

(a) Top View

(b) Cross-Section

FIGURE 19.1 Schematic of the otolith organ: (a) Top view showing the peripheral region with differential area *dA* where the model is developed. (b) Cross-section showing the layered structure where *dx* is the width of the differential area *dA* shown in the top view at the left.

Endolymph Fluid Layer The spatial coordinate in the vertical direction is y_f, and the velocity of the endolymph fluid is $u(y_f,t)$, a function of the fluid depth y_f and time t.

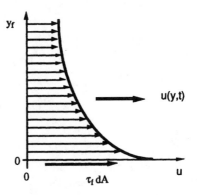

Otoconial Layer The otoconial layer with thickness b and at a height b above the gel layer vertical coordinate origin. The velocity of the otoconial layer is $v(t)$, which is a function of time only.

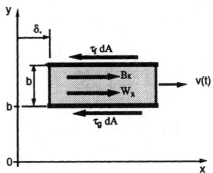

Gel Layer Gel layer of thickness b, vertical coordinate y_g, and horizontal coordinate δ_g the gel deflection. The gel deflection is a function of both y_g and time t. The velocity of the gel is $w(y_g, t)$, a function of y_g and t.

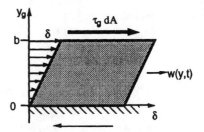

FIGURE 19.2 The free-body diagrams of each layer of the otolith with the forces that act on each layer. The interfaces are coupled by shear stresses of equal magnitude that act in opposite directions at each surface. The τ_g shear stress acts between the gel-otoconial layer, and the τ_f acts between the fluid-otoconial layer. The forces acting at these interfaces are the product of shear stress τ and area dA. The B_x and W_x forces are respectively the components of the buoyant and weight forces acting in the plane of the otoconial layer. See the nomenclature table for definitions of other variables.

In the equation of motion for the endolymph fluid, the force $\tau_f dA$ acts on the fluid, at the fluid–otoconial layer interface. This shear stress τ_f is responsible for driving the fluid flow. The linear Navier–Stokes equations for an incompressible fluid are used to describe this endolymph flow. Expressions for the pressure gradient, the flow velocity of the fluid measured with respect to an inertial reference frame, and the force due to gravity (body force) are substituted into the Navier–Stokes equation for flow in the x-direction yielding:

$$\rho_f \frac{\partial u}{\partial t} = \mu_f \frac{\partial^2 u}{\partial y_f}$$
(19.1)

with boundary and initial conditions: $u(0, t) = v(t)$; $u(\infty, t) = 0$; $u(y_f, 0) = 0$.

The gel layer is treated as a *Kelvin–Voight viscoelastic material* where the gel shear stress has both an elastic component and a viscous component acting in parallel. This viscoelastic material model is substituted into the momentum equation, and the resulting gel layer equation of motion is:

$$\rho_g \frac{\partial w}{\partial t} = G \int_0^t \left(\frac{\partial^2 w}{\partial y_g^2} \right) dt + \mu_g \frac{\partial^2 w}{\partial y_g^2} \tag{19.2}$$

with boundary and initial conditions: $w(b,t) = v(t)$; $w(0,t) = 0$; $w(y_g,0) = 0$; $\delta_g(y_g,0) = 0$. The elastic term in the equation is written in terms of the integral of velocity with respect to time, instead of displacement, so the equation is in terms of a single dependent variable, the velocity.

The otoconial layer equation was developed using Newton's second law of motion, equating the forces that act on the otoconial layer—fluid shear, gel shear, buoyancy, and weight—to the product of mass and inertial acceleration. The resulting otoconial layer equation is:

$$\rho_o b \frac{\partial v}{\partial t} + \left(\rho_o - \rho_f \right) \left[\frac{\partial V_s}{\partial t} - g_x \right] = \mu_f \left(\frac{\partial u}{\partial y_f} \bigg|_{y_f=0} \right) - G \int_0^t \left(\frac{\partial w}{\partial y_g} \bigg|_{y_g=b} \right) dt + \mu_g \left(\frac{\partial w}{\partial y_g} \bigg|_{y_g=b} \right) \tag{19.3}$$

with the initial condition $v(0) = 0$.

19.3   Nondimensionalization of the Motion Equations

The equations of motion are then nondimensionalized to reduce the number of physical and dimensional parameters and combine them into some useful nondimensional numbers. The following nondimensional variables, which are indicated by overbars, are introduced into the motion equations:

$$\bar{y}_f = \frac{y_f}{b} \quad \bar{y}_g = \frac{y_g}{b} \quad \bar{t} = \left(\frac{\mu_f}{\rho_o b^2} \right) t \quad \bar{u} = \frac{u}{V} \quad \bar{v} = \frac{v}{V} \quad \bar{w} = \frac{w}{V} \tag{19.4}$$

Several nondimensional parameters occur naturally as a part of the nondimensionalization process. These parameters are

$$R = \frac{\rho_f}{\rho_o} \quad \epsilon = \frac{Gb^2 \rho_o}{\mu_f^2} \quad M = \frac{\mu_g}{\mu_f} \quad \bar{g}_x = \left(\frac{\rho_o b^2}{V \mu_f} \right) g_x \tag{19.5}$$

where R is the density ratio, ϵ is a nondimensional elastic parameter, M is the viscosity ratio and represents a major portion of the system damping, and $\bar{g}_x$ is the nondimensional gravity.

The governing equations of motion in nondimensional form are then as follows. For the endolymph fluid layer:

$$R \frac{\partial \bar{u}}{\partial \bar{t}} = \frac{\partial^2 \bar{u}}{\partial \bar{y}_f^2} \tag{19.6}$$

with boundary conditions of $\bar{u}(0,t) = v(\bar{t})$ and $\bar{u}(\infty,\bar{t}) = 0$ and initial conditions of $\bar{u}(\bar{y}_f,0) = 0$. For the otoconial layer:

$$\frac{\partial \overline{v}}{\partial \overline{t}} + \left(1 - R\right)\left[\frac{\partial \overline{V}_s}{\partial \overline{t}} - \overline{g}_x\right] = \left(\left.\frac{\partial \overline{u}}{\partial \overline{y}_f}\right|_0\right) - \epsilon \int_0^t \left(\left.\frac{\partial \overline{w}}{\partial \overline{y}_g}\right|_1\right) dt - M\left(\left.\frac{\partial \overline{w}}{\partial \overline{y}_g}\right|_1\right) \tag{19.7}$$

with an initial condition of $\overline{v}(0) = 0$. For the gel layer:

$$R\frac{\partial \overline{w}}{\partial \overline{t}} = \epsilon \int_o^{\overline{t}} \left(\frac{\partial^2 \overline{w}}{\partial \overline{y}_g^2}\right) d\overline{t} + M\frac{\partial^2 \overline{w}}{\partial \overline{y}_g^2} \tag{19.8}$$

with boundary conditions of $\overline{w}(1,\overline{t}) = \overline{v}(\overline{t})$ and $\overline{w}(0,\overline{t}) = 0$ and initial conditions of $\overline{w}(\overline{y}_g,0) = 0$ and $\overline{\delta}_g(\overline{y}_g,0) = 0$.

These equations can be solved numerically for the case of a step change in velocity of the skull. This solution is shown in Figure 19.3 for a step change in the velocity of the head and a step change in the acceleration of the head, and can be found in Grant and Cotton [1991].

19.4   Otolith Transfer Function

A transfer function of otoconial layer deflection related to skull acceleration can be obtained from the governing equations. For details of this derivation see Grant and colleagues [1994].

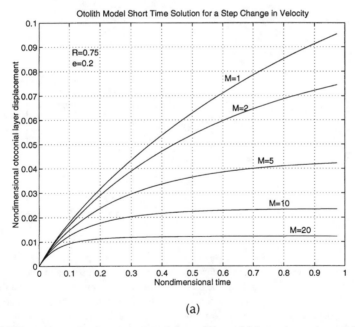

(a)

FIGURE 19.3   This figure shows the time response of the otolith model for various values of the nondimensional parameters. Parts (a) through (d) show the response to a step change in the velocity of the head. This response can be thought of as a transient response since it is an impulse in acceleration. The step change has a nondimensional skull velocity of magnitude one. Parts (e) and (f) show the response to a step change in the acceleration of the skull or simulates a constant acceleration stimulus. Again the magnitude of the nondimensional step change is one. (a) Short time response for various values of the damping parameter *M*. Note the effect on maximum displacement produced by this parameter.

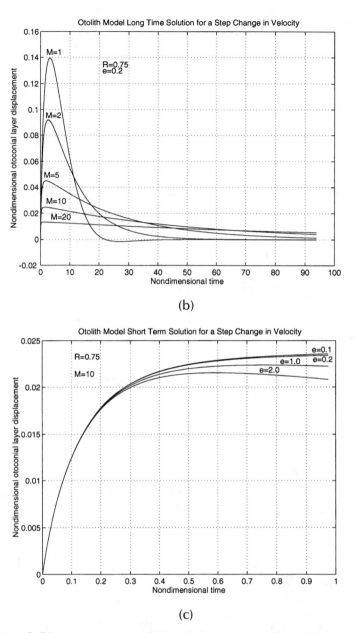

(b)

(c)

FIGURE 19.3 (continued) (b) Long time response for various M values. Note the small undershoot for $M = 1$, no conditions of an overdamped system were put on the model. As the system is highly overdamped this set of nondimensional parameters ($M = 1$, $R = 0.75$, and $\varepsilon = 0.2$) would not be an acceptable solution. (c) Short time response for various values of ε. Note that this elastic parameter does not have much effect on the otoconial layer maximum displacement for a step change in velocity solution. This is not true for a step change in acceleration, see part (f).

Starting with the nondimensional fluid and gel layer equations, taking the Laplace transform with respect to time and using the initial conditions give two ordinary differential equations. These equations can then be solved, using the boundary conditions. Taking the Laplace transform of the otoconial layer motion equation, combining with the two differential equation solutions, and integrating otoconial layer velocity to get deflection produces the transfer function for displacement reacceleration:

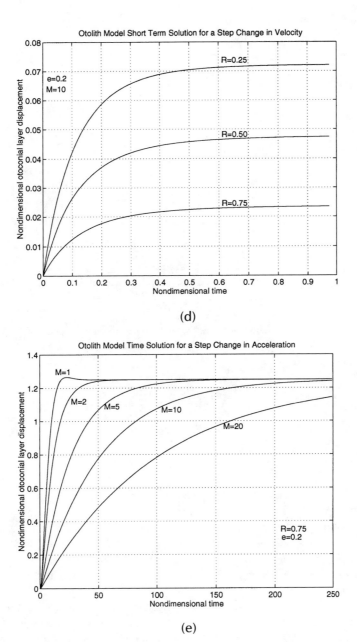

FIGURE 19.3 (continued) (d) Effect of the mass or density parameter R on response. R changes the maximum otoconial layer displacement. (e) Response to a step change in acceleration for various values of the damping parameter M. Note the overshoot for $M = 1$.

$$\frac{\delta_o}{A}(s) = \frac{(1-R)}{s\left[s+\sqrt{Rs}+\left(\frac{\epsilon}{s}+M\right)\sqrt{\frac{Rs}{\frac{\epsilon}{s}+M}}\coth\left(\sqrt{\frac{Rs}{\frac{\epsilon}{s}+M}}\right)\right]} \qquad (19.9)$$

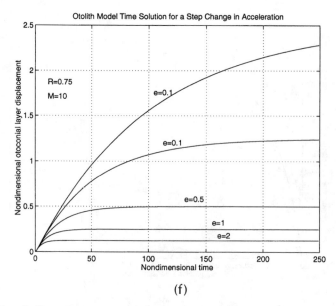

(f)

FIGURE 19.3 (continued) (f) Acceleration step change time response for various values of the elastic parameter ε. Note that this parameter affects the final displacement of the system for this type of constant stimulus (compare with part (c)).

where the overbars denoting nondimensional variables have been dropped, *s* is the Laplace transform variable, and a general acceleration term *A* is defined as:

$$A = -\left(\frac{\partial V_s}{\partial t} - g_x\right)$$
(19.10)

19.5 Otolith Frequency Response

This transfer function can now be studied in the frequency domain. It should be noted that these are linear partial differential equations and that the process of frequency domain analysis is appropriate. The range of values of ε = 0.01–0.2, M = 5–20, and R = 0.75 have been established by Grant and Cotton [1991] in a numerical finite difference solution of the governing equations. With these values established, the frequency response can be completed.

In order to construct a magnitude- and phase-vs.-frequency plot of the transfer function, the nondimensional time will be converted back to real time for use on the frequency axis. For the conversion to real time the following physical variables will be used: ρ_o = 1.35 g/cm³, b = 15 μm, μ_f = 0.85 mPa·s. The general frequency response is shown in Fig. 19.4. The flat response from DC up to the first corner frequency establishes this system as an accelerometer. These are the range-of-motion frequencies encountered in normal motion environments where this transducer is expected to function.

The range of flat response can be easily controlled with the two parameters ε and M. It is interesting to note that both the elastic term and the system damping are controlled by the gel layer, and thus an animal can easily control the system response by changing the parameters of this saccharide gel layer. The cross-linking of saccharide gels is extremely variable, yielding vastly different elastic and viscous properties of the resulting structure.

The otoconial layer transfer function can be compared to recent data from single-fiber neural recording. The only discrepancy between the experimental data and theoretical model is a low-frequency phase lead and accompanying amplitude reduction. This has been observed in most experimental single-fiber recordings and has been attributed to the hair cell.

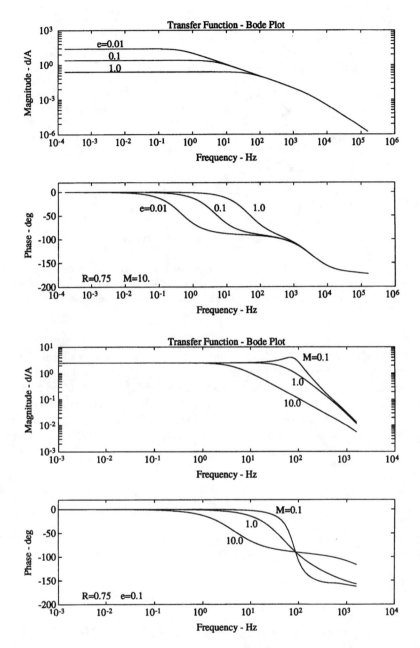

FIGURE 19.4 General performance of the octoconial layer transfer function shown for various values of the nondimensional elastic parameter E and M. For these evaluations the other parameter was held constant and $R = 0.75$. The value of $M = 0.1$ shows an underdamped response which is entirely feasible with the model formulation, since no restriction was incorporated which limits the response to that of an overdamped system.

19.6 Semicircular Canal Distributed Parameter Model

The membranous SCC duct is modeled as a section of a rigid torus filled with an incompressible newtonian fluid. The governing equations of motion for the fluid are developed from the Navier–Stokes equations. Refer to the nomenclature section for definition of all variables, and Fig. 19.5 for a cross-section of the SCC and utricle.

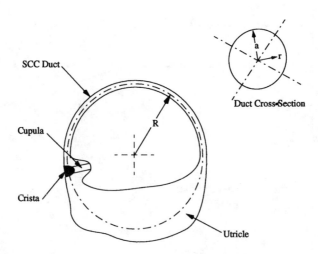

FIGURE 19.5   Schematic structure of the semicircular canal showing a cross-section through the canal duct and utricle. Also shown in the upper right corner is a cross-section of the duct. *R* is the radius of curvature of the semicircular canal, *a* is the inside radius to the duct wall, and *r* is the spatial coordinate in the radial direction of the duct.

We are interested in the flow of endolymph fluid with respect to the duct wall, and this requires that the inertial motion of the duct wall $R\Omega$ be added to the fluid velocity u measured with respect to the duct wall. The curvature of the duct can be shown to be negligible since $a \ll R$, and no secondary flow is induced; thus the curve duct can be treated as straight. Pressure gradients arise from two sources in the duct: (1) the utricle, (2) the cupula. The cupula when deflected exerts a restoring force on the endolymph. The cupula can be modeled as a membrane with linear stiffness $K = \Delta p / \Delta V$, where Δp is the pressure difference across the cupula and ΔV is the volumetric displacement, where:

$$\Delta V = 2\pi \int_0^t \int_0^a u(r,t)\, r dr\, dt \tag{19.11}$$

If the angle subtended by the membranous duct is denoted by β, the pressure gradient in the duct produced by the cupula is:

$$\frac{\partial p}{\partial z} = \frac{K\Delta V}{\beta R} \tag{19.12}$$

The utricle pressure gradient can be approximated [see Van Buskirk, 1977] by:

$$\frac{\partial p}{\partial z} = \frac{(2\pi - \beta)}{\beta} \rho R \alpha \tag{19.13}$$

When this information is substituted into the Navier–Stokes equation, the following governing equation for endolymph flow relative to the duct wall is obtained:

$$\frac{\partial u}{\partial t} + \left(\frac{2\pi}{\beta}\right) R\alpha = -\frac{2\pi K}{\rho \beta R}\int_0^t \int_0^a u\,(r dr)\, dt + v\frac{1}{r}\frac{\partial}{\partial r}\left(r\frac{\partial u}{\partial r}\right) \tag{19.14}$$

This equation can be nondimensionalized using the following nondimensional variables denoted by overbars:

$$\bar{r} = \frac{r}{a} \quad \bar{t} = \left(\frac{v}{a^2}\right)t \quad \bar{u} = \frac{u}{R\Omega}$$

(19.15)

where Ω is a characteristic angular velocity of the canal. In terms of the nondimensional variables, the governing equation for endolymph flow velocity becomes:

$$\frac{\partial \bar{u}}{\partial \bar{t}} + \left(\frac{2\pi}{\beta}\right)\alpha(\bar{t}) = -\epsilon \int_0^{\bar{t}} \int_0^1 \left(\bar{u}\bar{r}d\bar{r}\right) d\bar{t} + \frac{1}{\bar{r}}\frac{\partial}{\partial \bar{r}}\left(\bar{r}\frac{\partial \bar{u}}{\partial \bar{r}}\right)$$

(19.16)

where the nondimensional parameter ϵ is defined by:

$$\epsilon = \frac{2K\pi a^6}{\rho \beta R v^2}$$

(19.17)

The boundary conditions for this equation are as follows:

$$\bar{u}\left(1, \bar{t}\right) = 0 \quad \frac{\partial \bar{u}}{\partial \bar{r}}\left(0, \bar{t}\right) = 0$$

and the initial condition is $\bar{u}(\bar{r},0) = 0$.

19.7   Semicircular Canal Frequency Response

To examine the frequency response of the SCC, we will first get a solution to the nondimensional canal equation for the case of a step change in angular velocity of the skull. A step in angular velocity corresponds to an impulse in angular acceleration, and in dimensionless form this impulse is:

$$\alpha\left(t\right) = -\Omega\delta\left(t\right)$$

(19.18)

where $\delta(t)$ is the unit impulse function and Ω is again a characteristic angular velocity of the canal. The nondimensional volumetric displacement is defined as:

$$\phi = \int_0^{\bar{t}} \int_0^1 \left(\bar{u}\bar{r}d\bar{r}\right) d\bar{t}$$

(19.19)

the dimensional volumetric displacement is given by $\Delta V = (4\pi R\Omega a^4/v)\phi$ and the solution (for $\epsilon \ll 1$) is given by:

$$\phi = \sum_{n=1}^{\infty} \frac{2(2\pi/\beta)}{\lambda_\pi^4}$$

(19.20)

where λ_n represents the roots of the equation $J_0(x) = 0$, where J_0 is the Bessel function of zero order ($\lambda_1 = 2.405$, $\lambda_2 = 5.520$), and for infinite time $\phi = \pi/8\beta$. For details of the solution see Van Buskirk and Grant [1987] and Van Buskirk and coworkers [1978].

A transfer function can be developed from the previous solution for a step change in angular velocity of the canal. The transfer function of mean angular displacement of endolymph θ related to ω, the angular velocity of the head, is:

$$\frac{\theta}{\omega}(s) = \left(s\frac{(2\pi/\beta)\lambda_1^2}{8} \right)\left(\frac{1}{\left(s+\frac{1}{\tau_L}\right)\left(s+\frac{1}{\tau_s}\right)} \right) \tag{19.21}$$

where $\tau_L = 8\rho\nu\beta R/K\pi a^4$, $\tau_s = a^2/\lambda_1^2\nu$, and s is the Laplace transform variable.

The utility of the above transfer function is apparent when used to generate the frequency response of the system. The values for the various parameters are as follows: $a = 0.15$ mm, $R = 3.2$ mm, the dynamic viscosity of endolymph $\mu = 0.85$ mPa·s ($\nu = \mu/\rho$), $\rho = 1000$ kg/m^3, $\beta = 1.4\pi$, and $K = 3.4$ GPa/m^3. This produces values of the two time constants of $\tau_L = 20.8$ s and $\tau_s = 0.00385$ s. The frequency response of the system can be see in Fig. 19.6. The range of frequencies from 0.01 Hz to 30 Hz establishes the SCCs as angular velocity transducers of head motion. This range of frequencies is that encountered in

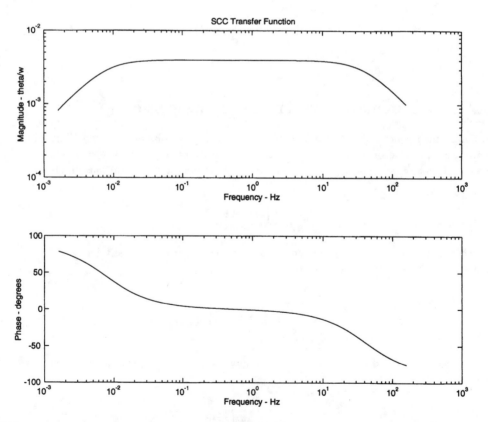

FIGURE 19.6 Frequency response of the human semicircular canals for the transfer function of mean angular displacement of endolymph fluid θ related to angular velocity of the head *w*.

everyday movement. Environments such as an aircraft flight can produce frequencies outside the linear range for these transducers.

Rabbit and Damino [1992] have modeled the flow of endolymph in the ampulla and its interaction with a cupula. This model indicates that the cupula in the mechanical system appears to add a high-frequency gain enhancement as well as phase lead over previous mechanical models. This is consistent with measurements of vestibular nerve recordings of gain and phase. Prior to this work this gain and phase enhancement were thought to be of hair cell origin.

Defining Terms

Endolymph: Fluid similar to intercellular fluid (high in potassium) which fills the membranous labyrinth, canals, utricle, and saccule.

Kelvin–Voight viscoelastic material: The simplest of solid materials which have both elastic and viscous responses of deformation. The viscous and elastic responses appear to act in parallel.

Otolith: Linear accelerometers of the vestibular system whose primary transduced signal is the sum of linear acceleration and gravity in the frequency range from DC (static) up to the maximum experienced by an animal.

Semicircular canals: Angular motion sensors of the vestibular system whose primary transduced signal is angular velocity in the frequency range of normal animal motion.

References

Grant JW, Cotton JR. 1991. A model for otolith dynamic response with a viscoelastic gel layer. *J Vestibular Res* 1:139.

Grant JW, Best WA, Lonegro R. 1984. Governing equations of motion for the otolith organs and their response to a step change in velocity of the skull. *J Biomech Eng* 106:203.

Grant JW, Huang CC, Cotton JR. 1994. Theoretical mechanical frequency response of the otolith organs. *J Vestibular Res* 4(2):137.

Lewis ER, Leverens EL, Bialek WS. 1985. *The Vertebrate Inner Ear*, Boca Raton, FL, CRC Press.

Rabbit RD, Damino ER. 1992. A hydroelastic model of macromachanics in the endolymphatic vestibular canal. *J Fluid Mech* 238:337.

Van Buskirk WC. 1977. The effects of the utricle on flow in the semicircular canals. *Ann Biomed Eng* 5:1.

Van Buskirk WC, Grant JW. 1973. *Biomechanics of the Semicircular Canals*, pp. 53–54, New York, Biomechanics Symposium of American Society of Mechanical Engineers.

Van Buskirk WC, Grant JW. 1987. Vestibular mechanics. In R. Skalak, S. Chien (Eds.), *Handbook of Bioengineering*, pp. 31.1–31.17, New York, McGraw-Hill.

Van Buskirk WC, Watts RG, Liu YU. 1976. Fluid mechanics of the semicircular canal. *J Fluid Mech* 78:87.

Nomenclature

Otolith Variables

x = coordinate direction in the plane of the otoconial layer
y_g = coordinate direction normal to the plane of the otolith with origin at the gel base
y_f = coordinate direction normal to the plane of the otolith with origin at the fluid base
t = time
$u(y_f, t)$ = velocity of the endolymph fluid measured with respect to the skull
$v(t)$ = velocity of the otoconial layer measured with respect to the skull
$w(y_g, t)$ = velocity of the gel layer measured with respect to the skull
$\delta_g(y_g, t)$ = displacement of the gel layer measured with respect to the skull
δ_o = displacement of the otoconial layer measured with respect to the skull
V_s = skull velocity in the x direction measured with respect to an inertial reference frame
V = a characteristic velocity of the skull in the problem (magnitude of a step change)

ρ_o = density of the otoconial layer
ρ_f = density of the endolymph fluid
τ_g = gel shear stress in the x-direction
μ_g = viscosity of the gel material
μ_f = viscosity of the endolymph fluid
G = shear modulus of the gel material
b = gel layer and otoconial layer thickness (assumed equal)
g_x = gravity component in the x-direction

Semicircular Canal Variables

$u(r,t)$ = velocity of endolymph fluid measured with respect to the canal wall
r = radial coordinate of canal duct
a = inside radius of the canal duct
R = radius of curvature of semicircular canal
ρ = density of endolymph fluid
v = endolymph kinematic viscosity
ω = angular velocity of the canal wall measured with respect to an inertial frame
α = angular acceleration of the canal wall measured with respect to an inertial frame
K = pressure-volume modulus of the cupula = $\Delta p/\Delta V$
Δp = differential pressure across the cupula
ΔV = volumetric displacement of endolymph fluid
β = angle subtended by the canal in radians ($\beta = \pi$ for a true semicircular canal)
λ_n = roots of $J_o(x) = 0$, where J_o is Bessel's function of order 0 ($\lambda_1 = 2.405$, $\lambda_2 = 5.520$)

Index